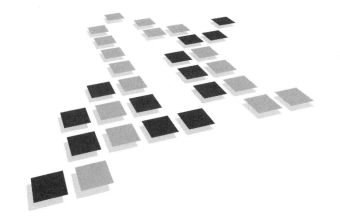

# An SPSS Companion
# to Political Analysis

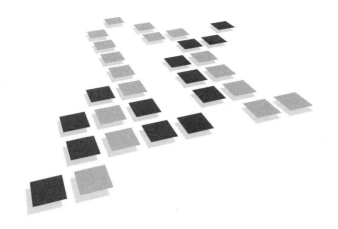

# An SPSS Companion to Political Analysis

### Fourth Edition

**Philip H. Pollock III**
*University of Central Florida*

Los Angeles | London | New Delhi
Singapore | Washington DC

CQ Press
2300 N Street, NW, Suite 800
Washington, DC 20037

Phone: 202-729-1900; toll-free, 1-866-4CQ-PRESS (1-866-427-7737)

Web: www.cqpress.com

*Credits:* Datasets on the CD: **GSS2008, GSS2008A_student, and GSS2008B_student:** selected variables obtained from General Social Surveys, 1972-2008 Cumulative Data File. James A. Davis, Tom W. Smith, and Peter V. Marsden. General Social Surveys, 1972-2008 [Cumulative File] [Computer file]. ICPSR25962-v2. Chicago: Storrs, CT: Roper Center for Public Opinion Research, University of Connecticut/Ann Arbor, MI: Inter-university Consortium for Political and Social Research [distributors], 2010-02-08. doi:10.3886/ICPSR25962. **NES2008, NES2008A_student, and NES2008B_student:** selected variables obtained from the American National Election Studies (ANES; www.electionstudies.org). The ANES 2008 Time Series Study [dataset]. Stanford University and the University of Michigan [producers]. These materials are based on work supported by the National Science Foundation under grants SES-0535334, SES-0720428, SES-0840550, and SES-0651271, Stanford University, and the University of Michigan. **World:** all variables obtained from Pippa Norris, John F. Kennedy School of Government, Harvard University, www.pippanorris.com, Democracy Crossnational Data, Release 3.0 Spring 2009, http://ksghome.harvard.edu/~pnorris/Data/Data.htm. **States:** selected variables obtained from Gerald C. Wright, Indiana University. See Robert S. Erikson, Gerald C. Wright, and John P. McIver, *Statehouse Democracy* (Cambridge, U.K.: Cambridge University Press, 1993).

Cover design: Auburn Associates Inc.
Typesetting: C&M Digitals (P) Ltd.

♾ The paper used in this publication exceeds the requirements of the American National Standard for Information Sciences—Permanence of Paper for Printed Library Materials, ANSI Z39.48-1992.

Printed and bound in the United States of America

15  14  13  12  11      1  2  3  4  5

ISBN 978-1-60871-687-6

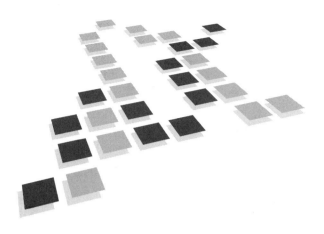

# Contents

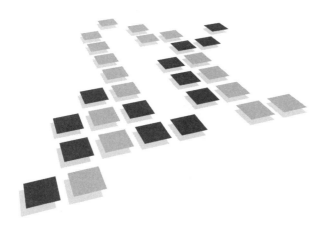

# Tables and Figures

## Tables

## Figures

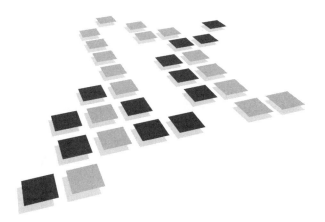

# Preface

Since I began teaching research methods over thirty years ago, I have introduced my students to the joys of doing their own political analysis. In my experience, students who appreciate the practical side of research are better prepared to contribute to class discussions of methodological concepts and problems. Moreover, students often develop a usable skill they can hone as they continue their academic careers or pursue employment opportunities. Whatever their professional goals, students need a solid foundation in basic political analysis techniques. They should learn to manipulate variables, explore patterns, and graph relationships. They also need a working knowledge of powerful yet easy-to-learn software, such as SPSS. This book instructs students in using SPSS to construct meaningful descriptions of variables and to perform substantive analysis of political relationships. The chapters cover all major topics in data analysis, from descriptive statistics to logistic regression. A final chapter describes several doable research projects, shows how to find or create analyzable data, and lays out a framework for a well-organized research paper.

In its essential features—multiple datasets, guided examples, screenshots, graphics instruction, and end-of-chapter exercises—this book continues in the tradition of previous editions.

## DATASETS AND EXERCISES

The SPSS datasets have been thoroughly revised and updated. The CD that accompanies this book contains four SPSS data files: selected variables from the 2008 General Social Survey (dataset GSS2008) and the 2008 National Election Study (NES2008), as well as datasets on the 50 states (States) and 191 countries of the world (World). As before, each chapter is written as a step-by-step tutorial, taking students through a series of guided examples and providing many annotated screenshots. Because of the revised and updated datasets, all the examples are fresh and all the screenshots are new.

This book contains fifty-six end-of-chapter exercises. The exercises are designed to give students opportunities to apply their new skills and to engage students in discovering the meaning of their findings and learning to interpret them. I have included exercises that reflect current scholarly debates in American political behavior and issues in comparative politics. The exercises test a full range of competencies, and most chapters include at least one more-challenging exercise. Even so, I continue to assume that students using this workbook have never heard of SPSS and have never used a computer to analyze data. After completing this book, students will have become competent SPSS users, and they will have learned a fair amount about substantive political science, too.

Any student who has access to SPSS—the full version or the student version—can use this book. When students analyze the States and World datasets using the student version, their results will be identical to full version output. Because NES2008 and GSS2008 exceed student version limitations, I created student-version compatible sets (NES2008A_student, NES2008B_student, GSS2008A_student, and GSS2008B_student) by drawing random samples from the full datasets. SPSS output from these sets will be similar to, but not the same as, output from the full-version datasets. To perform logistic regression, covered in Chapter 10, students will need access to the full version of SPSS.

## DIFFERENT RELEASES OF SPSS

SPSS 18 and 19 are featured here, but anyone running release 12 or later can profitably use this book. There are many commonalities across releases 12 through 19, including the graphic dialogs and the Chart Editor. Although SPSS now uses one editor for all graphics output, there are currently two ways to obtain unedited charts: the Legacy Dialogs and Chart Builder. Chart Builder was meant to be a one-stop, easy-to-use vehicle for constructing any chart, from the simple to the complex. However, the Legacy Dialogs offer superior flexibility and intuitiveness. Thus, for creating unedited charts this book uses the Legacy Dialogs, not Chart Builder. This edition carries forward the emphasis on elegant graphic display to complement and clarify empirical results. I have sought to instruct students in using the Chart Editor to emulate the techniques advocated by Edward R. Tufte and other experts on the visual display of data.

## CHAPTER ORGANIZATION

Chapter organization follows that of the previous edition. The "Getting Started" introduction describes the datasets, alerts students to differences between the full and student versions, and describes how to install the student version software. Chapter 1 introduces the SPSS Data Editor, discusses the output Viewer, and illustrates the print procedure. Chapter 2 covers central tendency and dispersion and guides students in using the Frequencies routine. Chapter 2 includes coverage of Case Summaries, which can be quite useful for providing insights into small datasets, such as States and World. Chapter 2 also shows how a frequency distribution, examined in conjunction with a bar chart or histogram, can enrich the description of a variable. Chapter 3 describes the main SPSS data transformation procedures, Recode and Compute. Chapter 3 also discusses Visual Binning (labeled Visual Banding through release 15), which is a powerful and efficient alternative to Recode, especially for collapsing interval variables into ordinal categories of roughly equal size. In Chapter 4, which covers Crosstabs and Compare Means, students learn bivariate analysis. Chapter 4 introduces line charts and bar charts, and students are given an initial tour of the Chart Editor. In Chapter 5 students use Crosstabs and Compare Means to obtain and interpret controlled comparisons. The chapter also discusses graphic support for controlled relationships. Chapter 6 uses One-Sample T Test and Independent-Samples T Test to demonstrate statistical significance for interval-level dependent variables. Chapter 6 features an expanded discussion of one-tailed versus two-tailed tests of significance—a persistent point of confusion for students. Chapter 7 covers chi-square and measures of association for nominal and ordinal variables. In Chapter 8 students work through an extended guided example to learn Correlate (Bivariate) and Regression (Linear). Chapter 8 discusses advanced editing using the Chart Editor. Chapter 9 shows how to create dummy variables (essentially an application of Recode, which students learned in Chapter 3), perform dummy variable regression analysis, and model interaction in multiple regression. Chapter 10 covers binary logistic regression, including a discussion of how to present logistic regression results in terms of probabilities. Chapter 11 guides students as they collect and analyze their own data.

These chapters are organized in the way that I typically teach my methods courses. I prefer to cover the logic of description and hypothesis testing before introducing inferential statistics and statistical significance. However, with a little rearranging of the chapters, this book will prove useful for instructors who do things differently. For example, after covering basic data transformations (Chapter 3) and discussing cross-tabulation analysis (Chapter 4), an instructor could assign Chapter 7, which covers chi-square and bivariate measures of association for categorical variables. Instructors who prefer using the regression approach to evaluating the statistical significance of mean differences might decide to skip Chapter 6 and move on to Chapters 8 and 9.

## ACCOMPANYING CORE TEXT

Instructors will find that this book makes an effective supplement to any of a variety of methods textbooks. However, it is a particularly suitable companion to my own core text, *The Essentials of Political Analysis*. The textbook's substantive chapters cover basic and intermediate methodological issues and ideas: measurement, explanations and hypotheses, univariate statistics and bivariate analysis, controlled relationships, sampling and inference, statistical significance, correlation and linear regression, and logistic regression.

Each chapter also includes end-of-chapter exercises. Students can read the textbook chapters, do the exercises, and then work through the guided examples and exercises in *An SPSS Companion to Political Analysis*. The idea is to get students in front of the computer, experiencing political research firsthand, fairly early in the academic term. An instructor's solutions manual, available for download online and free to adopters, provides solutions for all the textbook and workbook exercises.

## ACKNOWLEDGMENTS

I received more than a few friendly e-mails suggesting ways to improve this book. I am grateful for this advice. Many thanks, as well, to current and past reviewers for pointing me in the right direction: Holly Brasher, University of Alabama at Birmingham; Matthew Davis, University of Delaware; Jason Kehrberg, University of Kentucky; Thad Kousser, University of California, San Diego; Nancy Martorano, University of Dayton; Matthew Streb, Northern Illinois University; Brian Vargus, Indiana University–Purdue University Indianapolis; and Julian Westerhout, Illinois State University. I thank my University of Central Florida colleagues Bruce Wilson and Kerstin Hamann for helping me with ideas for exercises on comparative politics. I offer special thanks to Joe Simons-Rudolph of North Carolina State University for suggesting that I aim the exercises at a wider range of competencies and ground them more firmly in current scholarly debate. Thanks to Pete Furia of Wake Forest University and Bill Claggett of Florida State University for sharing their SPSS know-how with me. It was Claggett who reminded me of an ancient flaw in Compute: Multiply 0 times missing, and SPSS interprets the product as 0, not as missing. This quirk—and how to avoid it—is discussed in Chapter 9. Many encouraging people have helped me make this a better book. Any remaining errors, however, are mine.

I gratefully acknowledge the encouragement and professionalism of everyone associated with the College Division of CQ Press: Catherine Getzie, managing editor; Amy Marks, copy editor; Gwenda Larsen, senior production editor; and Nancy Loh, editorial assistant. I owe a special debt to Charisse Kiino, editorial director, without whose commitment this entire project would still be an incoherent jumble of datasets and syntax files lying fallow on my hard drive.

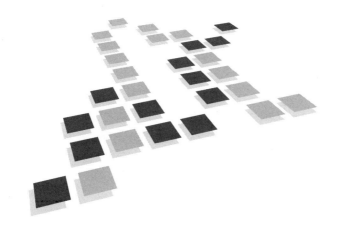

# An SPSS Companion
# to Political Analysis

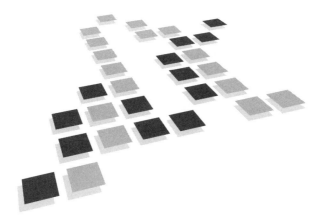

# Getting Started

To get started with this book you will need

- Access to a Microsoft Windows–based computer
- The CD that accompanies this workbook
- A USB flash drive or other portable media

As you have learned about political research and explored techniques of political analysis, you have studied many examples of other people's work. You may have read textbook chapters that present frequency distributions, or you may have pondered research articles that use cross-tabulation, correlation, or regression analysis to investigate interesting relationships between variables. As valuable as these learning experiences are, they can be enhanced greatly by performing political analysis firsthand—handling and modifying social science datasets, learning to use data analysis software, obtaining your own descriptive statistics for variables, setting up the appropriate analysis for interesting relationships, and running the analysis and interpreting your results.

This book is designed to guide you as you learn these valuable practical skills. In this volume you will gain a working knowledge of SPSS, a data analysis package used widely in academic institutions and business environments. SPSS has been in use for many years (it first appeared in 1968), and it contains a great variety of statistical analysis routines—from basic descriptive statistics to sophisticated predictive modeling. It is extraordinarily user friendly. In fact, although this book assumes that you have practical knowledge of the Windows operating system and that you know how to perform elemental file-handling tasks, it also assumes that you have never heard of SPSS and that you have never used a computer to analyze data of any kind. By the time you complete the guided examples and the exercises in this book, you will be well on your way to becoming an SPSS aficionado. The skills you learn will be durable, and they will serve you well as you continue your educational career or enter the business world.

This book's chapters are written in tutorial fashion. Each chapter contains several guided examples, and each includes exercises at the end. You will read each chapter while sitting in front of a computer, doing the analyses described in the guided examples, and analyzing the datasets that accompany this text. Each data analysis procedure is described in step-by-step fashion, and the book has many figures that show you what your computer screen should look like as you perform the procedures. Thus the guided examples allow you to develop your skills and to become comfortable with SPSS. The end-of-chapter exercises allow you to apply your new skills to different substantive problems.

This book will provide you with a solid foundation in data analysis. You will learn to obtain and interpret descriptive statistics (Chapter 2), to collapse and combine variables (Chapter 3), to perform cross-tabulation and mean analysis (Chapter 4), and to control for other factors that might be affecting your results (Chapter 5).

Techniques of statistical inference (Chapters 6 and 7) are covered, too. On the somewhat more advanced side, this book introduces correlation and linear regression (Chapter 8), and it teaches you how to use dummy variables and how to model interaction effects in regression analysis (Chapter 9). If you are running SPSS Full Version, Chapter 10 provides an introduction to logistic regression, an analytic technique that has gained wide currency in recent years. Chapter 11 shows you how to read data into SPSS, and it provides guidance on writing up your results.

## DATASET CD

The CD accompanying this book contains four datasets that you will analyze using SPSS. Insert the CD into your computer's drive. The CD contains two folders, one containing the datasets you will analyze if you are running SPSS Full Version, and one containing the datasets for SPSS Student Version (Figure I-1). If you are a Full Version user, open the Full Version folder (Figure I-2). If you are using Student Version, open the Student Version folder (Figure I-3). Even though the Student Version folder contains six datasets, the Student datasets are based on the four Full Version sets. (The major differences between versions are discussed below.)

1. **GSS2008** (for Student Version users: GSS2008A_student and GSS2008B_student). This dataset has selected variables from the 2008 General Social Survey, a random sample of 2,023 adults aged 18 years or older, conducted by the National Opinion Research Center and made available through the Inter-university Consortium for Political and Social Research (ICPSR) at the University of Michigan.[1] Some of the scales in GSS2008 were constructed by the author. These constructed variables are described in the appendix (Table A-1).

**Figure I-1**   Two Folders on CD: Full Version and Student Version

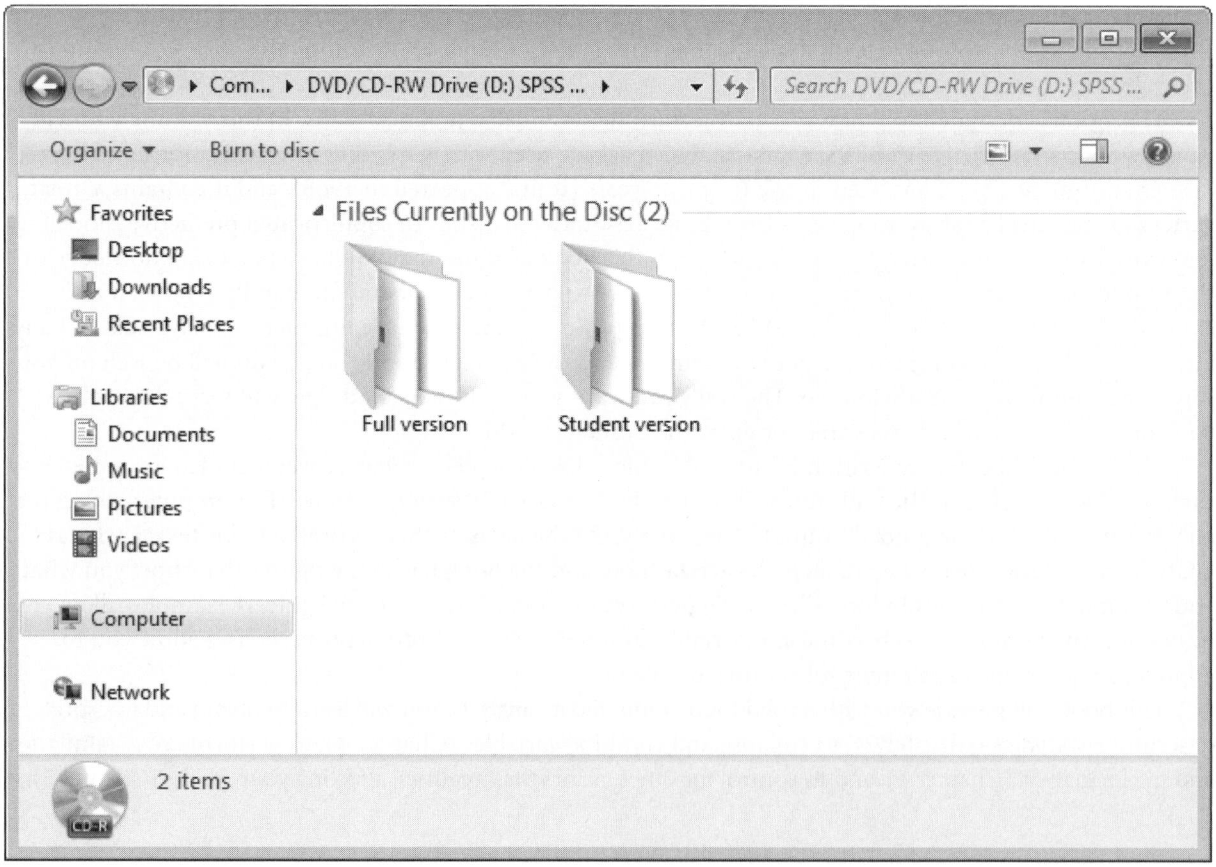

**Figure I-2**  Full Version Datasets

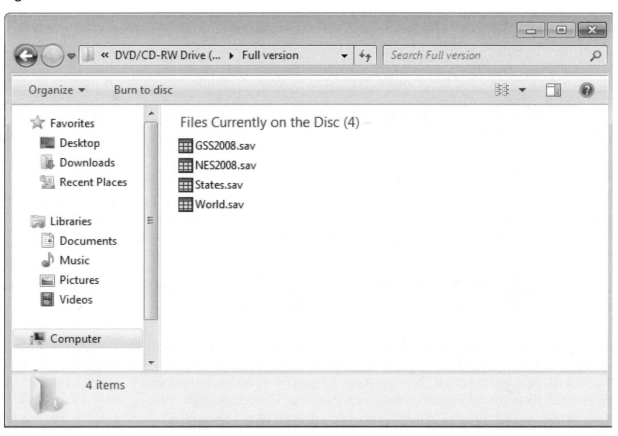

**Figure I-3**  Student Version Datasets

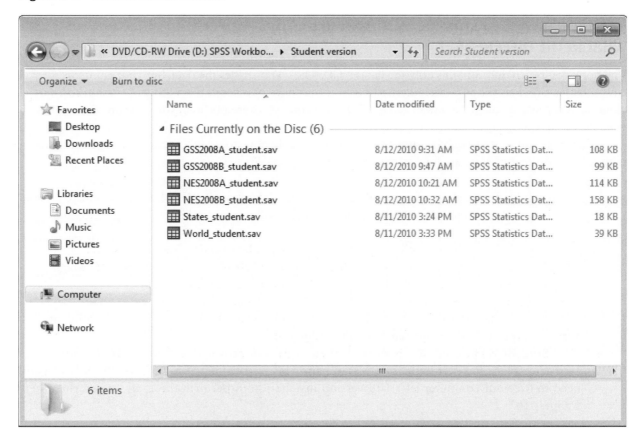

2. **NES2008** (NES2008A_student and NES2008B_student). This dataset includes selected variables from the 2008 National Election Study, a random sample of 2,323 citizens of voting age, conducted by the University of Michigan's Institute for Social Research and made available through ICPSR.[2] See the appendix (Table A-4).

3. **States** (States_student). This dataset includes variables on each of the 50 states. Most of these variables were compiled by the author from various sources. A complete description of States is found in the appendix (Table A-2).

4. **World** (World_student). This dataset includes variables on 191 countries of the world. These variables are based on data compiled by Pippa Norris, John F. Kennedy School of Government, Harvard University, and made available to the scholarly community through her Web site.[3] A complete description of World appears in the appendix (Table A-3).

As you work your way through this book, you will modify these datasets—recoding some variables, computing new variables, and otherwise tailoring the datasets to suit your purposes. You will need to make personal copies of the datasets and store them on a removable drive, such as a USB flash drive.

 **DO THIS NOW:** Before proceeding, copy the datasets from the accompanying CD to a personal USB drive or other portable media.[4]

When you begin each chapter's guided examples, or when you do the exercises, you will want to insert your personal media into the appropriate computer drive. SPSS will read the data from the drive. (Chapter 1 covers this operation in detail.) If you make any changes to a dataset, you can save the newly modified dataset directly to your drive. Alternatively, your computer lab's administrator may permit you to work on datasets that have been copied to the lab computer's desktop or to a folder designated for such a purpose. In any case, if you have modified a dataset during a data analysis session, it is important that you copy the dataset to your personal drive and take the datasets with you! (A comforting thought: The original datasets are still stored safely on the CD.)

## SPSS FULL VERSION AND SPSS STUDENT VERSION: WHAT IS THE DIFFERENCE?

The book you purchased contains only the dataset CD. The CD provides the data, but your campus computer lab provides the software. Alternatively, you may have purchased SPSS Student Version software along with the workbook. After you install the Student Version (covered below), you can run SPSS from your PC or laptop.

In terms of the guided examples and exercises in this book, how does SPSS Student Version compare with SPSS Full Version? Here are four facts worth knowing:

1. When a guided example or exercise calls for the States dataset or the World dataset, Student Version will analyze States_student or World_student. Output will be identical in every respect to Full Version output.

2. When a guided example or exercise calls for NES2008, Student Version users will analyze NES2008A_student (Chapter 1 through Chapter 6) or NES2008B_student (Chapter 7 through Chapter 9). Student Version output will not be the same as Full Version output.

3. When a guided example or exercise calls for GSS2008, Student Version users will analyze GSS2008A_student (Chapter 1 through Chapter 8) or GSS2008B_student (Chapter 9). Student Version output will not be the same as Full Version output.

4. SPSS Student Version will not perform logistic regression, covered in Chapter 10. If your instructor plans to cover logistic regression, and you are running Student Version, then you will need to use your computer lab's Full Version installation to perform the analyses described in Chapter 10.[5]

These similarities and differences are summarized in the table that follows:

| Version of SPSS | | | | |
|---|---|---|---|---|
| Full Version | | Student Version | | |
| Dataset | Chapters | Dataset | Chapters | Same output as Full Version? |
| GSS2008 | All | GSS2008A_student<br>GSS2008B_student | 1 through 8<br>9 | No<br>No |
| NES2008 | All | NES2008A_student<br>NES2008B_student | 1 through 6<br>7 through 9 | No<br>No |
| States | All | States_student | 1 through 9 | Yes |
| World | All | World_student | 1 through 9 | Yes |

## INSTALLING STUDENT VERSION

If you purchased SPSS Student Version, *now* would be a good time to install it on your PC or laptop. This software follows a standard installation protocol. Relevant instructions follow. (You will also find these instructions in the file Installation Instructions.pdf on the Student Version CD.[6])

1. Insert the Student Version CD into your CD drive. The AutoPlay feature presents a menu.
2. On the AutoPlay menu, click Install SPSS for Windows Integrated Student Version, and then follow the instructions that appear on the screen. (*Note:* If you already have another version of SPSS installed, then install the new version in a separate directory.)

You can also run the Setup program manually by following these steps:

1. From the Windows Start menu, choose Run.
2. In the Run dialog box, type D:\setup. (This assumes that drive D is your CD drive. If your CD drive is identified by a different letter, then substitute that drive's letter for D.) Follow the instructions that appear on the screen.

## NOTES

1. GSS2008 was created from the General Social Surveys 1972–2008 Cumulative Data File. James A. Davis, Tom W. Smith, and Peter V. Marsden. General Social Surveys, 1972–2008 [Cumulative File] [Computer file]. ICPSR25962-v2. Chicago: Storrs, CT: Roper Center for Public Opinion Research, University of Connecticut/Ann Arbor, MI: Inter-university Consortium for Political and Social Research [distributors], 2010-02-08. doi:10.3886/ICPSR25962.
2. The American National Election Studies (ANES; www.electionstudies.org). The ANES 2008 Time Series Study [dataset]. Stanford University and the University of Michigan [producers]. These materials are based on work supported by the National Science Foundation under grants SES-0535334, SES-0720428, SES-0840550, and SES-0651271, Stanford University, and the University of Michigan.
3. See www.pippanorris.com.
4. In the open folder window, click Edit → Select All. This selects all the datasets in the folder. Click Edit → Copy. Open your removable media. Click Edit → Paste. After the datasets have been copied, remove your portable media and the CD and put them in a safe place.
5. Student Version will not handle datasets containing more than 50 variables or having more than 1,500 cases. To get around these limitations, the author split NES2008 into two sets (NES2008A_student and NES2008B_student) and split GSS2008 into two sets (GSS2008A_student and GSS2008B_student), each of which contains fewer than 50 variables and no more than 1,500 cases. Cases were selected by taking random samples from the full datasets. The Full Version States and World datasets have more than 50 variables. States_student and World_student were created by deleting variables that are not used in this book's guided examples or exercises.
6. SPSS for Windows Integrated Student Version installation instructions.

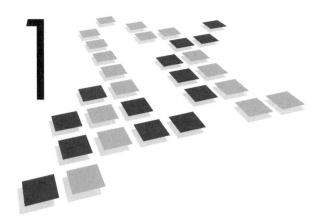

# Introduction to SPSS

Suppose you were hired by a telephone-polling firm to interview a large number of respondents. Your job is to find out and record three characteristics of each person you interview: age, gender, and the party with which the respondent identifies. The natural human tendency would be to record these attributes in words. For example, you might describe a respondent this way: "the respondent is 19 years old, is male, and identifies—but not strongly—with the Democratic Party." This would be a good thumbnail description, easily interpreted by another person. To SPSS, though, it would make no sense at all. Whereas people excel at recognizing and manipulating words, SPSS excels at recognizing and manipulating numbers. This is why researchers devise a *coding system*, a set of numeric identifiers for the different values of a variable. For one of the above variables, age, a coding scheme would be straightforward: Simply record the respondent's age in number of years, 19. In recording information about gender and party identification, however, a different set of rules is needed. For example, the National Election Study (NES) applies these codes for gender and party identification:

| Variable | Response | Code |
|---|---|---|
| Gender | Male | 1 |
| | Female | 2 |
| Party Identification | Strong Democrat | 0 |
| | Weak Democrat | 1 |
| | Independent Democrat | 2 |
| | Independent | 3 |
| | Independent Republican | 4 |
| | Weak Republican | 5 |
| | Strong Republican | 6 |

Thus the narrative profile "the respondent is 19 years old, is male, and identifies—but not strongly—with the Democratic Party" becomes "19 1 1" to SPSS. SPSS doesn't really care what the numbers stand for. As long as SPSS has numeric data, it will crunch the numbers—telling you the mean age of all respondents or the median of party identification. It is important, therefore, to provide SPSS with labels for each code so that the software's analytic work makes sense to the user. Accordingly, the SPSS Data Editor has two "views." The Data View shows the codes that SPSS recognizes and analyzes. The Variable View, among other useful features, shows the word labels that the researcher has assigned to the numeric codes.

## THE DATA EDITOR

Let's open the 2008 National Election Study dataset, NES2008, and see how this works. (If you are running SPSS Student Version, open NES2008A_student. You will use NES2008A_student through Chapter 6.) Insert into the computer's appropriate drive (for example, the USB port) the storage device containing your personal copies of the datasets. You created these copies in the Getting Started chapter. After it recognizes the drive, Windows will probably ask you whether you want to open its files. If this offer is not forthcoming, double-click the My Computer icon and then double-click the icon representing your computer's drive where the datasets are stored. Double-click the NES2008 icon. (Alternatively, you may want to open NES2008 after you have copied it onto the desktop. SPSS will run faster if the dataset is located on the computer's hard drive.)

SPSS opens the data file and displays the Data Editor (Figure 1-1). Notice the two tabs at the bottom of the window: Data View and Variable View. Let's look at the Data View first. (Make sure the Data View tab is clicked.) This shows how all the cases are organized for analysis. Information for each case occupies a separate row. The variables, given brief yet descriptive names, appear along the columns of the editor. You can tell that the first respondent in the dataset is 19 years old and is also coded 1 on age5, a 5-category ordinal measure of respondents' age groups. If you were to scroll to the right, you would also find that this respondent is male (coded 1 on the variable named gender) and is a weak Democrat (coded 1 on the variable named partyid7). To paint a more complete word-portrait of this respondent, however, you need to see how all the variables are coded. To reveal this information, click the Variable View tab (Figure 1-2). This view shows complete information on the meaning and measurement of each variable in the dataset. (You can adjust the width of a column by clicking, holding, and dragging the column border.)

The most frequently used variable information is contained in Name, Label, Values, and Missing. Name is the brief descriptor recognized by SPSS when it does analysis. Names can be up to 64 characters in length, although they need to begin with a letter (not a number). Plus, names must not contain any special characters, such as dashes or commas, although underscores are okay. You are encouraged to make good use of Label, a long descriptor (up to 256 characters are allowed), for each variable name. For example, when SPSS analyzes the variable camp_int, it will look in the Variable View for a label. If it finds one, then it will label the results of its analysis by using Label instead of Name. So camp_int shows up as "Campaign interest"—a

**Figure 1-1**   SPSS Data Editor: Data View

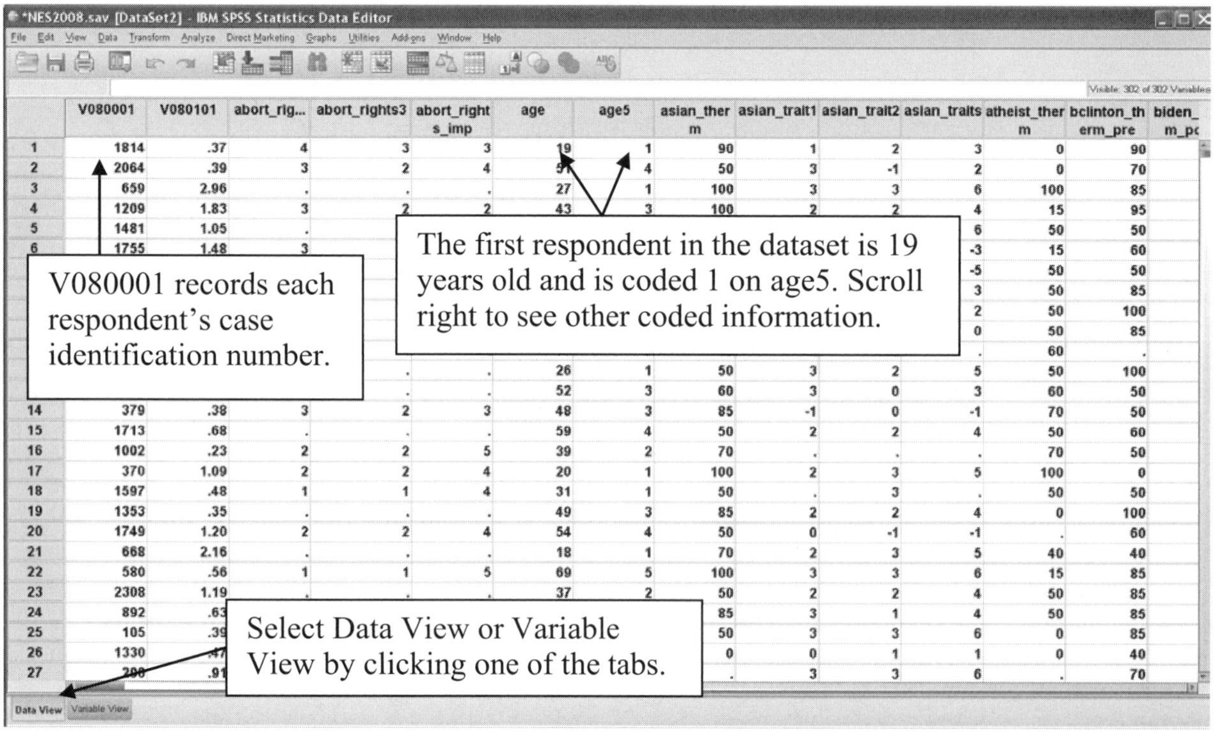

**Figure 1-2**   SPSS Data Editor: Variable View

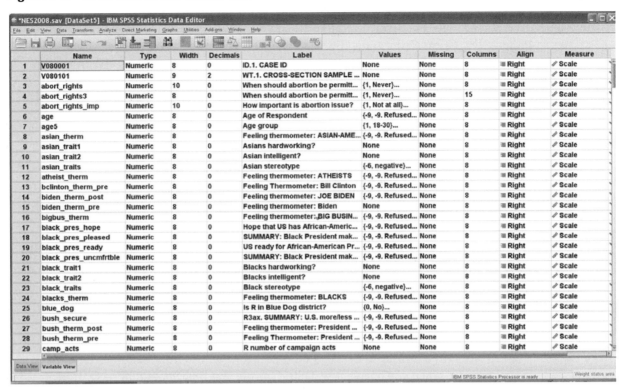

**Figure 1-3**   Value Labels Box

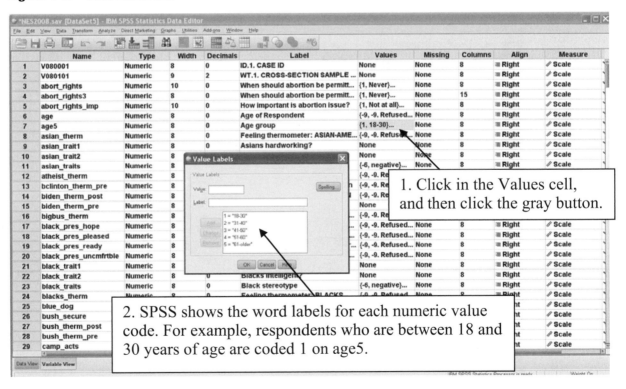

1. Click in the Values cell, and then click the gray button.

2. SPSS shows the word labels for each numeric value code. For example, respondents who are between 18 and 30 years of age are coded 1 on age5.

bit more descriptive than "camp_int." Just as Label permits a wordier description for Name, Values attaches word labels to the numeric variable codes. Consider age5, which, according to Label, records the respondent's age group. Click the mouse anywhere in the Values cell and then click the gray button that appears. A Value Labels window pops up, revealing the labels that SPSS will attach to the numeric codes of age5 (Figure 1-3). Unless you instruct it to do otherwise, SPSS will apply these labels to its analysis of age5. (Click the Cancel button in the Value Labels window to return to the Variable View.)

Finally, a word about Missing. Sometimes a dataset does not have complete information for some variables on a number of cases. In coding the data, researchers typically give a special numeric code to these missing values. In coding age, for example, the NES coders entered a value of "-9" for respondents who refused to reveal their ages. To avoid confusion, the author has set most missing values in the datasets to *system-missing*, which SPSS automatically removes from the analysis. However, when you use an existing variable to create a new variable, SPSS may not automatically transfer missing values on the existing variable to missing values on the new variable. Later in this volume we discuss how to handle such situations.

## A MUST-DO: SETTING OPTIONS FOR VARIABLE LISTS

Now you have a feel for the number-oriented side and the word-oriented side of SPSS. Before looking at how SPSS produces and handles output, you must do one more thing. To ensure that all the examples in this workbook correspond to what you see on your screen, you will need to follow the steps given in this section when you open each dataset for the first time.

 **DO THIS NOW:** In the main menu bar of the Data Editor, click Edit → Options. Make sure that the General tab is clicked. (See Figure 1-4.) If the radio button Display names *and* the radio button Alphabetical were already selected when you opened the Options menu, you are set to go. Click Cancel. If, however, Display names and/or Alphabetical were not already selected when you opened the Options menu, select them (as in Figure 1-4). Click Apply. Click OK, returning to the Data Editor. Again, you will need to do this for each dataset. When you open a new dataset for the first time, go to Edit → Options and ensure that Display names / Alphabetical are selected and applied.

**Figure 1-4**   Setting Options for Variable Lists

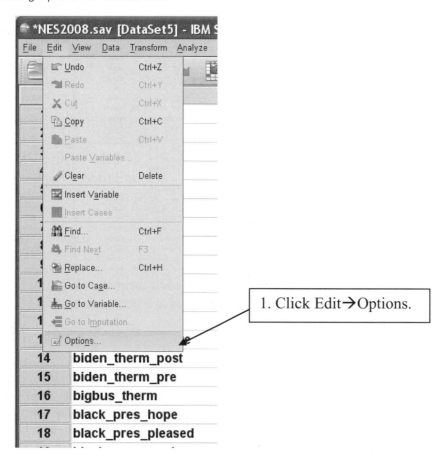

**Figure 1-4**  Setting Options for Variable Lists (continued)

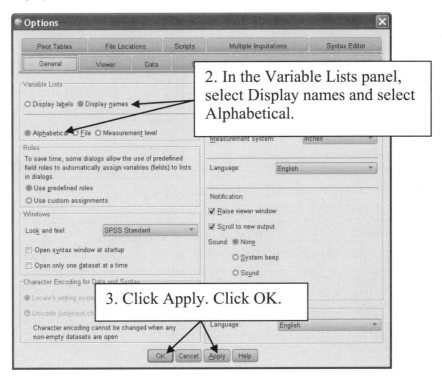

## THE VIEWER

Let's run through a quick analysis and see how SPSS handles variables and output. On the main menu bar, click Analyze → Descriptive Statistics → Frequencies. The Frequencies window appears (Figure 1-5). There are two panels. On the right is the (currently empty) Variable(s) panel. This is the panel where you enter the variables you want to analyze. On the left you see the names of all the variables in NES2008 in alphabetical order, just as you specified in the Options menu. Although the names are not terribly informative, complete coding information is just a (right) mouse click away. Suppose you want to analyze the variable named cong_approval. Scroll down the variable list until you find cong_approval. Put the mouse pointer on the variable, cong_approval, and right-click. Then click on Variable Information. As shown in Figure 1-6, SPSS retrieves and displays the label (R approve/dispp Congress handling job), name (cong_approval), and, most usefully, the value labels for the numeric codes. (To see all the codes, click the drop-down arrow in the Value Labels box.) Respondents who approve strongly are coded 1, those who approve not strongly are coded 2, and so on.

Return the mouse to the Frequencies window and click cong_approval into the Variable(s) panel. (Click on cong_approval and then click the arrow between the panels.) Click OK. SPSS runs the analysis and displays the results in the Viewer (Figure 1-7). The Viewer has two panes. In the Outline pane, SPSS keeps a running log of the analyses you are performing. The Outline pane references each element in the Contents pane, which reports the results of your analyses. In this book we are interested exclusively in the Contents pane. Reduce the size of the Outline pane by first placing the cursor on the Pane divider. Click and hold the left button of the mouse and then move the Pane divider over to the left-hand border of the Viewer. The Viewer should now look like Figure 1-8. The output for cong_approval shows you the frequency distribution, with value codes labeled. In Chapter 2 we discuss frequency analysis in more detail. Our immediate purpose is to become familiar with SPSS output.

Here are some key facts about the Viewer. First, the Viewer is a separate file, created by you during your analysis of the data. It is completely distinct from the data file. Whereas SPSS data files all have the file extension *.sav, Viewer files have the file extension *.spv. The output can be saved, under a name that you choose,

**Figure 1-5** Requesting Frequencies

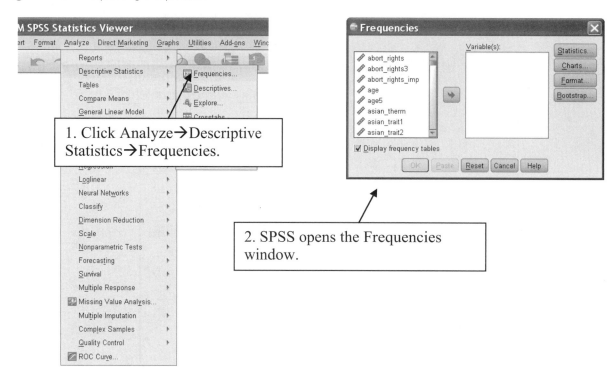

1. Click Analyze→Descriptive Statistics→Frequencies.

2. SPSS opens the Frequencies window.

**Figure 1-6** Retrieving Coding Information

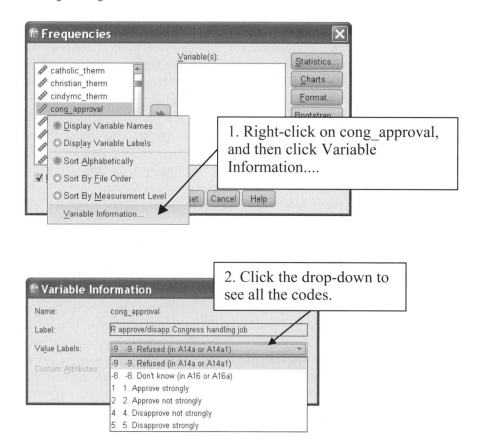

1. Right-click on cong_approval, and then click Variable Information....

2. Click the drop-down to see all the codes.

**Figure 1-7** SPSS Viewer: Outline Pane and Contents Pane

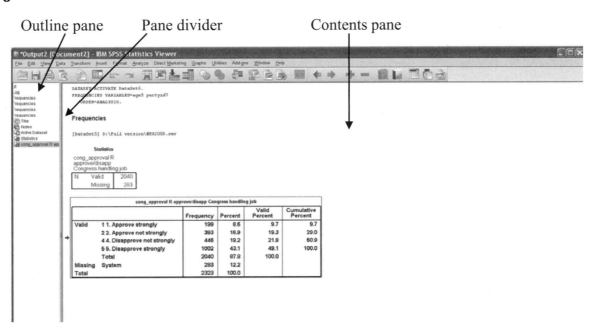

**Figure 1-8** SPSS Viewer: Outline Pane Minimized

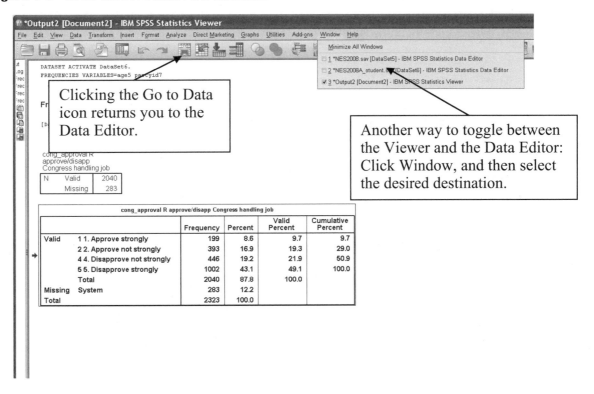

and then reopened later. Second, the output from each succeeding analysis does not overwrite the file. Rather, it appends new results to the Viewer file. If you were to run another analysis for a different variable, SPSS would dump the results in the Viewer below the analysis you just performed. Third, there are a number of ways to toggle between the Viewer and the Data Editor. (See Figure 1-8.) Clicking the Go to Data icon on the toolbar takes you to the Data Editor. Clicking Window on the main menu bar and checking "NES2008.

sav - IBM SPSS Statistics Data Editor" also returns you to the Data Editor. Clicking Window on the main menu bar in the Data Editor and checking "*.spv - SPSS Viewer" sends you to your output file. (And, of course, Windows accumulates icons for all open files along the bottom Taskbar.) Finally, you may select any part of the output file, print it, or copy and paste it into a word processing program.

Many of the exercises in this workbook will ask you to print the results of your SPSS analyses, so let's cover the print procedure. We'll also address a routine necessity: saving output.

### Selecting, Printing, and Saving Output

Printing desired results requires, first, that you select the output or portion of output you want to print. A quick and easy way to select a single table or chart is to place the cursor anywhere on the desired object and click once. Let's say you want to print the cong_approval frequency distribution. Place the cursor on the frequency table and click. A red arrow appears in the left-hand margin next to the table (Figure 1-9). Now click the Printer icon on the Viewer menu bar. The Print window opens. In the window's Print Range panel, the radio button next to "Selected output" should already be clicked. Clicking OK would send the frequency table to the printer. To select more than one table or graph, hold down the Control key (Ctrl) while selecting the desired output with the mouse. Thus, if you wanted to print the frequency table and the statistics table, first click on one of the desired tables. While holding down the Ctrl key, click on the other table. SPSS will select both tables.

To save your output, simply click the familiar Save icon on the Viewer menu bar (refer to Figure 1-9). Browse for an appropriate location—most likely the removable media containing your copy of the datasets. Invent a file name (but preserve the .spv extension), such as "chap1.spv," and click Save. SPSS saves all of the information in the Viewer to the file chap1.spv. Saving your output protects your work. Plus, the output file can always be reopened later. Suppose you are in the middle of a series of SPSS analyses and you want to stop and return later. You can save the Viewer file, as described here, and exit SPSS. When you return, you start SPSS and load a data file (like NES2008) into the Data Editor. In the main menu bar of the Data Editor, you click File → Open → Output, find your .spv file, and open it. Then you can pick up where you left off.

**Figure 1-9** Selecting, Printing, and Saving Output

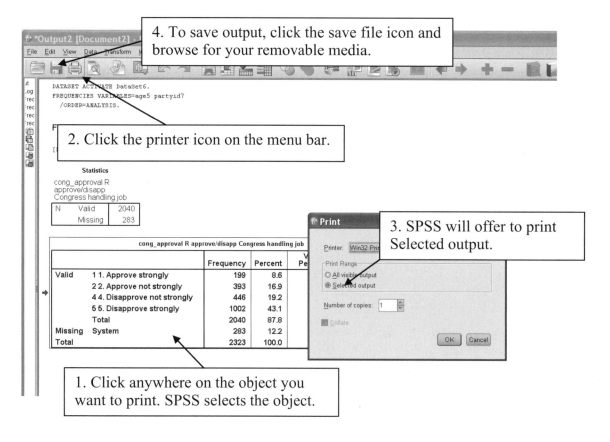

## EXERCISES

1. (Dataset: NES2008. Variables: contact_GOTV, income_r.) Earlier we spent some time using the Data View and the Variable View to describe the first respondent in the NES2008 dataset. In this exercise you will use your familiarity with the Data Editor to find out this respondent's income (income_r) and determine whether this respondent was contacted during the campaign and encouraged to vote (contact_GOTV).

   A. With NES2008 open, go to the Data View. What numeric code does the first respondent have on contact_GOTV? A code of (fill in the blank) _____. Go to the Variable View. Just as you did earlier in this chapter, find contact_GOTV and click in the Values cell. When asked whether anyone talked to him about registering or getting out to vote, this respondent (circle one)

      refused to answer.      responded "No."      responded "Yes."

   B. Return to the Data View. What is this respondent's code on the variable income_r? A code of (fill in the blank) _____. Go to the Variable View. This respondent's income is (circle one)

      $3,000–$4,999      $5,000–$7,499      $7,500–$9,999

2. Suppose that you have just opened World, States, or GSS2008 for the first time. The first thing you do is to click Edit → Options and consider the Variable Lists panel of the General tab. You must make sure that which two choices are selected and applied? (check two)

   ❑  Display labels              ❑  File

   ❑  Display names               ❑  Measurement level

   ❑  Alphabetical

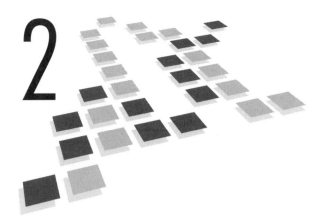

# Descriptive Statistics

**Procedures Covered**

Analyze → Descriptive Statistics → Frequencies

Analyze → Reports → Case Summaries

Analyzing descriptive statistics is the most basic—and sometimes the most informative—form of analysis you will do. Descriptive statistics reveal two attributes of a variable: its typical value (central tendency) and its spread (degree of dispersion or variation). The precision with which you can describe central tendency for any given variable depends on the variable's level of measurement. For nominal-level variables you can identify the *mode,* the most common value of the variable. For ordinal-level variables, those whose categories can be ranked, you can find the mode and the *median*—the value of the variable that divides the cases into two equal-size groups. For interval-level variables you can obtain the mode, median, and arithmetic *mean,* the sum of all values divided by the number of cases.

In this chapter you will use Analyze → Descriptive Statistics → Frequencies to obtain appropriate measures of central tendency, and you will learn to make informed judgments about variation. With the correct prompts, the Frequencies procedure also provides valuable graphic support—bar charts and (for interval variables) histograms. These tools are essential for distilling useful information from datasets having hundreds of anonymous cases, such the National Election Studies (NES2008) or General Social Survey (GSS2008). For smaller datasets with aggregated units, such as the States and World datasets, SPSS offers an additional procedure: Analyze → Reports → Case Summaries. Case Summaries lets you see first-hand how specific cases are distributed across a variable that you find especially interesting.

## INTERPRETING MEASURES OF CENTRAL TENDENCY AND VARIATION

Finding a variable's central tendency is ordinarily a straightforward exercise. Simply read the computer output and report the numbers. Describing a variable's degree of dispersion or variation, however, often requires informed judgment.[1] Here is a general rule that applies to any variable at any level of measurement: A variable has no dispersion if all the cases—states, countries, people, or whatever—fall into the same value of the variable. A variable has maximum dispersion if the cases are spread evenly across all values of the variable. In other words, the number of cases in one category equals the number of cases in every other category.

Central tendency and variation work together in providing a complete description of any variable. Some variables have an easily identified typical value and show little dispersion. For example, suppose you were to ask a large number of U.S. citizens what sort of economic system they believe to be the best: capitalism, communism, or socialism. What would be the modal response, or the economic system preferred by most people? Capitalism. Would there be a great deal of dispersion, with large numbers of people choosing the alternatives, communism or socialism? Probably not.

In other instances, however, you may find that one value of a variable has a more tenuous grasp on the label *typical.* And the variable may exhibit more dispersion, with the cases spread out more evenly across the variable's other values. For example, suppose a large sample of voting-age adults were asked, in the weeks preceding a presidential election, how interested they are in the campaign: very interested, somewhat interested, or not very interested. Among your own acquaintances you probably know a number of people who fit into each category. So even if one category, such as "somewhat interested," is the median, many people will likely be found at the extremes of "very interested" and "not very interested." In this instance the amount of dispersion in a variable—its degree of spread—is essential to understanding and describing it.

These and other points are best understood by working through some guided examples. For the next several analyses, you will use GSS2008. Open the dataset by double-clicking the GSS2008 icon. (If you are using *SPSS Student Version,* open GSS2008A_Student.) In the Data Editor, click Edit → Options and then click on the General tab. Just as you did with NES2008 in Chapter 1, make sure that the radio buttons in the Variable Lists area are set for Display names and Alphabetical. (If these options are already set, click Cancel. If they are not set, select them, click Apply, and then click OK. Now you are ready to go.)

## DESCRIBING NOMINAL VARIABLES

First, you will obtain a frequency distribution and bar chart for a nominal-level variable, zodiac, which records respondents' astrological signs. Click Analyze → Descriptive Statistics → Frequencies. Scroll down to the bottom of the left-hand list until you find zodiac. Click zodiac into the Variable(s) panel. To the right of the Variable(s) panel, click the Charts button (Figure 2-1). The Frequencies: Charts window appears. In

**Figure 2-1**   Obtaining Frequencies and a Bar Chart (nominal variable)

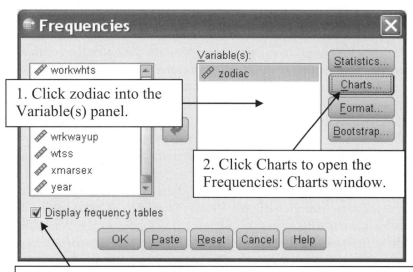

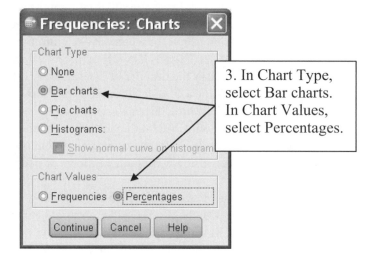

**Figure 2-2**  Frequencies Output (nominal variable)

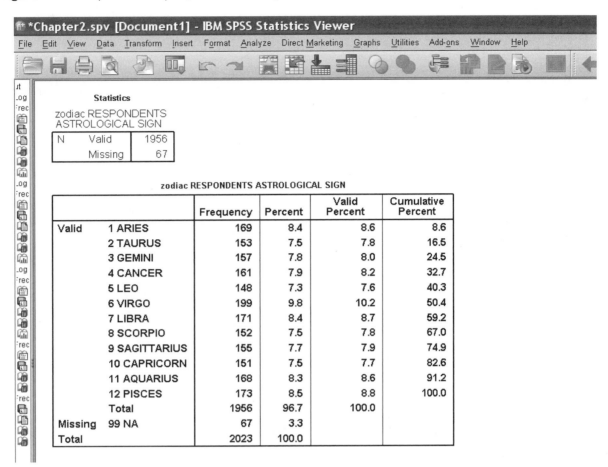

Chart Type, select Bar charts. In Chart Values, select Percentages. Click Continue, which returns you to the main Frequencies window. Click OK. SPSS runs the analysis.

SPSS has produced two items of interest in the Viewer: a frequency distribution of respondents' astrological signs and a bar chart of the same information. Examine the frequency distribution (Figure 2-2). The value labels for each astrological code appear in the leftmost column, with Aries occupying the top row of numbers and Pisces the bottom row. There are four numeric columns: Frequency, Percent, Valid Percent, and Cumulative Percent. What does each column mean? The Frequency column shows raw frequencies, the actual number of respondents having each zodiac sign. Percent is the percentage of *all* respondents, including missing cases, in each category of the variable. Ordinarily the Percent column can be ignored, because we generally are not interested in including missing cases in our description of a variable. Valid Percent is the column to focus on. Valid Percent tells us the percentage of nonmissing responses in each value of zodiac. Finally, Cumulative Percent reports the percentage of cases that fall in *or below* each value of the variable. For ordinal or interval variables, as you will see, the Cumulative Percent column can provide valuable clues about how a variable is distributed. But for nominal variables, which cannot be ranked, the Cumulative Percent column provides no information of value.

Now consider the Valid Percent column more closely. Scroll between the frequency distribution and the bar chart, which depicts the zodiac variable in graphic form (Figure 2-3). What is the mode, the most common astrological sign? For nominal variables, the answer to this question is (almost) always an easy call: Simply find the value with the highest percentage of responses. Virgo is the mode. Does this variable have little dispersion or a lot of dispersion? Again study the Valid Percent column and the bar chart. Apply the following rule: A variable has no dispersion if the cases are concentrated in one value of the

**Figure 2-3** Bar Chart (nominal variable)

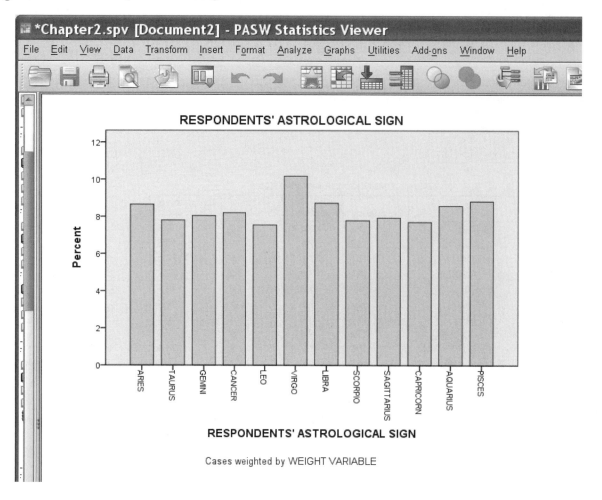

variable; a variable has maximum dispersion if the cases are spread evenly across all values of the variable. Are most of the cases concentrated in Virgo, or are there many cases in each value of zodiac? Because respondents show great heterogeneity in astrological signs, you would conclude that zodiac has a high level of dispersion.

## DESCRIBING ORDINAL VARIABLES

Next, you will analyze and describe two ordinal-level variables, one of which has little variation and the other of which is more spread out. Along the top menu bar of the Viewer, click Analyze → Descriptive Statistics → Frequencies. SPSS remembers the preceding analysis, so zodiac is still in the Variable(s) list. Click zodiac back into the left-hand list. Scroll through the list until you find these variables: helppoor and helpsick. Each of these is a 5-point ordinal scale. Helppoor asks respondents to place themselves on a scale between 1 ("The government should take action to help poor people") and 5 ("People should help themselves"). Helpsick, using a similar 5-point scale, asks respondents about government responsibility or individual responsibility for medical care. Click helppoor and helpsick into the Variable(s) list. SPSS retained your earlier settings for Charts, so accompanying bar charts will appear in the Viewer. Click OK.

SPSS runs the analysis for each variable and produces two frequency distributions, one for helppoor and one for helpsick, followed by two bar charts of the same information. First, let's focus on helppoor. To get a feel for this variable, scroll back and forth between the frequency distribution (Figure 2-4) and the bar chart (Figure 2-5). How would you describe the central tendency and dispersion of this variable? Because helppoor is an ordinal variable, you can report both its mode and its median. Its mode, clearly enough, is the response "Agree with both." What about the median? This is where the Cumulative

**Figure 2-4**   Frequencies Output (ordinal variables)

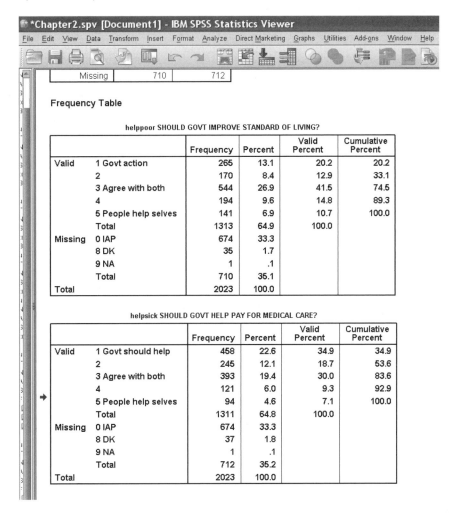

**Figure 2-5**   Bar Chart (ordinal variable with low dispersion)

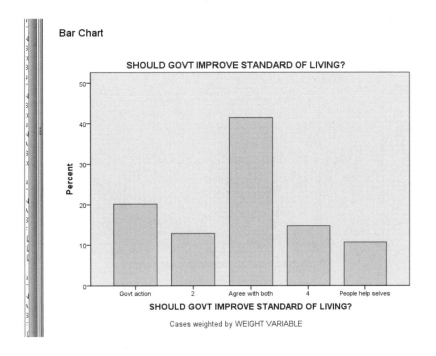

Percent column of the frequency distribution comes into play. *The median for any ordinal (or interval) variable is the category below which 50 percent of the cases lie.* Is the first category, "Govt action," the median? No, this code contains fewer than half the cases. How about the next higher category? No, again. The Cumulative Percent column still has not reached 50 percent. The median occurs in the "Agree with both" category.

Does helppoor have a high or low degree of dispersion? The dispersion of an ordinal variable can be evaluated in two complementary ways. One way is to take a close look at the bar graph. If helppoor had a high level of variation, the bars would have roughly equal heights, much like the zodiac variable that you analyzed earlier. If helppoor had no dispersion, then all the cases would fall into one category—only one bar would be showing in the graphic. Another way to evaluate variation is to compare the mode and the median. If the mode and the median fall in the same category, the variable has lower dispersion than if the mode and the median fall in different values of the variable. You sometimes have to exercise judgment in determining variation, but it seems clear that helppoor is a variable with a fairly low degree of dispersion. The fence-straddling response, "Agree with both," is prominent in the bar graph. What is more, both the mode and the median are within this response category.

Now turn your attention to the frequency distribution for helpsick (see Figure 2-4) and the accompanying bar chart (Figure 2-6). What is the mode? The most common response is "Govt should help." So "Govt should help" is the mode. Use the Cumulative Percent column to find the median. Is the first value, "Govt should help," the median? No, this value contains fewer than half the cases. Now go up one value to the category labeled "2" on the 5-point ordinal scale. Is this the median? Yes, it is. According to the Cumulative Percent column, more than 50 percent of the cases fall in or below this value. So "Govt should help" is the mode, but the median falls within response category 2, which lies a bit more toward the "People help selves" side of the variable.

Which variable, helppoor or helpsick, has higher variation? Notice that, unlike helppoor, helpsick has respondents' values that are more spread out, with sizable numbers of cases falling in the first response category ("Govt should help") and the middle category ("Agree with both"). Indeed, these two quite different responses are close rivals for the distinction of being the modal opinion on this issue. And, unlike helppoor, helpsick's mode and median are different, providing a useful field mark of higher variation. Thus helpsick has more variation—greater dispersion—than helppoor.

**Figure 2-6**    Bar Chart (ordinal variable with high dispersion)

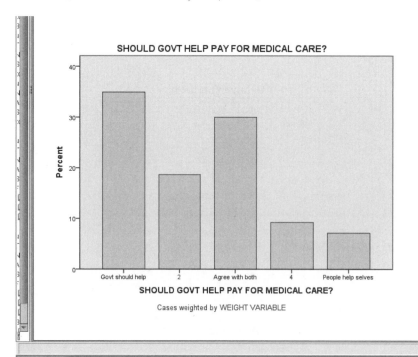

## DESCRIBING INTERVAL VARIABLES

Let's now turn to the descriptive analysis of interval-level variables. An interval-level variable represents the most precise level of measurement. Unlike nominal variables, whose values stand for categories, and ordinal variables, whose values can be ranked, the values of an interval variable *tell you the exact quantity of the characteristic being measured.* For example, age qualifies as an interval-level variable because its values impart each respondent's age in years.

Because interval variables have the most precision, they can be described more completely than can nominal or ordinal variables. For any interval-level variable, you can report its mode, median, and arithmetic average, or *mean.* In addition to these measures of central tendency, you can make more sophisticated judgments about variation. Specifically, you can determine if an interval-level distribution is *skewed.* What is skewness, and how do you know it when you see it?

Skewness refers to the symmetry of a distribution. If a distribution is not skewed, the cases tend to cluster symmetrically around the mean of the distribution, and they taper off evenly for values above and below the mean. If a distribution is skewed, by contrast, one tail of the distribution is longer and skinnier than the other tail. Distributions in which some cases occupy the higher values of an interval variable—distributions with a skinnier right-hand tail—are said to have a *positive skew.* By the same token, if the distribution has some cases at the extreme lower end—the distribution has a skinnier left-hand tail—then the distribution has a *negative skew.* Skewness has a predictable effect on the mean. A positive skew tends to "pull" the mean upward; a negative skew pulls it downward. However, skewness has less effect on the median. Because the median reports the middlemost value of a distribution, it is not tugged upward or downward by extreme values. *For badly skewed distributions, it is a good practice to use the median instead of the mean in describing central tendency.*

A step-by-step analysis of a GSS2008 variable, age, will clarify these points. Click Analyze → Descriptive Statistics → Frequencies. If helppoor and helpsick are still in the Variable(s) list, click them back into the left-hand list. Click age into the Variable(s) list. Click the Charts button. Make sure that Bar charts (under Chart Type) and Percentages (under Chart Values) are selected. Click Continue, which returns you to the main Frequencies window (Figure 2-7).

So far, this procedure is the same as in your analysis of zodiac, helppoor, and helpsick. When running a frequencies analysis of an interval-level variable, however, you need to do two additional things. One of these is a must-do. The other is a may-want-to-do. The must-do: Click the Statistics button in the Frequencies window, as shown in Figure 2-7. The Frequencies: Statistics window appears. In the Central Tendency panel, click the boxes next to Mean, Median, and Mode. In the Distribution panel, click Skewness. Click Continue, returning to the main Frequencies window. The may-want-to-do: *Un*check the box next to Display frequency tables, appearing at the foot of the left-hand list.[2] Click OK.

SPSS runs the analysis of age and dumps the requested statistics and bar chart into the Viewer (Figure 2-8). Most of the entries in the Statistics table are familiar to you: valid number of cases (N); number of missing cases; and mean, median, and mode. In addition, SPSS reports values for skewness and for something called standard error of skewness, which are as precise as they are mysterious. What do these numbers mean? When a distribution is perfectly symmetrical—no skew—it has skewness equal to 0. If the distribution has a skinnier right-hand tail—positive skew—then skewness will be a positive number. A skinnier left-hand tail, logically enough, returns a negative number for skewness. Just about all distributions will have some degree of skewness. How much is too much? That's where the standard error of skewness comes into play. Follow this simple rule: Divide skewness by its standard error. If the result has a magnitude (absolute value) of greater than 2, then the distribution is significantly skewed, and you should use the median as the best measure of central tendency. If you divide skewness by its standard error and get a number whose magnitude is 2 or less, the distribution is not significantly skewed, and you can use the mean as the best measure of central tendency.

The age variable plainly has a positive skew. Compare the mean age with the median age. Recall that a positive skew pulls the mean upward. Remember also that the mean is susceptible to skewness, and the median less so. Is the skewness more than twice its standard error? Yes, and then some. Obviously, a number of cases in the upper reaches of this variable have pulled the mean off the exact 50-50 center of the distribution. The bar chart confirms this suspicion. The skinnier right-hand tail is the telltale sign of a positive skew.[3]

**Figure 2-7** Requesting Statistics for an Interval Variable

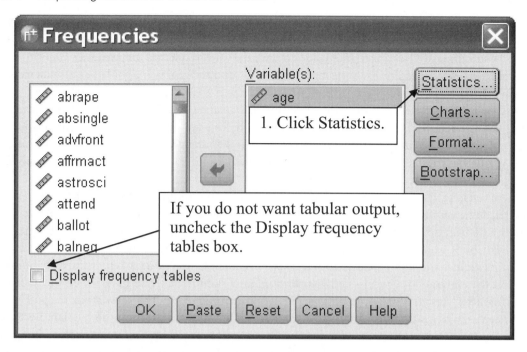

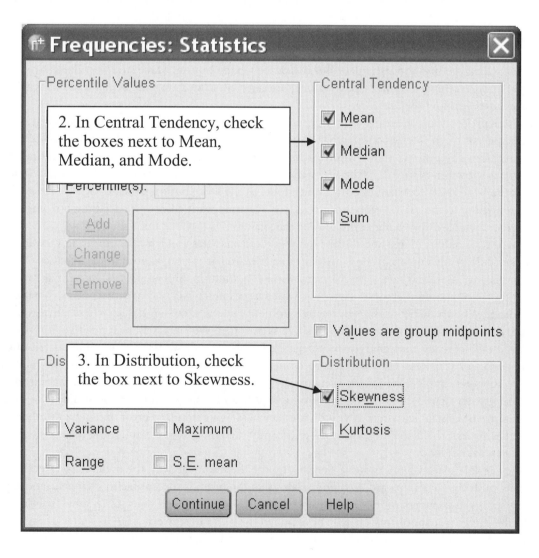

**Figure 2-8**   Statistics and Bar Chart (interval variable)

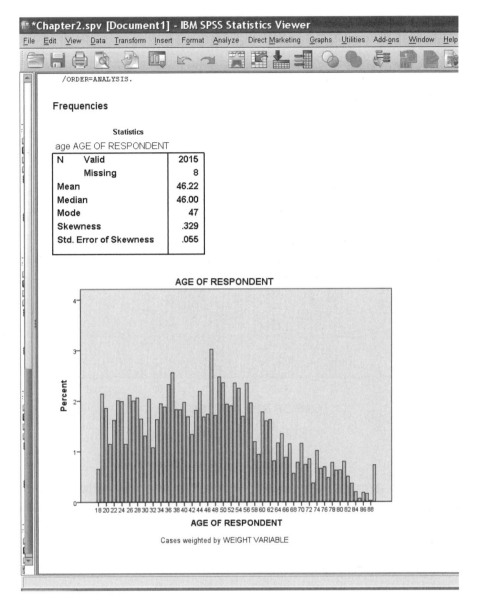

All the guided examples thus far have used bar charts for graphic support. For nominal and ordinal variables, a bar chart should always be your choice. For interval variables, however, you may want to ask SPSS to produce a histogram instead. What is the difference between a bar chart and a histogram? When is one preferred over the other? A bar chart displays each value of a variable and shows you the percentage (alternatively, the raw number) of cases that fall into each category. A histogram is similar, but instead of displaying each discrete value, it collapses categories into ranges (called bins), resulting in a compact display. Histograms are sometimes more readable and elegant than bar charts. Most of the time a histogram will work just as well as a bar chart in summarizing an interval-level variable. For interval variables with a large number of values, a histogram is the graphic of choice. (Remember: For nominal or ordinal variables, you always want a bar chart.)

So that you can become familiar with histograms, let's run the analysis of age once again—only this time we'll ask SPSS to produce a histogram instead of a bar chart. Click Analyze → Descriptive Statistics → Frequencies. Make sure age is still in the Variable(s) list. Click Statistics, and then uncheck all the boxes: Mean, Median, Mode, and Skewness. Click Continue. Click Charts, and then select the Histograms radio button in Chart Type. Click Continue. For this analysis, we do not need a frequency table. In the Frequencies window, uncheck the Display frequency tables box. (Refer to Figure 2-7.) Click OK.

**Figure 2-9**  Histogram (interval variable)

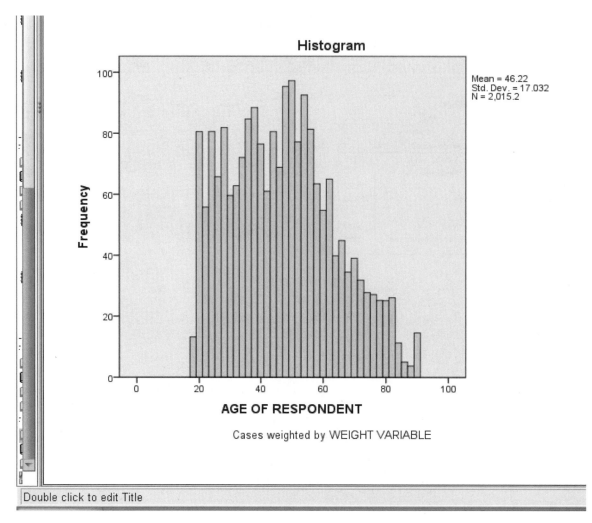

This is a bare-bones run. SPSS reports its obligatory count of valid and missing cases, plus a histogram for age (Figure 2-9). On the histogram's horizontal axis, notice the hash marks, which are spaced at 20-year intervals. SPSS has compressed the data so that each bar represents about 2 years of age rather than 1 year of age. Now scroll up the Viewer to the bar chart of age, which you produced in the preceding analysis. Notice that the histogram has smoothed out the nuance and choppiness of the bar chart, though it still captures the essential qualities of the age variable.

## OBTAINING CASE-LEVEL INFORMATION WITH CASE SUMMARIES

When you analyze a large survey dataset, as you have just done, you generally are not interested in how respondent x or respondent y answered a particular question. Rather, you want to know how the entire sample of respondents distributed themselves across the response categories of a variable. Sometimes, however, you gather data on particular cases because the cases are themselves inherently important. The States dataset (50 cases) and World dataset (191 cases) are good examples. With these datasets, you may want to push the descriptions beyond the relative anonymity of Frequencies analysis and find out where particular cases "are" on an interesting variable. Analyze → Reports → Case Summaries is ready-made for such elemental insights. Before beginning this guided example, close GSS2008 and open World.

Suppose you are interested in identifying the countries that have the most equitable distribution of wealth, as well as those in which wealth is more concentrated in the hands of a few. The World dataset contains gini08, the Gini coefficient for each country. The Gini coefficient measures wealth distribution on a scale that ranges from 0 (wealth is distributed equitably) to 100 (wealth is distributed inequitably). Exactly which countries are the most equitable? Which are the least equitable? Where does the United States fall on the list? Case Summaries can quickly answer questions like these. SPSS will sort the countries on the basis of a "grouping variable" (in this example, gini08) and then produce a report telling you which countries are in each group.

With the World dataset open, click Analyze → Reports → Case Summaries. The Summarize Cases window opens (Figure 2-10). You need to do three things here:

1.  Click the variable containing the cases' identities into the Variables window. In the World dataset, this variable is named country, an alphabetic descriptor of each country's name.
2.  Click the variable you are interested in analyzing, gini08, into the Grouping Variable(s) window.
3.  *Uncheck the Limit cases box.* This is important. If this box is left checked, SPSS will limit the analysis to the first 100 cases, which in many instances, such as the World dataset, will produce an incomplete analysis.

**Figure 2-10**   Summarize Cases Window (modified)

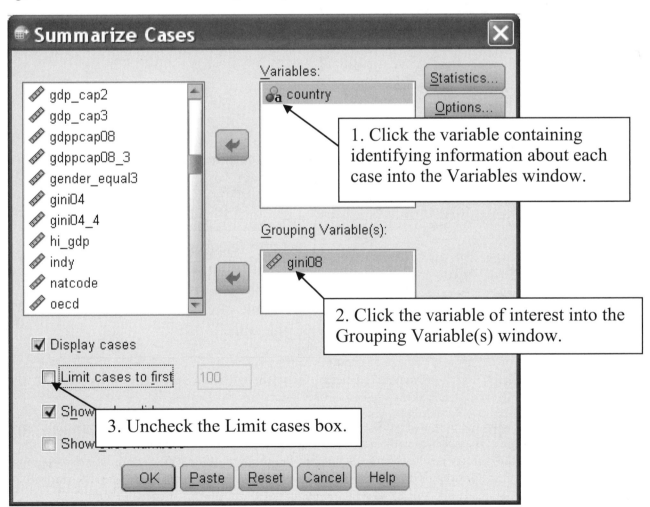

**Figure 2-11**  Case Summaries Output

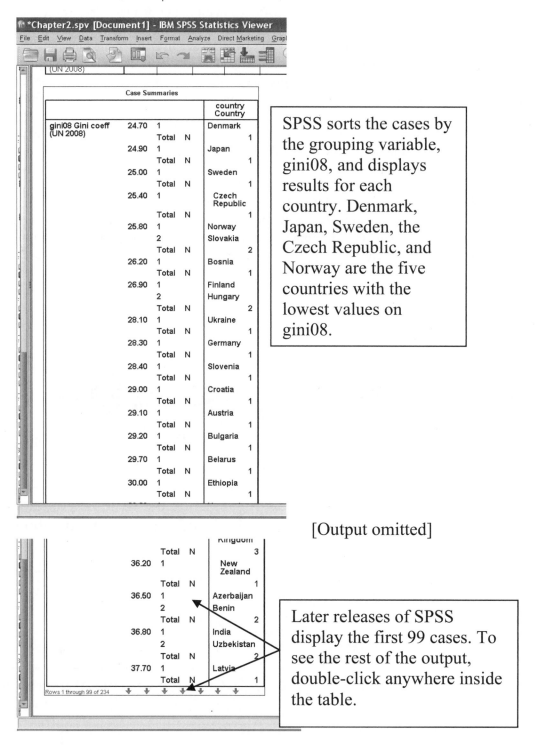

SPSS sorts the cases by the grouping variable, gini08, and displays results for each country. Denmark, Japan, Sweden, the Czech Republic, and Norway are the five countries with the lowest values on gini08.

[Output omitted]

Later releases of SPSS display the first 99 cases. To see the rest of the output, double-click anywhere inside the table.

Click OK and consider the output (Figure 2-11). SPSS sorts the cases on the grouping variable, gini08, and tells us which country is associated with each value of gini08. For example, Denmark, with a Gini coefficient of 24.70, is the country having the most equitable wealth distribution. Which countries rank highest on gini08? Scroll to the bottom of the tabular output. If you are using a recent release of SPSS, you will need to double-click inside the table to view the remainder of the output, as shown in Figure 2-11. Double-clicking the table reopens the output in a pivot table window (Figure 2-12). Once in the pivotal

**Figure 2-12**    Scrolling through Case Summaries Output

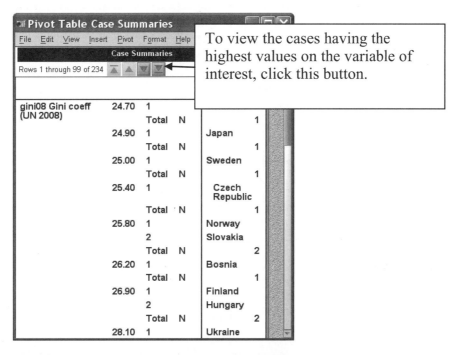

To view the cases having the highest values on the variable of interest, click this button.

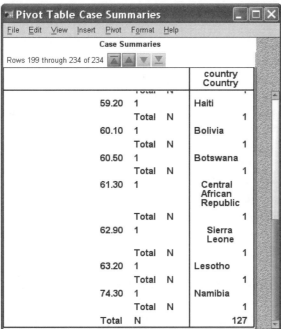

Namibia, Lesotho, Sierra Leone, Central African Republic, and Botswana have the highest values on gini08.

table window, you can scroll page-by-page or "fast forward" to the end of the output, as displayed in Figure 2-12. With a Gini coefficient of 74.30, Namibia is the country having the least equitable distribution of wealth.

## EXERCISES

1.  (Dataset: World. Variables: women09, country.) What percentage of members of the U.S. House of Representatives are women? In 2009 the number was 16.8 percent, according to the Inter-Parliamentary Union, an international organization of parliaments.[4] How does the United States compare to other democratic countries? Is 16.8 percent comparatively low, comparatively high, or average for a typical national legislature?

A. World contains women09, the percentage of women in the lower house of the legislature in each of 111 democracies. Perform a frequencies analysis on women09. In Statistics, obtain the mean, median, mode, and skewness. In the main Frequencies window, make sure that the Display frequency tables box is checked. (In Charts, Chart Type, make sure None is selected.) Fill in the table that follows:

| Statistics for women09 Percent women in lower house of parliament, 2009 | |
| --- | --- |
| Mean | ? |
| Median | ? |
| Mode | ? |
| Skewness | ? |
| Std. Error of Skewness | ? |

B. Based on the statistics you obtained, which is the better measure of central tendency to use, the mean or the median? (circle one)

Mean     Median

Briefly explain your answer. _____

_____

_____

C. Recall that 16.8 percent of U.S. House members are women. Which of the following statements *most accurately describes* the U.S. House? Using the appropriate measure of central tendency, the percentage of women in the U.S. House can be described as (circle one)

"below average."     "average."     "above average."

D. Suppose a women's advocacy organization vows to support female congressional candidates so that the U.S. House might someday "be ranked among the top one-fourth of democracies in the percentage of female members." According to the frequencies analysis, to meet this goal, women would need to constitute what percentage of the House? (circle one)

10.1 percent     22.1 percent     35.3 percent

E. Run Analyze → Reports → Case Summaries. Click country into the Variables box and women09 into the Grouping Variable(s) box. Make sure to uncheck the box next to Limit cases to first 100. Examine the output. (Reminder: If you are using a recent release of SPSS, you will need to double-click inside the table to view the remainder of the output. Refer to Figures 2-11 and 2-12.) Which five countries have the lowest percentages of women in their legislatures?

_____

_____

_____

_____

_____

Which five countries have the highest percentages of women in their legislatures?

_____

_____

_____

_____

2.  (Dataset: GSS2008. Variable: science_quiz.) The late Carl Sagan once lamented: "We live in a society exquisitely dependent on science and technology, in which hardly anyone knows anything about science and technology." This is a rather pessimistic assessment of the scientific acumen of ordinary Americans. Sagan seemed to be suggesting that the average level of scientific knowledge is quite low and that most people would fail even the simplest test of scientific facts.

    GSS2008 contains science_quiz, which was created from 10 questions testing respondents' knowledge of basic scientific facts. Values on science_quiz range from 0 (the respondent did not answer any of the questions correctly) to 10 (the respondent correctly answered all 10).[5]

    A.  Consider three possible scenarios for the distribution of science_quiz. All three scenarios assume that science_quiz has a median value of 6 on the 10-item scale. In scenario X, science_quiz has no skew. In scenario Y, science_quiz has a positive skew. In scenario Z, science_quiz has a negative skew. Below (and continued on page 32) are three graphic shells, labeled Scenario X, Scenario Y, and Scenario Z. For each scenario, sketch a curved line depicting what the distribution of science_quiz would look like if that scenario were accurate.

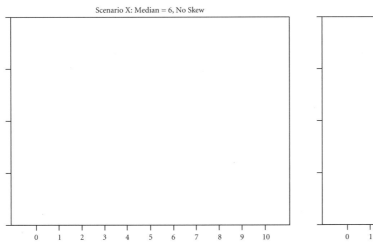

Scenario X: Median = 6, No Skew

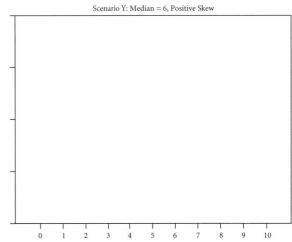

Scenario Y: Median = 6, Positive Skew

Scenario Z: Median = 6, Negative Skew

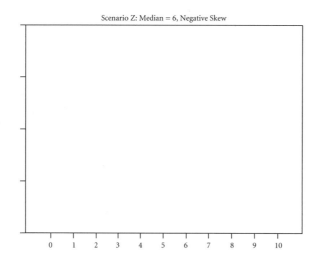

0   1   2   3   4   5   6   7   8   9   10

B. Run Frequencies to obtain a frequency distribution of science_quiz. In Charts, request a bar chart with percentages. Fill in the table that follows (in your output, be sure to focus on Valid Percent instead of Percent):

| science_quiz | Frequency | Valid Percent | Cumulative Percent |
|---|---|---|---|
| 0 | ? | ? | ? |
| 1 | ? | ? | ? |
| 2 | ? | ? | ? |
| 3 | ? | ? | ? |
| 4 | ? | ? | ? |
| 5 | ? | ? | ? |
| 6 | ? | ? | ? |
| 7 | ? | ? | ? |
| 8 | ? | ? | ? |
| 9 | ? | ? | ? |
| 10 | ? | ? | 100.0 |
| Total | ? | 100.0 | |

C. Examine the frequency distribution and the bar chart. Based on your analysis, which scenario does the distribution of science_quiz most closely approximate? (circle one)

Scenario X      Scenario Y      Scenario Z

Briefly explain your reasoning. _____

_____

_____

D. According to conventional academic standards, any science_quiz score of 5 or lower would be an F, a failing grade. A score of 6 would be a grade of D, a 7 would be a C, an 8 a B, and scores of 9 or 10 would be an A. Based on these standards, about what percentage of people got passing grades on science_quiz? (circle one)

About 30 percent      About 40 percent      About 50 percent      About 60 percent

What percentage got a C or better? (circle one)

About 30 percent    About 40 percent    About 50 percent    About 60 percent

E. Print the chart that you created for this exercise.

3. (Dataset: GSS2008. Variable: fem_role.) Two pundits are arguing about how the general public views the role of women in the home and in politics.

Pundit 1: "Our society has a sizable minority of traditionally minded individuals who think that the proper 'place' for women is taking care of the home and caring for children. This small but vocal group of traditionalists aside, the typical adult supports the idea that women belong outside the home and in the workplace."

Pundit 2: "Poppycock! It's just the opposite. The extremist 'women's liberation' crowd has distorted the overall picture. The typical view among most citizens is that women should be in the home, not in work."

A. GSS2008 contains fem_role, an interval-level variable that measures respondents' attitudes toward women in society and politics. Scores can range from 0 (women belong in the home) to 9 (women belong in work).

If pundit 1 is correct, fem_role will have (circle one)

a negative skew.    no skew.    a positive skew.

If pundit 2 is correct, fem_role will have (circle one)

a negative skew.    no skew.    a positive skew.

If pundit 1 is correct, fem_role's mean will be (circle one)

lower than its median.    the same as its median.    higher than its median.

If pundit 2 is correct, fem_role's mean will be (circle one)

lower than its median.    the same as its median.    higher than its median.

B. Perform a frequencies analysis of fem_role. Remember to obtain the mean, median, and mode, as well as skewness. Obtain a bar chart with percentages. Fill in the table that follows:

| Statistics for fem_role<br>Female role: home, work | |
| --- | --- |
| Mean | ? |
| Median | ? |
| Mode | ? |
| Skewness | ? |
| Std. Error of Skewness | ? |

C. Which of the following bar charts—X, Y, or Z—most closely resembles the bar chart you obtained? (circle one)

Bar chart X

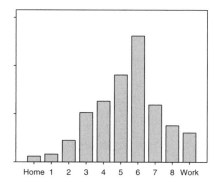

Bar chart Y

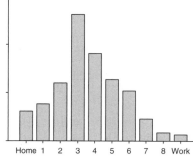

Bar chart Z

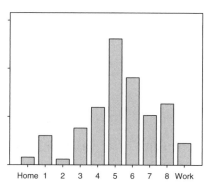

D. Based on your analysis, whose assessment is more accurate? (circle one)

Pundit 1's     Pundit 2's

Briefly explain your reasoning. _____

_____

_____

E. Print the output from this exercise.

4. (Dataset: GSS2008. Variable: attend.) The General Social Survey provides a rich array of variables that permit scholars to study religiosity among the adult population. GSS2008 contains attend, a 9-point ordinal scale that measures how often respondents attend religious services. Values can range from 0 ("Never attend") to 8 ("Attend more than once a week").

A. The shell of a bar chart is given below. The categories of attend appear along the horizontal axis. What would a bar chart of attend look like if this variable had maximum dispersion? Sketch inside the axes a bar chart that would depict maximum dispersion.

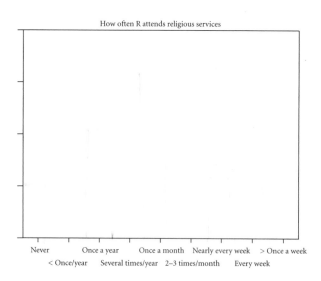

How often R attends religious services

Never        Once a year      Once a month   Nearly every week   > Once a week
   < Once/year    Several times/year   2–3 times/month    Every week

B. What would a bar chart of attend look like if this variable had no dispersion? Sketch inside the axes a bar chart that would depict no dispersion.

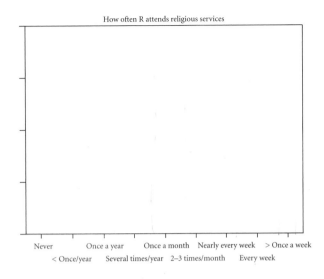

C. Obtain frequencies output and a bar chart for attend. In the main Frequencies window, make sure that the Display frequency tables box is checked. In Statistics, see that all the boxes are unchecked. In Charts, request a bar chart with percentages. Based on your examination of the frequency distribution,

the mode of attend is _____.

the median of attend is _____.

D. Based on your examination of the frequency distribution and bar chart, you would conclude that attend has (circle one)

low dispersion.     high dispersion.

E. Print the output from this exercise.

5. (Dataset: NES2008. Variables: enviro_emissions_r, illegal_citizenshp_r, medinsur_r.) We frequently describe public opinion by referring to how citizens distribute themselves on a political issue. *Consensus* is a situation in which just about everyone, 80–90 percent of the public, holds the same position, or very similar positions, on an issue. *Dissensus* is a situation in which opinion is spread out pretty evenly across all positions on an issue. *Polarization* refers to a configuration of opinion in which people are split between two extreme poles of an issue, with only a few individuals populating the more moderate, middle-of-the-road positions.

In this exercise you will decide whether consensus, dissensus, or polarization best describes public opinion, as measured by three NES2008 variables: whether the government should mandate reductions in power plant emissions (enviro_emissions_r), whether the government should make it possible for illegal immigrants to become U.S. citizens (illegal_citizenshp_r), and whether the government should provide medical insurance for everyone (medinsur_r). For each issue, respondents' opinions are measured on a 7-point scale. On the citizenship-process question, for example, respondents could choose any position between 1 ("favor a great deal") and 7 ("oppose a great deal").

A. Open NES2008. Perform a frequencies analysis on enviro_emissions_r, illegal_citizenshp_r, and medinsur_r. Obtain bar charts with percentages. Refer to the Valid Percent column of the frequency distributions. In the table that follows, write the appropriate valid percent next to each question mark (?):

| R's self-placement on 7-point scale | Lower emission standards (enviro_emissions_r) | Citizenship process for illegals (illegal_citizenshp_r) | Government vs. private medical insurance (medinsur_r) |
|---|---|---|---|
| | Valid percent | Valid percent | Valid percent |
| 1 | ? | ? | ? |
| 2 | ? | ? | ? |
| 3 | ? | ? | ? |
| 4 | ? | ? | ? |
| 5 | ? | ? | ? |
| 6 | ? | ? | ? |
| 7 | ? | ? | ? |
| Total | 100.0 | 100.0 | 100.0 |

B.  Examine the percentages in part A. Examine the bar charts that you created. Of the three issues, which one *most closely approximates* consensus (circle one):

Lower emission standards (enviro_emissions_r)

Citizenship process for illegals (illegal_citizenshp_r)

Government vs. private medical insurance (medinsur_r)

Briefly explain your reasoning. _____

_____

_____

_____

C.  Of the three issues, which one *most closely approximates* dissensus (circle one):

Lower emission standards (enviro_emissions_r)

Citizenship process for illegals (illegal_citizenshp_r)

Government vs. private medical insurance (medinsur_r)

Briefly explain your reasoning. _____

_____

_____

_____

D.  Of the three issues, which one *most closely approximates* polarization (circle one):

Lower emission standards (enviro_emissions_r)

Citizenship process for illegals (illegal_citizenshp_r)

Government vs. private medical insurance (medinsur_r)

Briefly explain your reasoning. _____

_____

_____

_____

E. Print the bar chart for the variable you chose in part D.

6. (Dataset: NES2008. Variables: cong_approval, cong_incumb_approval.) Pedantic pontificator believes he has discovered how voters evaluate the performance of House incumbents: "I call it my 'guilt by association' theory. When voters disapprove of the way Congress has been handling its job, they transfer that negative evaluation to their House incumbent. My theory is eminently plausible and surely correct. The distribution of opinions about House incumbents will be very similar to the distribution of opinion about Congress as a whole."

NES2008 contains cong_approval, which gauges respondent approval or disapproval of "the way the U.S. Congress has been handling its job." The dataset also has cong_incumb_approval, which measures approval or disapproval of the way each respondent's House incumbent "has been handling his or her job."

A. To test pedantic pontificator's theory, perform a frequencies analysis of cong_approval and cong_incumb_approval. Obtain bar charts with percentages. Refer to the Valid Percent column of the frequency distributions. In the table that follows, write the appropriate valid percent next to each question mark (?):

|  | Approve/disapprove Congress handling job | Approve/disapprove House incumbent |
|---|---|---|
|  | Valid percent | Valid percent |
| 1. Approve strongly | ? | ? |
| 2. Approve not strongly | ? | ? |
| 4. Disapprove not strongly | ? | ? |
| 5. Disapprove strongly | ? | ? |
| Total | 100.0 | 100.0 |

B. Consider the tabular and graphic evidence. Does pedantic pontificator's theory appear to be correct or incorrect? (Circle one.)

Correct        Incorrect

Explain your reasoning. _____

_____

_____

_____

7. (Dataset: States. Variables: defexpen.) Here is the conventional political wisdom: Well-positioned members of Congress from a handful of states are successful in getting the federal government to spend revenue in their states—defense-related expenditures, for example. The typical state, by contrast, receives far fewer defense budget dollars.

A. Suppose you had a variable that measured the amount of defense-related expenditures in each state. The conventional wisdom says that, when you look at how all 50 states are distributed on this variable, a few states would have a high amount of defense spending. Most states, however, would have lower values on this variable.

   If the conventional wisdom is correct, the distribution of defense-related expenditures will have (circle one)

   a negative skew.     no skew.     a positive skew.

   If the conventional wisdom is correct, the mean of defense-related expenditures will be (circle one)

   lower than its median.     the same as its median.     higher than its median.

B. States contains the variable defexpen, defense expenditures per capita for each of the 50 states. Perform a frequencies analysis of defexpen. In Statistics, obtain the mean and median, as well as skewness. (You do not need to obtain the mode for this exercise.) In the main Frequencies window, uncheck the Display frequency tables box. In Charts, request a histogram. Examine the results in the Viewer. Record the mean, median, skewness, and standard error of skewness in the table that follows:

| Statistics for defexpen Federal defense expenditures per capita | |
| --- | --- |
| Mean | ? |
| Median | ? |
| Skewness | ? |
| Std. Error of Skewness | ? |

C. Which is the better measure of central tendency? (circle one)

   Mean     Median

Briefly explain your answer. _____

_____

_____

D. Based on your analysis, would you say that the conventional wisdom is accurate or inaccurate? (check one)

   ❑  The conventional wisdom is accurate.

   ❑  The conventional wisdom is inaccurate.

E. Print the histogram you produced in part B.

8. (Dataset: States. Variables: blkpct08, hispanic08, state.) Two demographers are arguing over how best to describe the racial and ethnic composition of the "typical" state.

   Demographer 1: "The typical state is 10.5 percent black and 9.9 percent Hispanic."

   Demographer 2: "The typical state is 7.4 percent black and 6.8 percent Hispanic."

   A. Run frequencies for blkpct08 (the percentage of each state's population that is African American) and hispanic08 (the percentage of each state's population that is Hispanic). In Statistics, obtain the mean and median, as well as skewness. (You do not need to obtain the mode for this exercise.) In Charts, Chart Type, select None. In the main Frequencies window, uncheck the Display frequency tables box. Record the appropriate statistics for each variable in the table that follows:

   | | blkpct08 Percent black (2008) | hispanic08 Percent Hispanic (2008) |
   |---|---|---|
   | Mean | ? | ? |
   | Median | ? | ? |
   | Skewness | ? | ? |
   | Std. Error of Skewness | ? | ? |

   B. Based on your analysis, which demographer is more accurate? (circle one)

      Demographer 1      Demographer 2

      Briefly explain your reasoning. _____

      _____

      _____

      _____

   C. Run Case Summaries. Click state into Variables and click hispanic08 into Grouping Variable(s).

      Which five states have the lowest percentages of Hispanics?

      _____

      _____

      _____

      _____

      _____

      Which five states have the highest percentages of Hispanics?

      _____

      _____

_____

_____

_____

That concludes the exercises for this chapter. Before exiting SPSS, be sure to save your output file.

## NOTES

1.  In this chapter we use the terms *dispersion, variation,* and *spread* interchangeably.
2.  For interval-level variables that have a large number of categories, as does age, a frequency distribution can run to several output pages and is not very informative. Unchecking the Display frequency tables box suppresses the frequency distribution. A general guide: If the interval-level variable you are analyzing has 15 or fewer categories, go ahead and obtain the frequency distribution. If it has more than 15 categories, suppress the frequency distribution.
3.  For demographic variables that are skewed, median values rather than means are often used to give a clearer picture of central tendency. One hears or reads reports, for example, of median family income or the median price of homes in an area.
4.  See Inter-Parliamentary Union web site (www.ipu.org/english/home.htm).
5.  Science_quiz was created by summing the number of correct responses to the following questions (all are in true-false format, except for earthsun): The center of the Earth is very hot (General Social Survey variable, hotcore); It is the father's gene that decides whether the baby is a boy or a girl (boyorgrl); Electrons are smaller than atoms (electron); The universe began with a huge explosion (bigbang); The continents on which we live have been moving their locations for millions of years and will continue to move in the future (condrift); Human beings, as we know them today, developed from earlier species of animals (evolved); Does the Earth go around the Sun, or does the Sun go around the Earth? (earthsun); All radioactivity is man-made (radioact); Lasers work by focusing sound waves (lasers); Antibiotics kill viruses as well as bacteria (viruses).

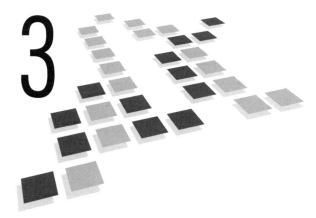

# 3

# Transforming Variables

| Procedures Covered | |
|---|---|
| | Transform → Recode into Different Variables |
| | Transform → Visual Binning |
| | Transform → Compute Variable |
| | Transform → Recode into Same Variables |

**P**olitical researchers sometimes must modify the variables they want to analyze. Generally speaking, such *variable transformations* become necessary or desirable in two common situations. Often a researcher wants to collapse a variable, combining its values or codes into a smaller number of useful categories. The researcher can do so through the Recode transformation feature or the Visual Binning procedure. In other situations a dataset may contain several variables that provide similar measures of the same concept. In these instances the researcher may want to combine the codes of different variables, creating a new and more precise measure. The Compute transformation feature is designed for this task.

In this chapter you will learn how to use the Recode, Visual Binning, and Compute commands. The chapter contains four guided examples, all of which use NES2008. The variables you modify or create in this chapter (and in this chapter's exercises) will become permanent variables in the datasets. After you complete each guided example, be sure to save the dataset.

## USING RECODE

With Recode, you can manipulate any variable at any level of measurement—nominal, ordinal, or interval. But you should exercise vigilance and care. Follow these three guidelines:

1. Before using Recode, obtain a frequency distribution of the variable you intend to manipulate.
2. After using Recode, check your work.
3. Properly label the new variable and its values.

Open NES2008, and let's work through the first example.

### Recoding a Categorical Variable
NES2008 contains marital, a demographic variable that measures marital status in six categories:

| Marital Status | Code |
|---|---|
| Married | 1 |
| Widowed | 2 |
| Divorced | 3 |
| Separated | 4 |
| Never married | 5 |
| Partnered, not married | 6 |

A Frequencies analysis of marital produced the following result:

| | | Frequency | Percent | Valid Percent | Cumulative Percent |
|---|---|---|---|---|---|
| Valid | 1 1. Married | 1159 | 49.9 | 50.1 | 50.1 |
| | 2 2. Divorced | 299 | 12.9 | 12.9 | 63.0 |
| | 3 3. Separated | 68 | 2.9 | 2.9 | 65.9 |
| | 4 4. Widowed | 181 | 7.8 | 7.8 | 73.8 |
| | 5 5. Never married | 571 | 24.6 | 24.7 | 98.4 |
| | 6 6. Partnered, not married {VOL} | 36 | 1.6 | 1.6 | 100.0 |
| | Total | 2314 | 99.6 | 100.0 | |
| Missing | System | 9 | .4 | | |
| Total | | 2323 | 100.0 | | |

Now, think about research questions for which you might want to make fine distinctions among people—comparing, for example, the 299 divorced individuals with the 571 individuals who never married. Much of the time, however, you might be after a simpler comparison—the 50.1 percent of the valid cases who are married (code 1) and the remaining 49.9 percent of the sample who are unmarried (codes 2 through 6). How would you collapse the codes of marital into two categories and still preserve the potentially useful values of the original variable?

On the main menu bar, click Transform and consider the array of choices (Figure 3-1). Notice that SPSS presents two recoding options: Recode into Same Variables and Recode into Different Variables. When you recode a variable into the same variable, SPSS replaces the original codes with the new codes. The original information is lost. When you recode a variable into a different variable, SPSS uses the original codes to create a new variable. The original variable is retained. In some situations (discussed later) you will want to pick Recode into Same Variables. Most of the time, however, you should use the second option, Recode into Different Variables.

Click Recode into Different Variables. The Recode into Different Variables window opens (Figure 3-2). Scroll down the left-hand variable list and find marital. Click marital into the Input Variable → Output Variable box. SPSS puts marital into the box, with this designation: "marital →?" This is SPSS-speak for "What do you want to name the new variable you are creating from marital?" Click in the Name box and type "married" (without quotation marks). Let's take this opportunity to give the new variable, married, a descriptive label. Click in the Label box and type "Is R married?" Click the Change button. The Recode into Different Variables window should now look like Figure 3-3.

**Figure 3-1**    Transform Drop-down Menu

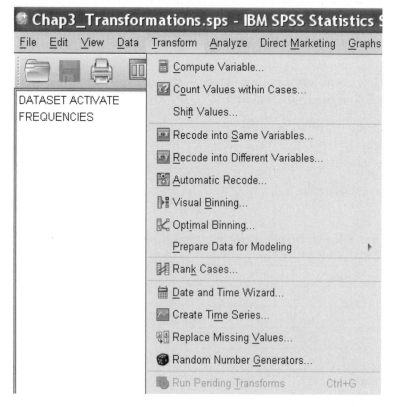

**Figure 3-2**  Recode into Different Variables Window

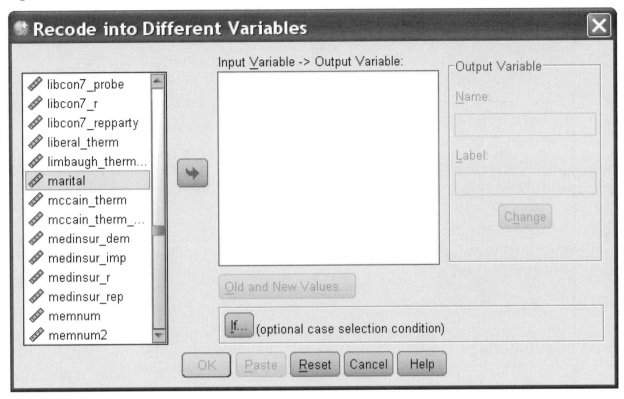

**Figure 3-3**  Recoding a Categorical Variable

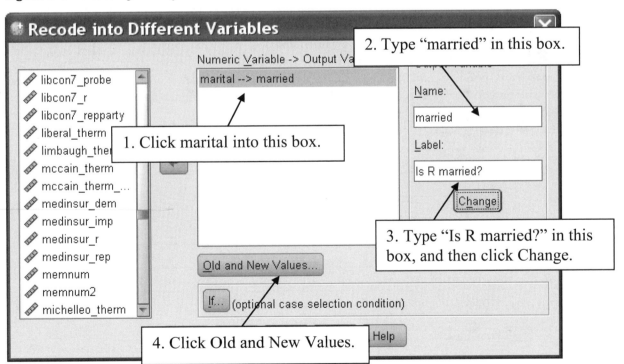

**Figure 3-4**   Recode into Different Variables: Old and New Values Window (default)

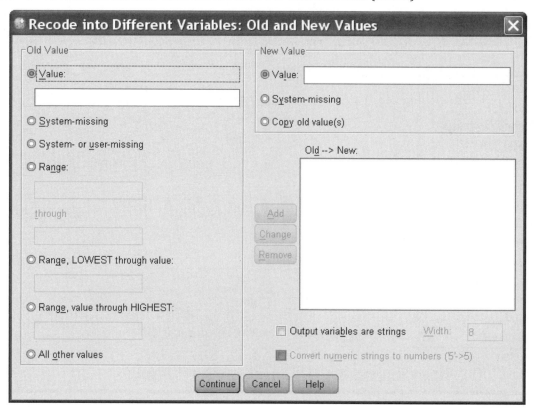

Now let's do the recoding. Click Old and New Values. The Recode into Different Variables: Old and New Values window pops up (Figure 3-4). There are two main panels. In the left-hand, Old Value panel, we will tell SPSS how to combine the original codes for marital. In the right-hand, New Value panel, we will assign codes for the new variable, which we have named married.

How do we want things to end up? Let's say that we want the new variable, married, to have two codes: code 1 for married respondents and code 0 for unmarried respondents. Plus we need to make sure that any respondents who have missing values on marital also have missing values on married. So we need to instruct SPSS to follow this recoding protocol:

| Marital status | Old value (marital) | New value (married) |
| --- | --- | --- |
| Married | 1 | 1 |
| Widowed | 2 | 0 |
| Divorced | 3 | 0 |
| Separated | 4 | 0 |
| Never married | 5 | 0 |
| Partnered, not married | 6 | 0 |
| | Missing | Missing |

Make sure that the top radio button in the Old Value panel is selected (the default), click the cursor in the box under "Value," and type "1." Move the cursor directly across to the right-hand, New Value panel, make sure the top radio button is selected (again, the default setting), and type "1" in the Value box. Click the Add button. In the Old → New box, SPSS records your instruction with "1 → 1," meaning "All respondents coded 1 on marital will be coded 1 on married." Now return to the left-hand, Old Value panel and select the uppermost Range button, the one simply labeled "Range." The two boxes beneath "Range" are activated. In the upper Range box, type "2." In the lower Range box, type "6." Move the cursor to the New Value panel and type "0" in the Value box. Click Add. SPSS responds, "2 thru 6 → 0," letting you know that all respondents coded 2, 3, 4, 5, or 6 on marital will be coded 0 on married. One last loose end: In the Old Value panel, click the radio button next to "System- or user-missing." In the New Value panel, click the radio button next to "System-missing." Click Add. SPSS records your instruction as "MISSING → SYSMIS," meaning that any respondents having missing values on marital will be assigned missing values on married. The Recode into Different Variables: Old and New Values window should now look like Figure 3-5. Click Continue, returning to the main Recode into Different Variables window. Click OK. SPSS runs the recode.

**Figure 3-5**   Collapsing a Categorical Variable into Fewer Categories

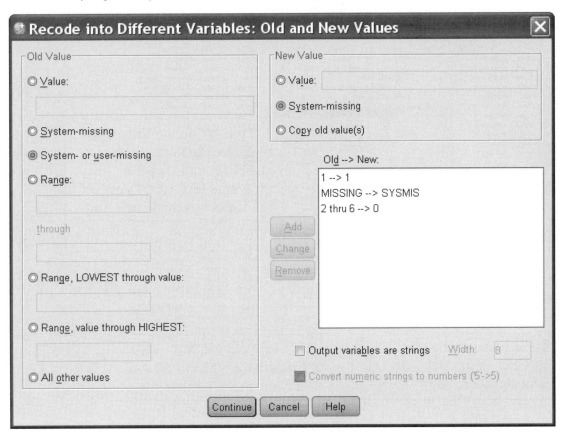

Did the recode work correctly? This is where the check-your-work rule takes effect. Run Frequencies on married to ensure that you did things right:

**married Is R married?**

| | | Frequency | Percent | Valid Percent | Cumulative Percent |
|---|---|---|---|---|---|
| Valid | .00 | 1155 | 49.7 | 49.9 | 49.9 |
| | 1.00 | 1159 | 49.9 | 50.1 | 100.0 |
| | Total | 2314 | 99.6 | 100.0 | |
| Missing | System | 9 | .4 | | |
| Total | | 2323 | 100.0 | | |

The frequency table displays the label for the newly minted variable. More important, the valid percentages check out: 49.9 percent coded 0 and 50.1 percent coded 1. The recode worked as planned. However, we still need to make sure that the numeric codes are labeled properly: "No" for numeric code 0 and "Yes" for numeric code 1. To complete the recoding process, one more step is required.

In the Data Editor, make sure that the Variable View tab is clicked. Scroll down to the bottom of the Data Editor, where you will find married (Figure 3-6). (SPSS always puts newly created variables on the bottom row of the Variable View.) While we are doing the essential work of assigning value labels, we will also tidy up the formatting of the variable we just created. Click in the Decimals cell, which shows "2," and change this value to "0." Next, click in the Values cell (which currently says "None"), and then click on the button that appears. The Value Labels window presents itself. In the box next to "Value," type "0." In the box next to "Label," type "No." Click Add. Repeat the process for code 1, typing "1" in the Value box and "Yes" in the Label box. Click Add. Click OK. Looks good.

**Figure 3-6** Assigning Value Labels to a Recoded Variable

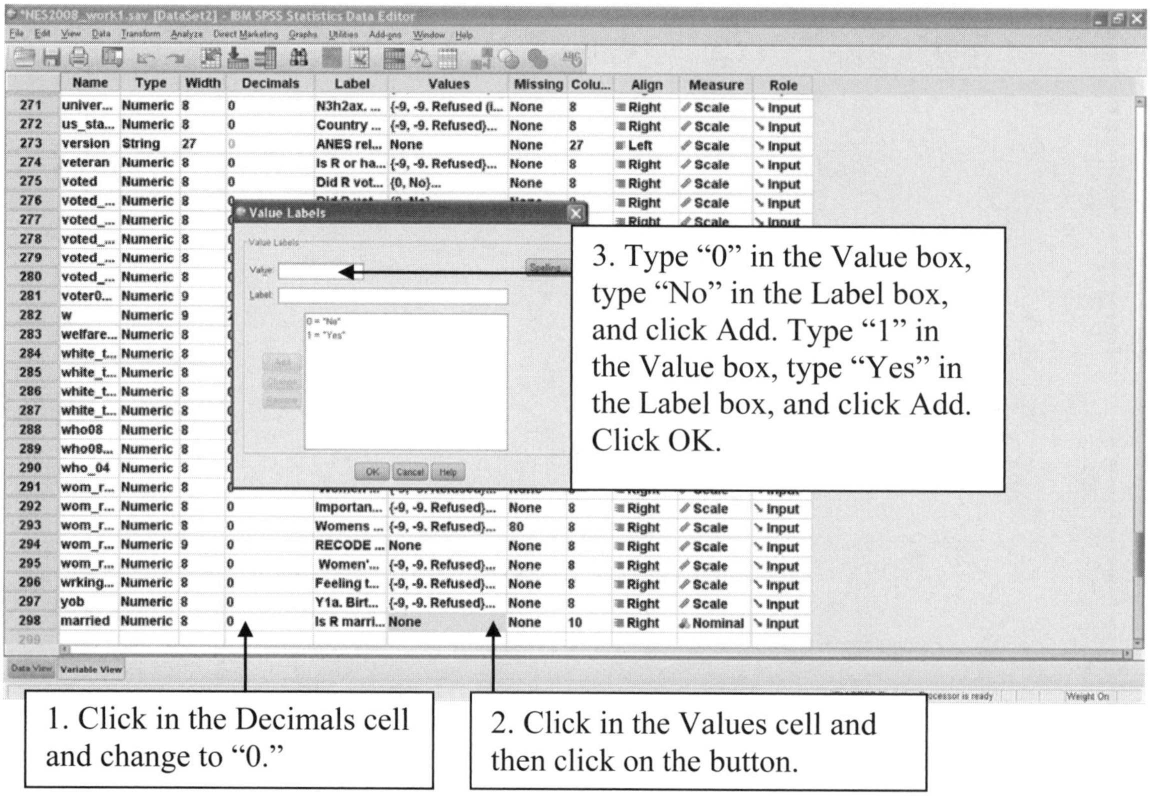

You have just invested your time in recoding an original variable into a new variable and, in the process, made NES2008 better and more usable. Before going on to the next example, make sure you save the dataset.

### Recoding an Interval-level Variable

Collapsing the values of a categorical variable, as you have just done, is perhaps the most common use of the Recode transformation feature. The original variable may be nominal level, such as marital. Or it may be ordinal level. For example, it might make sense to collapse four response categories such as "strongly agree," "agree," "disagree," and "strongly disagree" into two, "agree" and "disagree." At other times the original variable is interval level, such as age or income. In such cases you could use Recode to create a new variable having, say, three or four ordinal-level categories. Let's pursue this route.

NES2008 contains the variable yob, which records the year of birth for each respondent. Our goal here is to collapse yob into three theoretically useful categories: respondents born before 1950, those born between 1950 and 1965, and those born after 1965. How do we proceed? First, of course, we need a frequency distribution for yob. Click Analyze → Descriptive statistics → Frequencies, click yob into the Variable(s) list, and run the analysis. (In the main Frequencies window, make sure the Display frequency tables box is checked.) Consider the frequency output (Figure 3-7).

**Figure 3-7**  Frequency Distribution of an Interval-level Variable

| yob Y1a. Birthdate Year | | Frequency | Percent | Valid Percent | Cumulative Percent |
|---|---|---|---|---|---|
| Valid | 1915 | 2 | .1 | .1 | .1 |
| | 1917 | 7 | .3 | .3 | .4 |
| | 1918 | 5 | .2 | .2 | .6 |
| | 1919 | 3 | .1 | .1 | .7 |
| | 1920 | 2 | .1 | .1 | .8 |
| | 1921 | 7 | .3 | .3 | 1.1 |
| | 1922 | 7 | .3 | .3 | 1.4 |
| | 1923 | 7 | .3 | .3 | 1.7 |
| | 1924 | 12 | .5 | .5 | 2.3 |
| | 1925 | 7 | .3 | .3 | 2.6 |
| | 1926 | 9 | .4 | .4 | 3.0 |
| | 1927 | 14 | .6 | .6 | 3.6 |
| | 1928 | 19 | .8 | .8 | 4.4 |
| | 1929 | 13 | .5 | .6 | 5.0 |
| | 1930 | 26 | 1.1 | 1.2 | 6.2 |
| | 1931 | 13 | .6 | .6 | 6.7 |
| | 1932 | 16 | .7 | .7 | 7.4 |
| | 1933 | 13 | .5 | .6 | 8.0 |
| | 1934 | 15 | .6 | .6 | 8.6 |
| | 1935 | 21 | .9 | .9 | 9.5 |
| | 1936 | 22 | .9 | .9 | 10.5 |
| | 1937 | 21 | .9 | .9 | 11.4 |
| | 1938 | 24 | 1.0 | 1.1 | 12.4 |
| | 1939 | 20 | .9 | .9 | 13.3 |
| | 1940 | 18 | .8 | .8 | 14.1 |
| | 1941 | 25 | 1.1 | 1.1 | 15.2 |
| | 1942 | 32 | 1.4 | 1.4 | 16.6 |
| | 1943 | 25 | 1.1 | 1.1 | 17.7 |
| | 1944 | 33 | 1.4 | 1.5 | 19.1 |
| | 1945 | 25 | 1.1 | 1.1 | 20.2 |
| | 1946 | 35 | 1.5 | 1.5 | 21.8 |
| | 1947 | 28 | 1.2 | 1.2 | 23.0 |
| | 1948 | 33 | 1.4 | 1.5 | 24.5 |
| | 1949 | 42 | 1.8 | 1.8 | 26.3 |
| | 1950 | 40 | 1.7 | 1.7 | 28.0 |
| | 1951 | 50 | 2.2 | 2.2 | 30.3 |
| | 1952 | 28 | 1.2 | 1.2 | 31.5 |
| | 1953 | 36 | 1.6 | 1.6 | 33.1 |
| | 1954 | 43 | 1.9 | 1.9 | 34.9 |
| | 1955 | 50 | 2.2 | 2.2 | 37.1 |
| | 1956 | 38 | 1.6 | 1.7 | 38.8 |
| | 1957 | 43 | 1.9 | 1.9 | 40.7 |
| | 1958 | 49 | 2.1 | 2.1 | 42.8 |
| | 1959 | 46 | 2.0 | 2.0 | 44.9 |
| | 1960 | 55 | 2.4 | 2.4 | 47.3 |
| | 1961 | 57 | 2.4 | 2.5 | 49.7 |
| | 1962 | 51 | 2.2 | 2.2 | 52.0 |
| | 1963 | 48 | 2.1 | 2.1 | 54.1 |
| | 1964 | 35 | 1.5 | 1.5 | 55.6 |
| | 1965 | 31 | 1.3 | 1.4 | 56.9 |
| | 1966 | 42 | 1.8 | 1.9 | 58.8 |
| | 1967 | 40 | 1.7 | 1.7 | 60.5 |
| | 1968 | 40 | 1.7 | 1.7 | 62.3 |
| | 1969 | 34 | 1.5 | 1.5 | 63.8 |
| | 1970 | 34 | 1.5 | 1.5 | 65.3 |
| | 1971 | 43 | 1.9 | 1.9 | 67.1 |
| | 1972 | 38 | 1.6 | 1.7 | 68.8 |
| | 1973 | 33 | 1.4 | 1.5 | 70.3 |
| | 1974 | 37 | 1.6 | 1.6 | 71.9 |
| | 1975 | 30 | 1.3 | 1.3 | 73.2 |
| | 1976 | 42 | 1.8 | 1.9 | 75.1 |
| | 1977 | 47 | 2.0 | 2.0 | 77.1 |
| | 1978 | 40 | 1.7 | 1.8 | 78.9 |
| | 1979 | 35 | 1.5 | 1.5 | 80.4 |
| | 1980 | 53 | 2.3 | 2.3 | 82.7 |
| | 1981 | 46 | 2.0 | 2.0 | 84.8 |
| | 1982 | 56 | 2.4 | 2.4 | 87.2 |
| | 1983 | 36 | 1.6 | 1.6 | 88.8 |
| | 1984 | 48 | 2.1 | 2.1 | 90.9 |
| | 1985 | 49 | 2.1 | 2.2 | 93.0 |
| | 1986 | 30 | 1.3 | 1.3 | 94.4 |
| | 1987 | 34 | 1.4 | 1.5 | 95.8 |
| | 1988 | 39 | 1.7 | 1.7 | 97.5 |
| | 1989 | 23 | 1.0 | 1.0 | 98.5 |
| | 1990 | 33 | 1.4 | 1.5 | 100.0 |
| | Total | 2280 | 98.1 | 100.0 | |
| Missing | System | 43 | 1.9 | | |
| Total | | 2323 | 100.0 | | |

Let's use this distribution to get an idea of what the recoded variable should look like. To do this, focus on the Cumulative Percent column. What percentage of the sample falls into the oldest category—born before 1950? That's easy: 26.3 percent. What percentage of the sample falls *in or below* the middle category, people born between 1950 and 1965? Well, 56.9 percent of the sample was born in 1965 or earlier, so 56.9 percent of the sample should fall into the first two categories of the recoded variable. These two numbers, 26.3 percent and 56.9 percent, will help us to verify that our recode was performed properly.

Now do the recode. Click Transform → Recode into Different Variables. Click Reset to clear the panels. Click yob into the Input Variable → Output Variable box. Type "yob3" in the Name box. Type "Three generations" in the Label box and click Change (Figure 3-8). Click Old and New Values. Let's first create the oldest category for yob3. In the Old Value panel, select the radio button next to "Range, LOWEST through value"; doing so activates the box. Type "1949" in the box. In the New Value panel, type "1" in the Value box and click Add. SPSS translates the instruction as "Lowest thru 1949 → 1," lumping all respondents between the lowest value of yob (1914) and a value of 1949 on yob into code 1 of the new variable, yob3. In the Old Value panel, select the "Range" button and type "1950" in the upper box and "1965" in the lower box. Type "2" in the Value box in the New Value panel and click Add. That's the middle "baby boomer" age group, now coded 2 on yob3. In the Old Value panel, select the radio button next to "Range, value through HIGHEST" and type "1966" in the box. Type "3" in the Value box in the New Value panel and click Add. That puts the youngest generation into code 3 on yob3. Complete the recode by clicking the System- or user-missing button in the Old Value panel and the System-missing button in the New Value panel. Click Add. The Recode into Different Variables: Old and New Values window should now look like Figure 3-9.

**Figure 3-8**  Recoding an Interval-level Variable

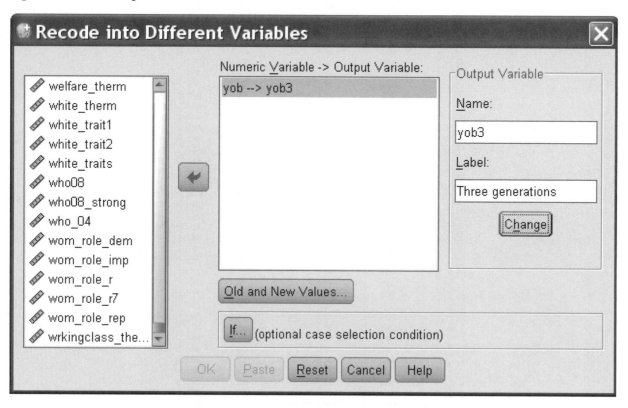

**Figure 3-9** Collapsing an Interval-level Variable into Categories

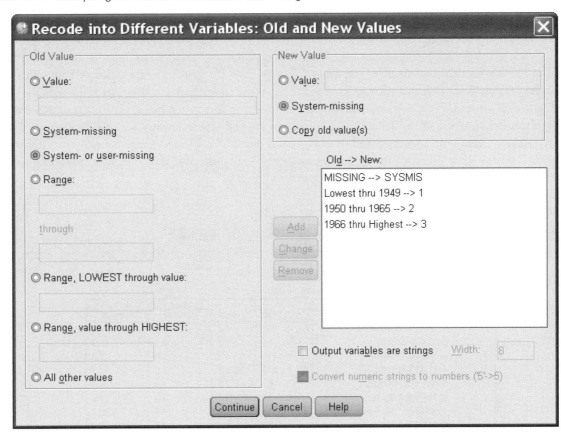

Click Continue. Click OK. Check your work by running Frequencies on yob3 and examining the output:

**yob3 Three generations**

|         |        | Frequency | Percent | Valid Percent | Cumulative Percent |
|---------|--------|-----------|---------|---------------|--------------------|
| Valid   | 1.00   | 600       | 25.8    | 26.3          | 26.3               |
|         | 2.00   | 698       | 30.1    | 30.6          | 56.9               |
|         | 3.00   | 982       | 42.3    | 43.1          | 100.0              |
|         | Total  | 2280      | 98.1    | 100.0         |                    |
| Missing | System | 43        | 1.9     |               |                    |
| Total   |        | 2323      | 100.0   |               |                    |

The cumulative percent markers, 26.3 percent and 56.9 percent, are just where they are supposed to be. Yob3 checks out. Before proceeding, scroll to the bottom of the Variable View in the Data Editor and make two changes to yob3. First, change Decimals to 0. Second, click in the Values cell and label yob3's values as follows:

| Value | Value label |
|-------|-------------|
| 1     | Before 1950 |
| 2     | 1950-1965   |
| 3     | After 1965  |

## USING VISUAL BINNING

For nominal and ordinal variables, Recode is fast and easy to use. However, for interval variables, as we have just seen, it can be a bit more cumbersome. For collapsing interval variables, SPSS's more obscure (and underappreciated) Visual Binning procedure provides an attractive alternative to Recode. Visual Binning is as good as Recode for creating variables, such as yob3, for which the researcher has selected theoretically meaningful cutpoints for defining the categories of the new variable. And it is superior to Recode for situations in which you want to quickly collapse an interval-level variable into a handful of values, each containing roughly equal numbers of cases. Let's work through a guided example using NES2008's income_r, a measure of each respondent's income. You will use Visual Binning to create a three-category ordinal, income_r3. Income_r3 will break income_r into thirds: the lowest-income third ("Low"), the middle-income third ("Middle"), and the highest-income third ("High").

### Collapsing an Interval-level Variable with Visual Binning

An obligatory Frequencies run on income_r produces the following output:

income_r R income

| | | Frequency | Percent | Valid Percent | Cumulative Percent |
|---|---|---|---|---|---|
| Valid | 1. None or less than $2,999 | 194 | 8.4 | 8.9 | 8.9 |
| | 2. $3,000 -$4,999 | 66 | 2.8 | 3.0 | 11.9 |
| | 3. $5,000 -$7,499 | 113 | 4.9 | 5.2 | 17.1 |
| | 4. $7,500 -$9,999 | 110 | 4.7 | 5.0 | 22.1 |
| | 5. $10,000 -$10,999 | 52 | 2.3 | 2.4 | 24.5 |
| | 6. $11,000-$12,499 | 71 | 3.1 | 3.3 | 27.8 |
| | 7. $12,500-$14,999 | 111 | 4.8 | 5.1 | 32.8 |
| | 8. $15,000-$16,999 | 76 | 3.3 | 3.5 | 36.3 |
| | 9. $17,000-$19,999 | 87 | 3.8 | 4.0 | 40.3 |
| | 10. $20,000-$21,999 | 75 | 3.2 | 3.4 | 43.7 |
| | 11. $22,000-$24,999 | 104 | 4.5 | 4.8 | 48.5 |
| | 12. $25,000-$29,999 | 150 | 6.5 | 6.9 | 55.4 |
| | 13. $30,000-$34,999 | 155 | 6.7 | 7.1 | 62.5 |
| | 14. $35,000-$39,999 | 134 | 5.8 | 6.1 | 68.6 |
| | 15. $40,000-$44,999 | 108 | 4.7 | 5.0 | 73.6 |
| | 16. $45,000-$49,999 | 74 | 3.2 | 3.4 | 76.9 |
| | 17. $50,000-$59,999 | 140 | 6.0 | 6.4 | 83.3 |
| | 18. $60,000-$74,999 | 109 | 4.7 | 5.0 | 88.3 |
| | 19. $75,000-$89,999 | 78 | 3.4 | 3.6 | 91.9 |
| | 20. $90,000-$99,999 | 38 | 1.7 | 1.8 | 93.7 |
| | 21.$100,000-$109,999 | 34 | 1.4 | 1.5 | 95.2 |
| | 22. $110,000-$119,999 | 19 | .8 | .9 | 96.1 |
| | 23. $120,000-$134,999 | 24 | 1.0 | 1.1 | 97.2 |
| | 24. $135,000-$149,999 | 15 | .6 | .7 | 97.9 |
| | 25. $150,000 and over | 47 | 2.0 | 2.1 | 100.0 |
| | Total | 2185 | 94.0 | 100.0 | |
| Missing | System | 138 | 6.0 | | |
| Total | | 2323 | 100.0 | | |

   If you were using Recode to collapse this variable into three equally sized groups, you could create the "Low" group by collapsing codes 1 through 8 (cumulative percent, 36.3), the "Middle" group by combining codes 9 through 14 (cumulative percent, 68.6), and the "High" group by collapsing codes 15 through 25. That's a fair amount of Recode drudgery. Visual Binning will accomplish the same task with fewer clicks and less typing.

   Click Transform → Visual Binning, opening the Visual Binning window. Scroll the variable list, find income_r, and click it into the Variables to Bin panel, as shown in Figure 3-10. Click Continue. To light up the panels, click on income_r in the Scanned Variable List (Figure 3-11). There are three panels to the right of the Scanned Variable List: the Name panel (top), a graphic display of the selected variable (middle), and

**Figure 3-10**  Visual Binning Opening Window

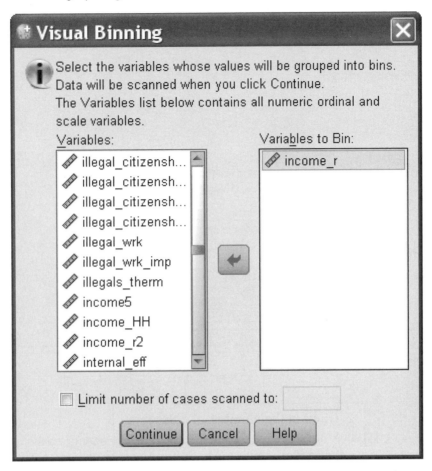

**Figure 3-11**  Visual Binning Continuation Window

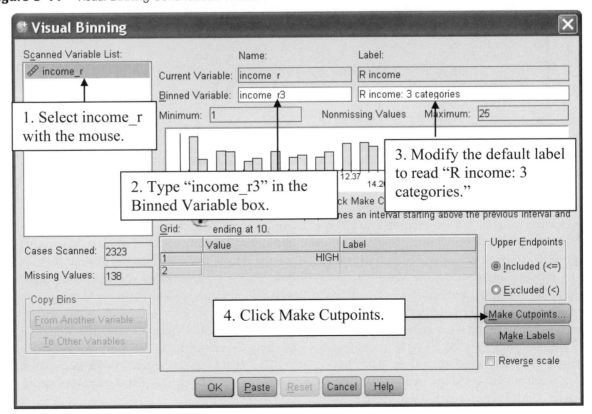

**Figure 3-12**   Visual Binning: Make Cutpoints Window

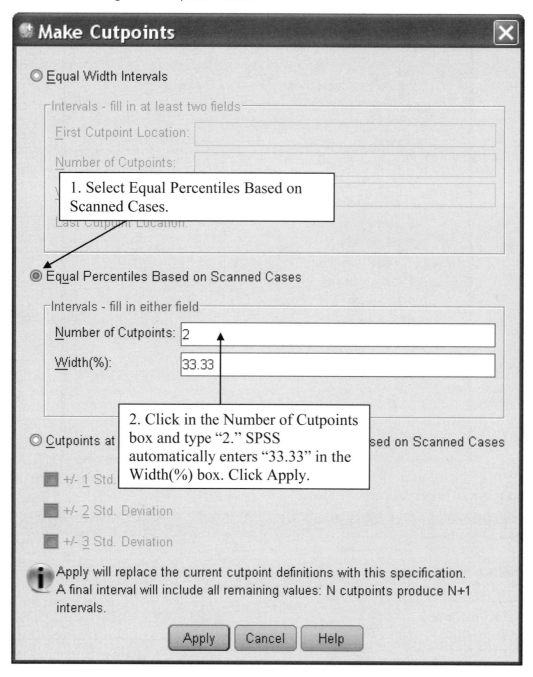

the Grid panel (bottom). Click in the Binned Variable box in the Name panel. This is where you provide a name for the variable you are about to create. Type "income_r3." Click in the box beneath Label, where SPSS has supplied a default name, "Respondent income (Binned)." Modify the label to read, "R income: 3 categories." Later you will attend to the Grid panel, but first you need to create income_r3's categories. Click Make Cutpoints and consider the Make Cutpoints window (Figure 3-12). Because you want equal-sized groups, select the radio button next to Equal Percentiles Based on Scanned Cases. And because you want three groups, click in the Number of Cutpoints box and type "2." Why 2? Here is the rule: If you wish to create a variable having k categories, then you must request k–1 cutpoints. (Reassuringly, after you type "2," SPSS automatically puts "33.3" in the Width(%) box.) Click Apply, returning to the continuation window (Figure 3-13). Now, notice the values "8.0," "14.0," and "HIGH" that SPSS has entered in the Grid panel. Earlier, when we inspected the frequency distribution of income_r, we knew that these cutpoints would divide respondents into nearly equal groups.[1]

**Figure 3-13** Labeling Values of Collapsed Variable in the Visual Binning Continuation Window

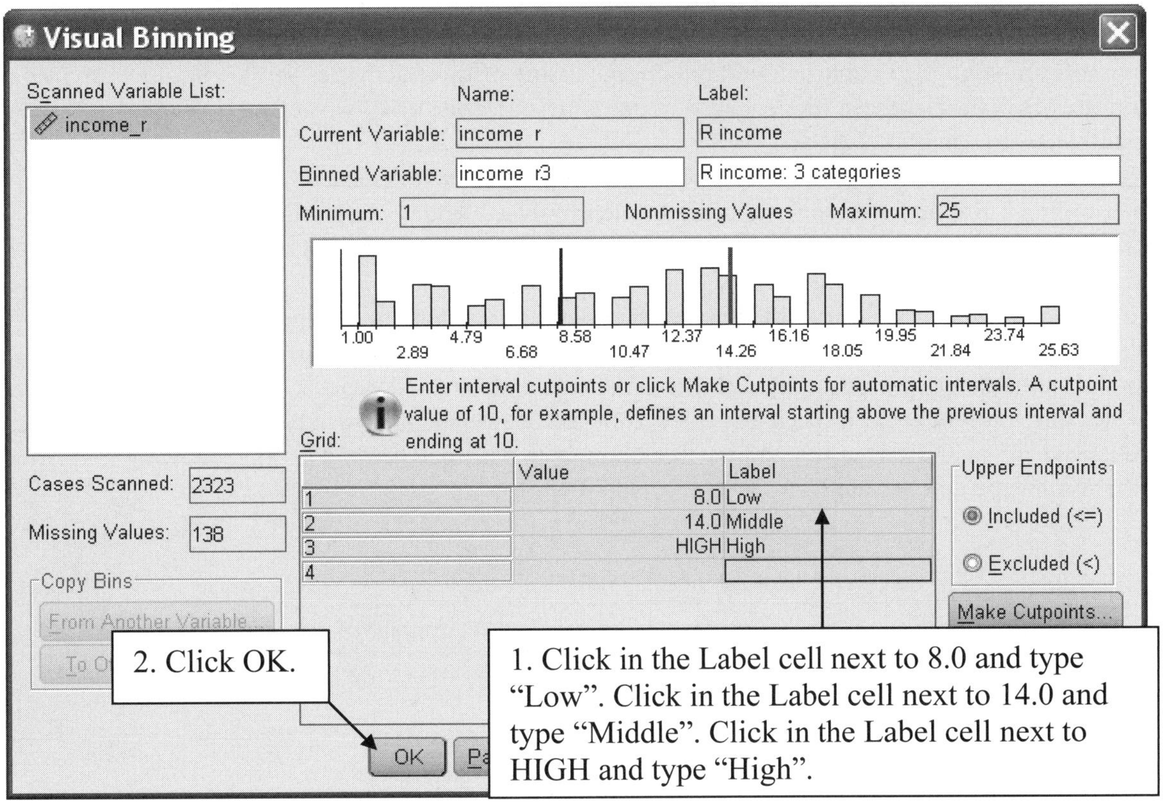

Let's finish the job by typing labels in the Values cells next to each cutpoint number, as shown in Figure 3-13: "Low" next to "8.0," "Middle" next to "14.0," and "High" next to "HIGH."[2] All set. Click OK. (Click OK again when SPSS issues the warning, "Binning specifications will create 1 variables.") Check your work by running Frequencies on income_r3:

income_r3 R income: 3 categories

|  |  | Frequency | Percent | Valid Percent | Cumulative Percent |
|---|---|---|---|---|---|
| Valid | Low | 793 | 34.1 | 36.3 | 36.3 |
|  | Middle | 706 | 30.4 | 32.3 | 68.6 |
|  | High | 686 | 29.5 | 31.4 | 100.0 |
|  | Total | 2185 | 94.0 | 100.0 |  |
| Missing | System | 138 | 6.0 |  |  |
| Total |  | 2323 | 100.0 |  |  |

This is a nice-looking three-category ordinal. Save the dataset, and let's move to the next topic.

## USING COMPUTE

Although SPSS permits the creation of new variables through a dizzying variety of complex transformations, the typical use of Compute is pretty straightforward. By and large, Compute is typically used to create a simple *additive index* from similarly coded variables. Consider a simple illustration. Suppose you have three variables, each of which measures whether or not a respondent engaged in each of the following activities during an election campaign: tried to convince somebody how to vote, put a campaign bumper sticker on his or her car, or gave money to one of the candidates or parties. Each variable is coded identically: 0 if the respondent did not engage in the activity and 1 if he or she did. Now, each of these variables is interesting in its own right, but you might want to add them together, creating an overall measure of campaigning: People who did not engage in any of these activities would end up with a value of 0 on the new variable; those who engaged in one activity, a code of 1; two activities, a code of 2; and all three activities, a code of 3.

Here are some suggested guidelines to follow in using Compute to create a simple additive index. First, before running Compute, make sure that each of the variables is coded identically. In the preceding illustration, if the "bumper sticker" variable were coded 1 for no and 2 for yes, and the other variables were coded 0 and 1, the resulting additive index would be incorrect. Second, make sure that the variables are all coded in the same *direction*. If the "contribute money" variable were coded 0 for yes and 1 for no, and the other variables were coded 0 for no and 1 for yes, the additive index would again be incorrect.[3] Third, after running Compute, obtain a frequency distribution of the newly created variable. Upon examining the frequency distribution, you may decide to use Recode to collapse the new variable into more useful categories. Suppose, for example, that you add the three campaign acts together and get the following frequency distribution for the new variable:

| Additive index: Number of campaign acts | | |
|---|---|---|
| Value label | Value | Percentage of sample |
| Engaged in none | 0 | 60 |
| Engaged in one | 1 | 25 |
| Engaged in two | 2 | 13 |
| Engaged in three | 3 | 2 |
| Total | | 100 |

It looks like a Recode run may be in order—collapsing respondents coded 2 or 3 into the same category.

These points are best understood firsthand. The 2008 National Election Study asked respondents whether certain internationalist policies should be important U.S. foreign policy goals: helping to bring a democratic form of government to other nations (NES2008 variable, goal_democ), promoting and defending human rights in other countries (goal_humanrights), combating world hunger (goal_hunger), and strengthening the United Nations and other international organizations (goal_UN). For each of these variables, respondents who said that the goal was "very important" are coded 1; those giving other responses are coded 0. We are going to add these variables together, using the expression, "goal_democ + goal_humanrights + goal_hunger + goal_UN." Think about this expression for a moment. Perhaps a respondent takes a dim view of internationalism and thinks that none of the four goals is very important. What would be his or her score on an additive index? It would be $0 + 0 + 0 + 0 = 0$. Another, international-minded respondent might say that all four are very important. For that respondent, $1 + 1 + 1 + 1 = 4$. Thus we know from the get-go that the values of the new variable will range from 0 to 4.

Let's get SPSS to compute a new variable, which we will name intism, by summing the codes of goal_democ, goal_humanrights, goal_hunger, and goal_UN. Click Transform → Compute, invoking the Compute Variable window (Figure 3-14). A box labeled "Target Variable" is in the window's upper left-hand corner. This is where we name the new variable. Click in the Target Variable box and type "intism" (as shown in Figure 3-15). The large box on the right side of the window, labeled "Numeric Expression," is where we tell SPSS which variables to use and how to combine them. Scroll down the left-hand variable list until you find goal_democ. Click goal_democ into the Numeric Expression box. Using the keyboard (or the calculator pad beneath the Numeric Expression box), type or click a plus sign (+) to the right of goal_democ. Returning to the variable list, click goal_humanrights into the Numeric Expression box. Repeat this process for the remaining variables, until the Numeric Expression box reads, "goal_democ + goal_human-rights + goal_hunger + goal_UN" (see Figure 3-15). Before we create intism, let's give it a descriptive label. Click the Type & Label button, which opens the Compute Variable: Type and Label window, as shown in Figure 3-15. Type "Internationalism support" in the Label box and click Continue. You are ready to run the compute. Click OK. SPSS does its work. What does the new variable, intism, look like? To find out, run Frequencies on intism:

**intism Internationalism support**

|  |  | Frequency | Percent | Valid Percent | Cumulative Percent |
|---|---|---|---|---|---|
| Valid | .00 | 525 | 22.6 | 25.0 | 25.0 |
|  | 1.00 | 553 | 23.8 | 26.3 | 51.3 |
|  | 2.00 | 538 | 23.1 | 25.6 | 76.9 |
|  | 3.00 | 348 | 15.0 | 16.6 | 93.4 |
|  | 4.00 | 138 | 6.0 | 6.6 | 100.0 |
|  | Total | 2102 | 90.5 | 100.0 |  |
| Missing | System | 221 | 9.5 |  |  |
| Total |  | 2323 | 100.0 |  |  |

**Figure 3-14**  Compute Variable Window

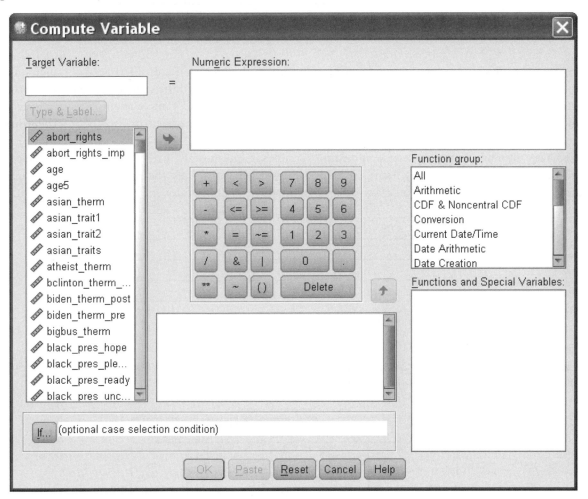

**Figure 3-15** Computing a New Variable

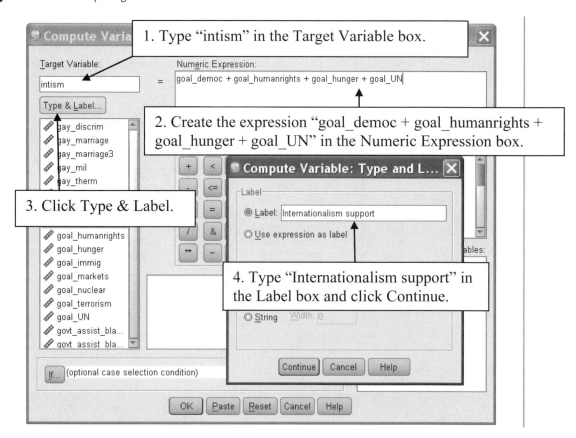

Notice the high variation in respondents' opinions about the importance of international policies. Very nearly 25 percent of the sample falls into each of the first three values, and the remaining 25 percent occupy the highest two codes. In fact, it might make sense to collapse codes 3 and 4 into a single category, which would create a four-category ordinal-level measure of support for internationalism: low (code 0), medium-low (code 1), medium-high (code 2), and high (codes 3 and 4).

This is a situation in which Recode into Same Variables is appropriate.[4] Click Transform → Recode into Same Variables. In the Recode into Same Variables window, click intism into the Numeric Variables box, as shown in Figure 3-16. Click Old and New Values. Follow this recoding protocol:

| Old value | New value |
| --- | --- |
| 0 | 0 |
| 1 | 1 |
| 2 | 2 |
| Range 3 through 4 | 3 |
| System- or user-missing | System-missing |

The Recode into Same Variables: Old and New Values window should look like Figure 3-17. Click Continue. Click OK. Again run Frequencies on intism to check the recode:

**intism Internationalism support**

| | | Frequency | Percent | Valid Percent | Cumulative Percent |
| --- | --- | --- | --- | --- | --- |
| Valid | .00 | 525 | 22.6 | 25.0 | 25.0 |
| | 1.00 | 553 | 23.8 | 26.3 | 51.3 |
| | 2.00 | 538 | 23.1 | 25.6 | 76.9 |
| | 3.00 | 486 | 20.9 | 23.1 | 100.0 |
| | Total | 2102 | 90.5 | 100.0 | |
| Missing | System | 221 | 9.5 | | |
| Total | | 2323 | 100.0 | | |

**Figure 3-16**   Recoding a New Variable

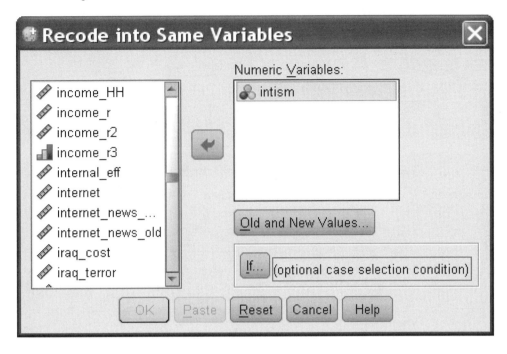

**Figure 3-17**   Collapsing a New Variable into Categories

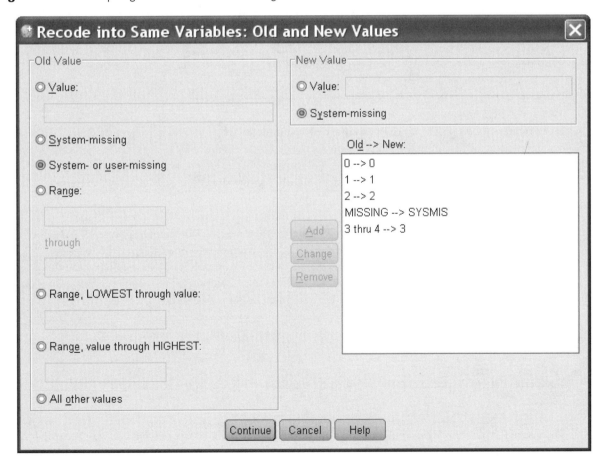

A flawless recode is a thing of beauty. Scroll to the bottom of the Variable View of the Data Editor and perform the usual housekeeping tasks with intism. First, change Decimals to 0. Second, click in the Values cell and assign these value labels:

| Value | Value label |
|-------|-------------|
| 0 | Low |
| 1 | Med-low |
| 2 | Med-high |
| 3 | High |

Before proceeding with the exercises, be sure to save the dataset.

## EXERCISES

1. (Dataset: GSS2008. Variable: polviews.) GSS2008 contains polviews, which measures political ideology—the extent to which individuals "think of themselves as liberal or conservative." Here is how polviews is coded:

| Value | Value label |
|-------|-------------|
| 1 | Extremely liberal |
| 2 | Liberal |
| 3 | Slightly liberal |
| 4 | Moderate |
| 5 | Slightly conservative |
| 6 | Conservative |
| 7 | Extremely conservative |

   A. Run Frequencies on polviews. The percentage of respondents who are either "extremely liberal," "liberal," or "slightly liberal" is (fill in the blank) _____ percent.

   The percentage of respondents who are either "slightly conservative," "conservative," or "extremely conservative" is (fill in the blank) _____ percent.

   B. Use polviews and Recode into Different Variables to create a new variable named polview3. Give polview3 this label: "Ideology: 3 categories." Collapse the three liberal codes into one category (coded 1 on polview3), put the moderates into their own category (coded 2 on polview3), and collapse the three conservative codes into one category (coded 3 on polview3). (Don't forget to recode missing values on polviews into missing values on polview3.) Run Frequencies on polview3.

   The percentage of respondents who are coded 1 on polview3 is (fill in the blank) _____ percent.

   The percentage of respondents who are coded 3 on polview3 is (fill in the blank) _____ percent.

   Make sure that the three percentages you wrote down in part B match the percentages you recorded in part A. The numbers may be very slightly different, such as 33.92 versus 33.93, and still be considered a match. If the two sets of numbers match, proceed to part C. If they do not match, you performed the recode incorrectly. Review this chapter's discussion of recode, and try the recode again.

C. In the Variable View of the Data Editor, change Decimals to 0, and then click in the Values cell and supply the appropriate labels: "Liberal" for code 1, "Moderate" for code 2, and "Conservative" for code 3. Run Frequencies on polview3. Print the Frequencies output.

2. (Dataset: GSS2008. Variables: mslm_col, mslm_lib, mslm_spk.) GSS2008 contains three variables that gauge tolerance toward "anti-American Muslim clergymen"—whether they should be allowed to teach in college (mslm_col), whether their books should be removed from the library (mslm_lib), and whether they should be allowed to preach hatred of the United States (mslm_spk). For each variable, a less-tolerant response is coded 0, and a more-tolerant response is coded 1.

A. Imagine creating an additive index from these three variables. The additive index would have scores that range between what two values?

Between a score of _____ and a score of _____.

B. Suppose a respondent takes the more-tolerant position on two questions and the less-tolerant position on the third question. What score would this respondent have?

A score of _____

C. Use Compute to create an additive index from mslm_col, mslm_lib, and mslm_spk. Name the new variable muslim_tol. Give muslim_tol this label: "Tolerance twrd Muslim clergy." Run Frequencies on muslim_tol. Referring to your output, fill in the table that follows:

| muslim_tol Tolerance twrd Muslim clergy | | |
|---|---|---|
| Score on muslim_tol | Frequency | Valid Percent |
| ? | ? | ? |
| ? | ? | ? |
| ? | ? | ? |
| ? | ? | ? |
| Total | ? | 100.0 |

D. Use Recode into Same Variables to collapse muslim_tol into three categories, following this recoding protocol:

| Old value | New value |
|---|---|
| 0 | 0 |
| Range: 1 through 2 | 1 |
| 3 | 2 |
| System- or user-missing | System-missing |

In the Variable View of the Data Editor, apply these value labels: 0, "Low"; 1, "Middle"; 2, "High."

E. Run a frequencies analysis of muslim_tol. Examine the output to ensure that the recode worked properly. Print the Frequencies output.

3. (Dataset: GSS2008. Variable: income06.) In this chapter you learned to use Visual Binning by collapsing an NES2008 measure of income into three roughly equal ordinal categories. In this exercise, you will use Visual Binning to collapse a very similar variable from GSS2008, income06. Just as you did with income_r, you will collapse income06 into income06_3, a three-category ordinal measure of income.

A.  Refer to this chapter's visual binning guided example and retrace the steps. Here is new information you will need:

| Variable to bin | income06 |
|---|---|
| Binned variable name | income06_3 |
| Binned variable label | Total income: 3 categories |
| Number of cutpoints | 2 |
| Labels for Value cells | Low, Middle, High |

B.  Run Frequencies on income08_3. Refer to your output. Fill in the table that follows:

| income06_3 Total family income: 3 categories | | |
|---|---|---|
| income06_3 | Frequency | Valid Percent |
| 1 Low | ? | ? |
| 2 Middle | ? | ? |
| 3 High | ? | ? |
| Total | ? | 100.0 |

By performing the exercises in this chapter, you have added three variables to GSS2008: polview3, muslim_tol, and income06_3.

That concludes the exercises for this chapter. Before exiting SPSS, be sure to save your output file.

## NOTES

1.  Suppose you wanted to collapse yob into yob3 using Visual Binning instead of Recode. You would follow these steps: 1. Click Transform → Visual Binning and scan yob. 2. In the Scanned Variable List of the continuation window, select yob. 3. In the Name panel, supply the name "yob3" and the label "Three generations." 4. In the Value cells of the Grid panel, type "1949" in the topmost cell and "1965" in the next lower cell. (SPSS automatically supplies the word "HIGH" in the lowest of the three cells.) 5. In the Label cells of the Grid panel, supply value labels for each value ("Before 1950" goes with "1949," "1950-1965" goes with "1965," and "After 1965" goes with "HIGH"). 6. Click OK. Yob3 will be created and labeled correctly.
2.  The SPSS-supplied numbers in the Value cells of the Grid panel are not the numeric codes that SPSS assigns to the categories. The numeric codes are numbered sequentially, beginning with 1. So income_r3 has numeric codes 1 ("Low"), 2 ("Middle"), and 3 ("High").
3.  Survey datasets are notorious for reverse-coding. Survey designers do this so that respondents don't fall into the trap of response-set bias, or automatically giving the same response to a series of questions.
4.  Recode into Same Variables is an appropriate choice because the original variables are not being replaced or destroyed in the process. If the recode goes badly and intism gets fouled up, you can always use the original variables to compute intism again.

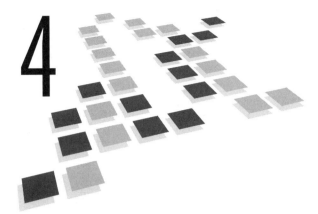

# 4

# Making Comparisons

Procedures Covered

Analyze → Descriptive Statistics → Crosstabs

Analyze → Compare Means → Means

Graphs → Legacy Dialogs → Line

Graphs → Legacy Dialogs → Bar

All hypothesis testing in political research follows a common logic of comparison. The researcher separates subjects into categories of the independent variable and then compares these groups on the dependent variable. For example, suppose you think that gender (independent variable) affects opinions about gun control (dependent variable) and that women are more likely than men to favor gun control. You would divide subjects into two groups on the basis of gender, women and men, and then compare the percentage of women who favor gun control with the percentage of men who favor gun control. Similarly, if you hypothesize that Republicans have higher incomes than do Democrats, you would divide subjects into partisanship groups (independent variable), Republicans and Democrats, and compare the average income (dependent variable) of Republicans with that of Democrats.

Although the logic of comparison is always the same, the appropriate method depends on the level of measurement of the independent and dependent variables. In this chapter you will learn to address two common hypothesis-testing situations: those in which both the independent and the dependent variables are categorical (nominal or ordinal) and those in which the independent variable is categorical and the dependent variable is interval level. You will also learn to add visual support to your hypothesis testing by creating and editing bar charts and line charts.

## CROSS-TABULATION ANALYSIS

Cross-tabulations are the workhorse vehicles for testing hypotheses for categorical variables. When setting up a cross-tabulation, you must observe the following three rules:

1. Put the independent variable on the columns and the dependent variable on the rows.
2. Always obtain column percentages, not row percentages.
3. Test the hypothesis by comparing the percentages of subjects who fall into the same category of the dependent variable.

Consider this hypothesis: In a comparison of individuals, older generations will take more interest in political campaigns than will younger generations. The dataset NES2008 contains the variable camp_int, which measures respondents' levels of interest in the 2008 campaigns: "high" (coded 1), "moderate" (coded 2), or "low" (coded 3). This will serve as the dependent variable. One of the variables that you created in Chapter 3, yob3, is the independent variable. You will recall that yob3 classifies individuals by year of birth: 1949 or before (coded 1), 1950–1965 (coded 2), and 1966–1990 (coded 3). Open NES2008, and let's test the hypothesis.

In the Data Editor, click Analyze → Descriptive Statistics → Crosstabs. The Crosstabs window appears, sporting four panels. For now, focus on the two upper right-hand panels: Row(s) and Column(s). (The oddly labeled Layer 1 of 1 panel comes into play in Chapter 5.) This is where we apply the first rule for a properly constructed cross-tabulation: The independent variable defines the columns, and the dependent variable defines the rows. Because camp_int is the dependent variable, click it into the Row(s) panel, as shown in Figure 4-1. Find yob3 in the left-hand variable list and click it into the Column(s) panel.

**Figure 4-1**   Crosstabs Window (modified)

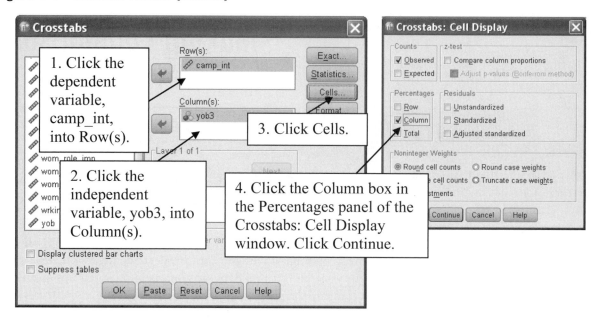

Now for the second rule of cross-tab construction: Always obtain column percentages. On the right-hand side of the Crosstabs window, click the Cells button (refer to Figure 4-1). SPSS displays the available options for Counts, Percentages, and Residuals. Left to its own defaults, SPSS will produce a cross-tabulation showing only the number of cases ("observed" counts) in each cell of the table. That's fine. But to follow the second rule, we also want column percentages—the percentage of each category of the independent variable falling into each category of the dependent variable. Click the Column box in the Percentages panel. Click Continue, which returns you to the Crosstabs window. That's all there is to it. Click OK.

SPSS runs the analysis and displays the results in the Viewer—a case processing summary followed by the requested cross-tabulation:

**camp_int Campaign interest * yob3 Three generations Crosstabulation**

| | | Statistics | yob3 Three generations | | | Total |
| --- | --- | --- | --- | --- | --- | --- |
| | | | 1 Before 1950 | 2 1950-1965 | 3 After 1965 | |
| camp_int Campaign interest | 1 High | Count | 357 | 372 | 366 | 1095 |
| | | % within yob3 Three generations | 59.5% | 53.3% | 37.3% | 48.0% |
| | 2 Moderate | Count | 172 | 235 | 401 | 808 |
| | | % within yob3 Three generations | 28.7% | 33.7% | 40.8% | 35.4% |
| | 3 Low | Count | 71 | 91 | 215 | 377 |
| | | % within yob3 Three generations | 11.8% | 13.0% | 21.9% | 16.5% |
| Total | | Count | 600 | 698 | 982 | 2280 |
| | | % within yob3 Three generations | 100.0% | 100.0% | 100.0% | 100.0% |

By convention, SPSS identifies its Crosstabs output with the label of the dependent variable, followed by an asterisk (*), and then the label of the independent variable. In fact, when SPSS runs Crosstabs, it produces a set of side-by-side frequency distributions of the dependent variable—one for each category of the independent variable—plus an overall frequency distribution for all analyzed cases. Accordingly, the table has four columns of numbers: one for respondents born before 1950, one for those born between 1950 and 1965, one for those born after 1965, and a total column showing the distribution of all cases across the dependent variable. And, as requested, each cell shows the number (count) and column percentage.

What do you think? Does the cross-tabulation fit the hypothesis? The third rule of cross-tabulation analysis is easily applied. Focusing on the "high" value of the dependent variable ("Campaign interest"), we see a clear pattern in the hypothesized direction. A comparison of respondents in the "Before 1950" column with those in the "1950–1965" column reveals a decrease in the "very much" percentage, from 59.5 percent to 53.3 percent. A comparison of the "1950–1965" and "After 1965" columns reveals yet another drop in campaign interest, from 53.3 percent to 37.3 percent. Yes, the analysis supports the hypothesis.

## MEAN COMPARISON ANALYSIS

We now turn to another common hypothesis-testing situation: when the independent variable is categorical and the dependent variable is interval level. The logic of comparison still applies—divide cases on the independent variable and compare values of the dependent variable—but the method is different. Instead of comparing percentages, we now compare means.

To illustrate, let's say that you are interested in explaining this dependent variable: attitudes toward Sarah Palin. Why do some people have positive feelings toward her whereas others harbor negative feelings? Here is a plausible idea: Partisanship (independent variable) will have a strong effect on attitudes toward Sarah Palin (dependent variable). The hypothesis: In a comparison of individuals, Republicans will have more favorable attitudes toward Sarah Palin than will Democrats.

NES2008 contains palin_therm_pre, a 100-point feeling thermometer. Each respondent was asked to rate Palin on this scale, from 0 (cold or negative) to 100 (warm or positive). This is the dependent variable. (The "pre" suffix means the question was asked in the survey's preelection wave. Several of the thermometer scales, including Mrs. Palin's, were readministered in the postelection wave as well. These variables have a "post" suffix. For this example, you will analyze the preelection Palin thermometer.) NES2008 also has partyid7, which measures partisanship in seven categories, from Strong Democrat (coded 0) to Strong Republican (coded 6). The intervening codes capture gradations between these poles: Weak Democrat (coded 1), Independent-Democrat (coded 2), Independent (coded 3), Independent-Republican (coded 4), and Weak Republican (coded 5). This is the independent variable. If the hypothesis is correct, we should find that strong Democrats have the lowest mean scores on palin_therm_pre and that mean scores increase systematically across categories of partyid7, hitting a peak among respondents who are strong Republicans. Is this what happens?

Click Analyze → Compare Means → Means. The Means window pops into view. Scroll down the left-hand variable list until you find palin_therm_pre, and then click it into the Dependent List panel, as shown in Figure 4-2. Now scroll to partyid7 and click it into the Independent List panel. In the Means window, click Options. The Means: Options window (also shown in Figure 4-2) permits you to select desired statistics from the left-hand Statistics panel and click them into the right-hand Cell Statistics panel. Alternatively, you can remove statistics from Cell Statistics by clicking them back into the left-hand panel. Unless instructed otherwise, SPSS will always report the mean, number of cases, and standard deviation of the dependent variable for each category of the independent variable. Because at present we are not interested in obtaining the standard deviation, select it with the mouse and click it back into the left-hand Statistics panel. Our mean comparison table will report only the mean value of palin_therm_pre and the number of cases for each category of partyid7. Click Continue, returning to the Means window. Click OK.

**Figure 4-2** Means Window (modified)

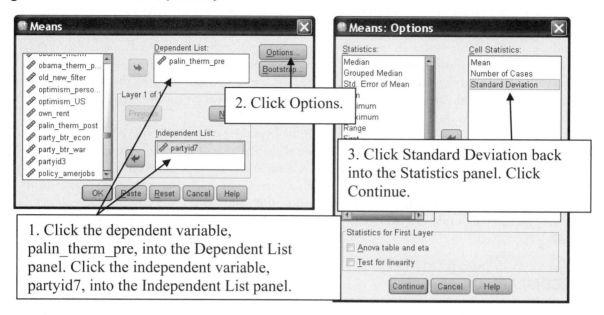

Compared with cross-tabulations, mean comparison tables are models of minimalism:

palin_therm_pre Feeling thermometer: Palin

| partyid7 Summar... | Mean | N |
|---|---|---|
| 0 StrDem | 30.89 | 375 |
| 1 WkDem | 42.37 | 291 |
| 2 IndDem | 41.35 | 335 |
| 3 Indep | 53.74 | 186 |
| 4 IndRep | 65.99 | 230 |
| 5 WkRep | 70.98 | 261 |
| 6 StrRep | 82.11 | 284 |
| Total | 53.40 | 1961 |

The label of the dependent variable, "Feeling thermometer: Palin," appears along the top of the table. The independent variable, partyid7, defines the leftmost column, which shows all seven categories, from Strong Democrat at the top to Strong Republican at the bottom. Beside each category, SPSS has calculated the mean of palin_therm_pre and reported the number of respondents falling into each value of partisanship. (The bottom row, "Total," gives the mean for the whole sample.)

Among strong Democrats the mean for palin_therm_pre is rather chilly—about 31 degrees. Do ratings of Palin increase as attachment to the Democratic Party weakens and identification with the Republican Party strengthens? Notice that the mean increases sharply among weak Democrats (who average about 42 degrees), remains essentially unchanged among independent-Democratic leaners (about 41), and then continues to increase predictably. Strong Republicans, who average 82 degrees on the thermometer, have the warmest response to Palin. Indeed, on average, the dependent variable increases by over 50 degrees between strong Democrats at one pole and strong Republicans at the other extreme. On the whole, then, the data support the hypothesis.

## GRAPHING RELATIONSHIPS

We have already seen that bar charts and histograms can be a great help in describing the central tendency and dispersion of a *single* variable. SPSS graphic procedures are also handy for illustrating relationships *between* variables. It will come as no surprise that SPSS supports a large array of graphic styles. To get a flavor of this variety, click Graphs → Legacy Dialogs and consider the choices (Figure 4-3). The legacy charts, as the name implies, use interfaces developed in earlier releases of SPSS. Even so, the Legacy Dialogs are still the best way to create graphics in SPSS. In this chapter you will learn to use Bar and Line. (In Chapter 8 you will work with Scatter/Dot.) A bar chart is useful for summarizing the relationship between two categorical variables. A line chart adds clarity to the relationship between a categorical independent variable and an interval-level dependent variable. Line charts are elegant and parsimonious, and they can be used to display the relationship between two categorical variables as well.

**Figure 4-3**   Graphs Drop-down Menu

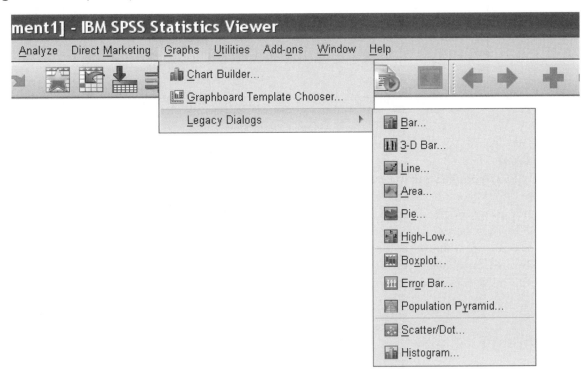

To get an idea of how SPSS produces a line chart, let's begin by creating one of our own, using the results from the palin_therm_pre-partyid7 example. Turn your attention to Figure 4-4, an empty graphic "shell." The horizontal axis, called the *category axis,* displays values of the independent variable, party identification. Each partisanship category is represented by a hash mark, from Strong Democrat on the left to Strong Republican on the right. The vertical axis records the mean values of the dependent variable, Sarah Palin thermometer ratings. Now, with a pen or pencil, make a dot directly above each category of the independent variable, recording the mean of palin_therm_pre for each partisan category. Above the Strong Democrat hash mark, for example, place a dot at 31 on the summary axis. Go to the right along the category axis until you reach the hash mark for Weak Democrat and make a dot directly above the hash mark, at about 42. Do the same for the remaining partisan groups, placing a dot vertically above each hash mark at the mean value of palin_therm_pre. (Don't worry about being precise. Just get the dots close to the mean values.) Using a straight edge, connect the dots. You've created a line chart for the relationship, a visual summary that is easy to interpret and present.

**Figure 4-4**   Line Chart Shell: Mean Values of Sarah Palin Thermometer, by Party Identification

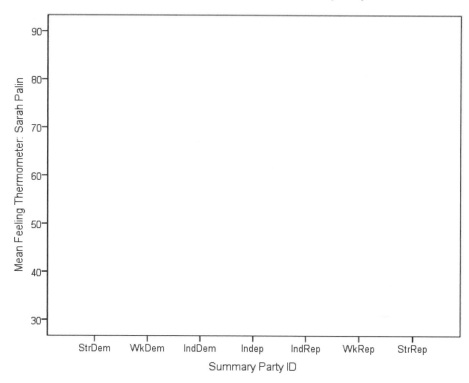

### Using Line Chart

Now let's get SPSS to do the work for us. Click Graphs → Legacy Dialogs → Line. The Line Charts window opens (Figure 4-5). Make sure that the icon next to "Simple" is clicked and that the radio button next to "Summaries for groups of cases" is selected.[1] Click Define. The Define Simple Line window appears (Figure 4-6). The two topmost boxes—the (currently inactive) Variable box in the Line Represents panel and the Category Axis box—are where we tailor the line chart to our specifications. (There are two additional boxes in the Panel by area, one labeled "Rows" and one labeled "Columns," as shown in Figure 4-6. For our purposes in this book, these boxes may be safely ignored.)

**Figure 4-5**   Line Charts Window (default)

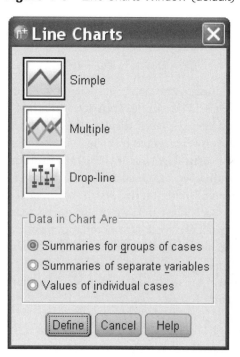

**Figure 4-6**   Define Simple Line Window (default)

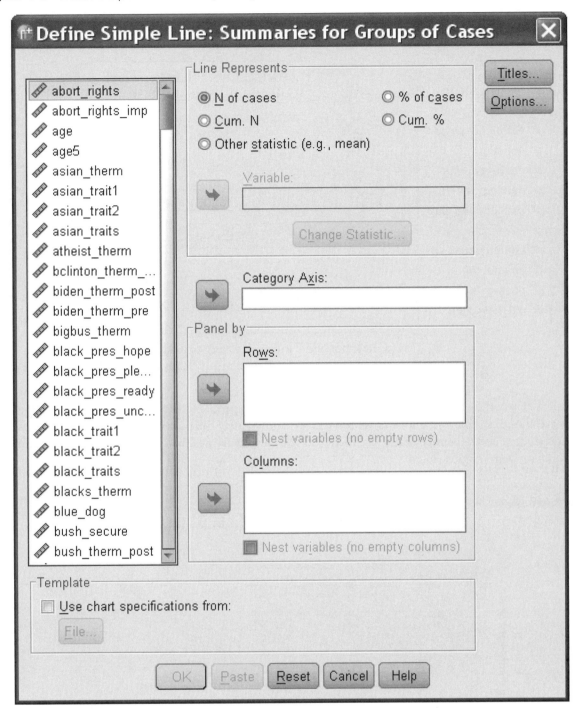

In SPSS idiom, "category axis" means x-axis or horizontal axis—the axis that represents values of the independent variable. Because, in the current example, partyid7 is the independent variable, scroll down to partyid7 and click it into the Category Axis box. We also want to graph the mean values of palin_therm_pre for each category of partyid7. To do this, SPSS requires instruction.[2] In the Line Represents panel, select the Other statistic radio button, as shown in Figure 4-7. The Variable box is activated. Now scroll the left-hand variable list until you find palin_therm_pre, and then click palin_therm_pre into the Variable box. SPSS moves palin_therm_pre into the Variable box and gives it the designation "MEAN(palin_therm_pre)." In Line Chart, whenever you request Other statistic and click a variable into the Variable box (as we have just done), SPSS assumes that you want to graph the mean values of the requested variable (as, in this case, we do).[3] So this default serves our current needs. All set. Click OK.

**Figure 4-7** Define Simple Line Window (modified)

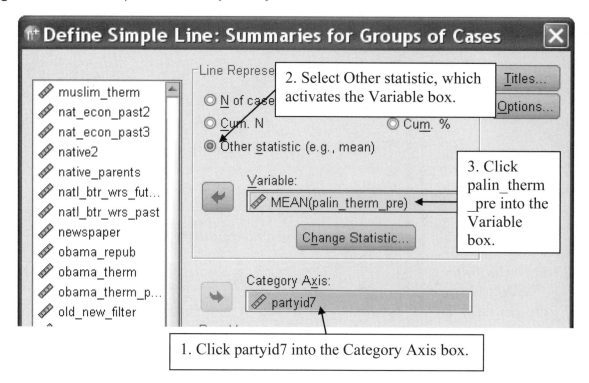

A line chart of the palin_therm_pre-partyid7 relationship appears in the Viewer (Figure 4-8). Line charts are at once simple and informative. You can see immediately the positive linear relationship between the independent and dependent variables. Note, too, the curious flatness between weak Democrats and independent-Democratic leaners.

**Figure 4-8** Line Chart Output

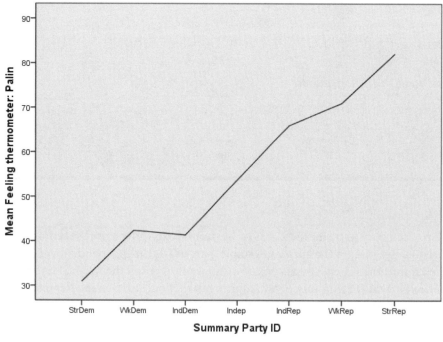

### *Using Bar Chart*

In most respects, Bar Chart is similar to Line Chart. However, a key difference between the two exists. In this guided example, you will obtain a bar chart of the relationship you analyzed earlier between campaign interest (the dependent variable camp_int) and generational cohort (the independent variable yob3). In the next section we will take an excursion into the Chart Editor, which allows you to enhance the appearance and content of the charts you create.

Click Graphs → Legacy Dialogs → Bar. The Bar Charts window gives you the same set of choices as the Line Charts window. Ensure that the same choices are selected: Simple and Summaries for groups of cases. Click Define. The Define Simple Bar window opens, and it, too, is identical to the Define Simple Line window in every detail. Because yob3 is the independent variable, it goes in the Category Axis box. Scroll to yob3 and click it over, as shown in Figure 4-9. So far, this is the same as before. At this point, however, the peculiarities of Bar Chart require that we refamiliarize ourselves with specific coding information about the dependent variable, camp_int. Why so? As a substantive matter, we want to depict the percentage of respondents in each category of yob3 who had a "high" level of interest in the 2008 campaigns. To accomplish this, Bar Chart will need specific coding information. Find camp_int in the left-hand variable list, place the cursor pointer on it, and then *right*-click (refer to Figure 4-9). Click on Variable Information and review the numeric codes. Respondents having a "high" level of interest are coded 1, those with "moderate" interest are coded 2, and those with "low" interest are coded 3 (Figure 4-9). Commit this fact to short-term memory: Respondents having "high" interest are coded 1 on the dependent variable, camp_int.

**Figure 4-9**   Reviewing Numeric Codes

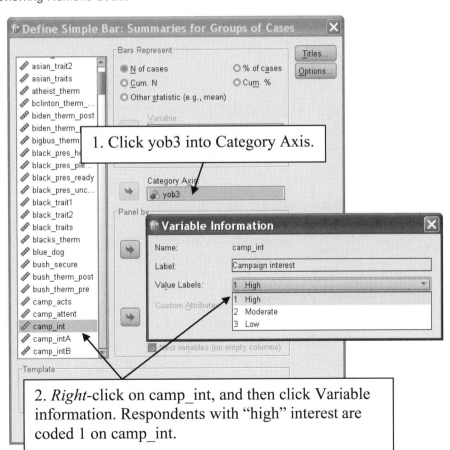

Now return to the Bars Represent panel. Select the Other statistic radio button, and click camp_int into the Variable box, as shown in Figure 4-10. The designation "MEAN(camp_int)" appears in the Variable box, as shown in Figure 4-10. Just as it did in Line Chart, SPSS assumes that we are after the mean of camp_int. This default is fine for mean comparisons, but in this case it won't do. Click the Change Statistic button.[4]

The Statistic window presents itself (Figure 4-10). The radio button for the default, Mean of values, is currently selected. However, we are interested in obtaining the percentage of cases in code 1 ("high" interest) on camp_int. How do we get SPSS to cooperate with this request? Click the radio button at the bottom on the left, the one labeled "Percentage inside," as shown in Figure 4-11. The two boxes, one labeled "Low" and the other labeled "High," go active. Our request is specific and restrictive: We want the percentage of respondents in code 1 only. Expressed in terms that SPSS can understand, we want the percentage of cases "inside" a coded value of 1 on the low side and a coded value of 1 on the high side. Click the cursor in the Low box and type a "1." Click the cursor in the High box and type a "1."[5] The Statistic window should now look like Figure 4-11. Click Continue, returning to the Define Simple Bar window. The Define Simple Bar window should now look like Figure 4-12. Click OK.

**Figure 4-10**   Define Simple Bar Window and Statistic Window (default)

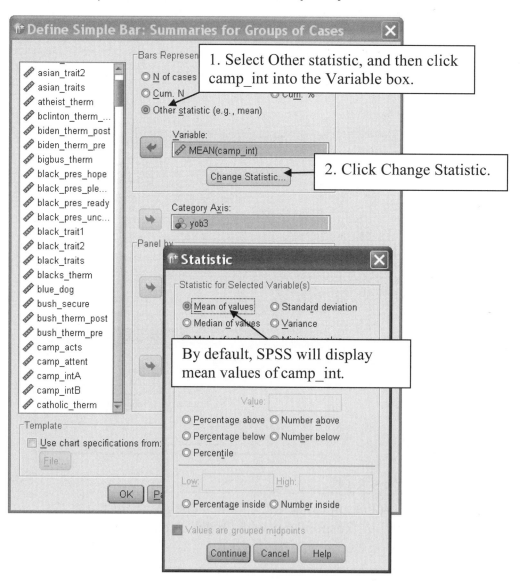

**Figure 4-11**  Statistic Window (modified)

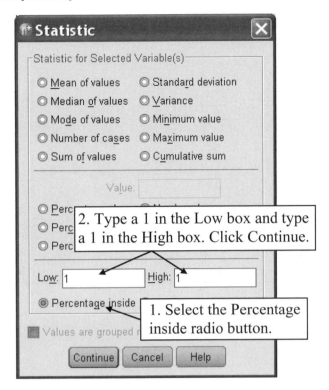

**Figure 4-12**  Define Simple Bar Window (modified)

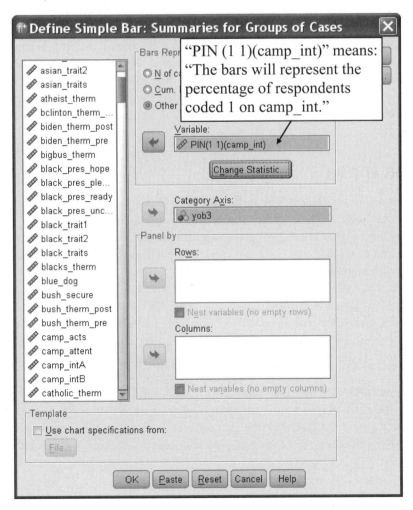

**Figure 4-13**   Bar Chart Output

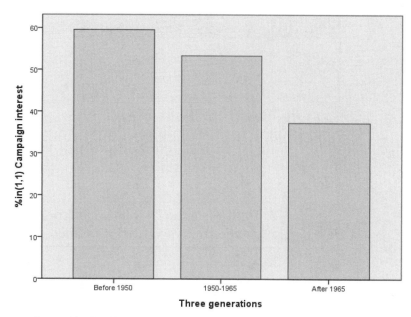

Our special instructions have paid off. SPSS displays a bar chart of the relationship between generation and campaign interest (Figure 4-13). The category axis is labeled nicely, and the heights of the bars clearly depict this pattern: As the generations change from older to younger, the percentage of people with high interest in the campaign declines. At least *we* know what the bars represent, because we did the analysis. An interested observer, however, might do a double-take at the title on the vertical axis, "%in(1,1) Campaign interest." SPSS is relentlessly literal. We asked it to graph the percentages of people between code 1 and code 1 on camp_int, so that is how SPSS has titled the axis. This chart is not ready for prime time. We need to give the vertical axis a more descriptive title, and perhaps make other appearance-enhancing changes.

## USING THE CHART EDITOR

SPSS permits the user to modify the content and appearance of any tabular or graphic object it produces in the Viewer. The user invokes the Editor, makes any desired changes, and then returns to the Viewer. The changes made in the Chart Editor are recorded automatically in the Viewer. In this section we describe how to retitle the vertical axis of the bar chart you just created. We'll also change the color of the bars. (The default color is rather uninspired, and it doesn't print well.)

In the Viewer, place the cursor anywhere on the bar chart and double-click. SPSS opens the Chart Editor (Figure 4-14). As with any editing software, the Chart Editor recognizes separate elements within an object. It recognizes some elements as text. These elements include the axis titles and the value labels for the categories of yob3. It recognizes other elements as graphic, such as the bars in the bar chart. First we will edit a text element, the title on the vertical axis. Then we will modify a graphic element, the color of the bars.

Place the cursor anywhere on the title "%in(1,1) Campaign interest" and single-click. SPSS selects the axis title. With the cursor still placed on the title, single-click again. SPSS moves the text into editing mode inside the chart (Figure 4-15). Delete the current text. In its place type the title "Percent 'High' Campaign Interest." Now click on one of the bars. (As soon as you click off the axis title, it returns to its

**Figure 4-14**  Chart Editor

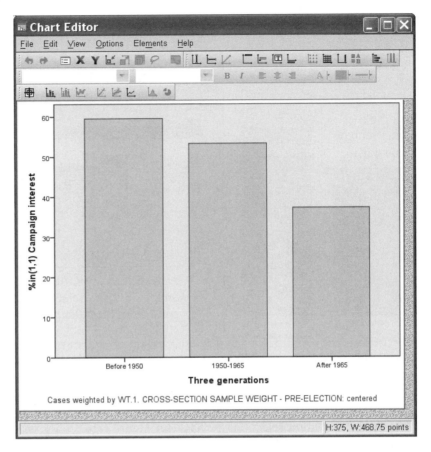

**Figure 4-15**  Bar Chart Axis Title Ready for Editing

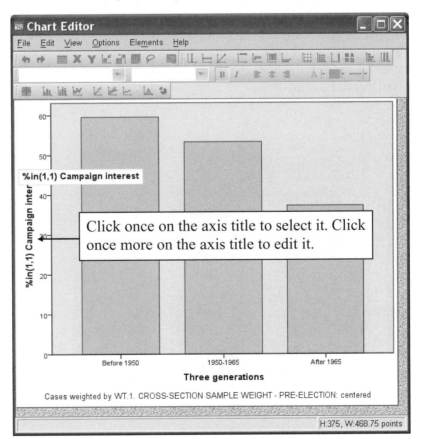

**Figure 4-16**    Using the Properties Window to Change the Bar Color

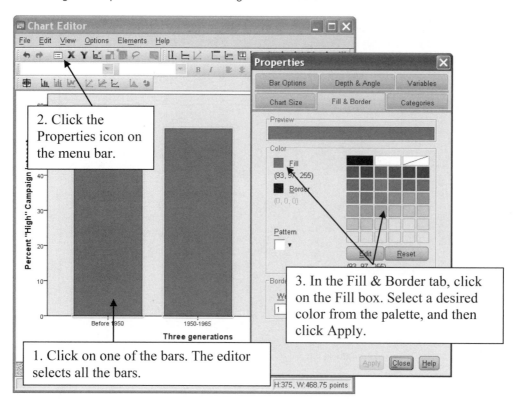

rightful position in the chart.) The editor selects all the bars (see Figure 4-16). Click on the Properties icon located near the upper-left corner of the Chart Editor window. This opens the Properties window, the most powerful editing tool in the Chart Editor's arsenal. (*Special note:* If you plan to do a lot of editing, it is a good idea to open the Properties window soon after you enter the Chart Editor. Each time you select a different text or graphic element with the mouse, the Properties window changes, displaying the editable properties of the selected element.) Click on the Fill & Border tab, as shown in Figure 4-16. In the Color panel, click in the Fill box. In the color palette, click on a desirable hue, and then click Apply. SPSS makes the change. Okay. Close the Properties window and exit the Chart Editor. The finished product appears in the Viewer (Figure 4-17).

**Figure 4-17**    Edited Bar Chart

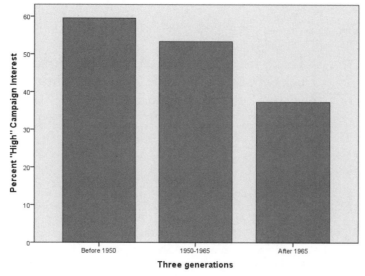

Cases weighted by WT.1. CROSS-SECTION SAMPLE WEIGHT - PRE-ELECTION: centered

## EXERCISES

1.  (Dataset: NES2008. Variables: medinsur_r, partyid7.) Here is a widely observed difference between Democrats and Republicans: Democrats favor government-funded medical insurance and Republicans prefer private insurance plans. Is this difference borne out by the data? Dataset NES2008 contains the variable medinsur_r, a 7-point scale that measures respondents' opinions on this issue. Respondents indicate their opinions by choosing any position on this scale, from 1 (government plan) at one end to 7 (private plan) at the other end. This is the dependent variable. Use the 7-point party identification scale (partyid7) as the independent variable.

    A.  If Democrats are more likely than Republicans to favor government-funded medical insurance, then Democrats will have (check one)

        ❏ a higher mean on medinsur_r than do Republicans.

        ❏ about the same mean on medinsur_r as do Republicans.

        ❏ a lower mean on medinsur_r than do Republicans.

    B.  Using Analyze → Compare Means → Means, obtain a mean comparison that shows the mean score on the dependent variable, medinsur_r, for each category of the independent variable, partyid7. Write the results in the table that follows:

| Govt/private medical insurance scale: self-placement | | |
|---|---|---|
| Party ID | Mean | N |
| Strong Democrat | ? | ? |
| Weak Democrat | ? | ? |
| Independent-Dem | ? | ? |
| Independent | ? | ? |
| Independent-Rep | ? | ? |
| Weak Republican | ? | ? |
| Strong Republican | ? | ? |
| Total | ? | ? |

    C.  Does your analysis support the idea that Democrats are more likely than Republicans to support government-funded medical insurance? (circle one)

    Yes    No

    Briefly explain your answer. _____

    _____

    _____

    _____

    _____

    D.  Suppose you overheard someone making this claim: "People might say they're 'independent-Democrats,' but when it comes to opinions about medical insurance, they're closer to 'independent' than to 'Democrat.'" Based on your analysis, is this claim correct? (circle one)

    Yes    No

Briefly explain your answer. _____

_____

_____

_____

E. Obtain a line chart of the relationship. Remember to put the independent variable, partyid7, on the Category Axis and the dependent variable, medinsur_r, in the Variable box of the Line Represents panel. Print the line chart.

2. (Dataset: NES2008. Variables: gay_marriage3, relig_attend3, gender, libcon3_r.) Should gay couples be legally permitted to marry? This controversial issue has gained center stage in recent elections. In 2004, amendments regarding the legal definition of marriage appeared on the ballot in 11 states. In 2008, three states considered such measures, and a fourth considered whether unmarried couples (including same-sex couples) should be prohibited from adopting or serving as foster parents. One can imagine several characteristics that divide people on this issue. People who are more religiously observant might be less likely to favor gay marriage than are the less observant. Men may be less likely than women to favor it. Or conservatives might be less open to the idea than are liberals.

Dataset NES2008 contains gay_marriage3, which is coded 1 (yes, gay marriage should be allowed), 2 (no, gay marriage should not be allowed, but civil unions should be permitted), or 3 (no, gay marriage should not be allowed). This is the dependent variable that you will use to test each of the following hypotheses:

Hypothesis 1: In a comparison of individuals, people who have low levels of attendance at religious services will be more likely to favor gay marriage than will people who have higher levels of religious attendance. (The independent variable is relig_attend3.)

Hypothesis 2: In a comparison of individuals, women are more likely than men to favor gay marriage. (The independent variable is gender.)

Hypothesis 3: In a comparison of individuals, liberals are more likely than conservatives to favor gay marriage. (The independent variable is libcon3_r.)

When using SPSS to obtain a series of cross-tabulations having the same dependent variable but different independent variables, only one Crosstabs run is required. In the Crosstabs window, click gay_marriage3 into the Row(s) panel. Click relig_attend3, gender, and libcon3_r into the Column(s) panel. (Make sure to request column percentages in the Crosstabs Cell Display.) Run the analysis. In the following spaces, record the percentages who say "Yes" when asked about gay marriage:

| | Level of religious attendance | | |
|---|---|---|---|
| | High | Middle | Low |
| Percentage "Yes" | ? | ? | ? |
| | R gender | | |
| | Male | | Female |
| Percentage "Yes" | ? | | ? |
| | R self-placement lib-con scale | | |
| | Liberal | Moderate | Conservative |
| Percentage "Yes" | ? | ? | ? |

B. These findings (circle one)

support Hypothesis 1.     do not support Hypothesis 1.

Briefly explain your reasoning. _____

_____

_____

_____

C. These findings (circle one)

support Hypothesis 2.     do not support Hypothesis 2.

Briefly explain your reasoning. _____

_____

_____

_____

D. These findings (circle one)

support Hypothesis 3.     do not support Hypothesis 3.

Briefly explain your reasoning. _____

_____

_____

_____

E. Obtain a bar chart of the relationship between gay_marriage3 and libcon3_r. You will want the vertical axis to depict the percentage of respondents in the "Yes" category of gay_marriage3. Remember that those who support gay marriage are coded 1 on the dependent variable. Using the Chart Editor, give the vertical axis a more descriptive title, such as "Percentage Favoring Gay Marriage." Change the default bar color to a color of your choosing. Print the bar chart you created.

3. (Dataset: NES2008. Variables: who08, r_econ_past3.) What factors determine how people vote in presidential elections? Political scientists have investigated and debated this question for many years. A particularly powerful and elegant perspective emphasizes voters' *retrospective* evaluations. According to this view, for example, voters whose financial situations have gotten better during the year preceding the election are likely to reward the candidate of the incumbent party. Voters whose economic situations have worsened, by contrast, are likely to punish the incumbent party by voting for the candidate of the party not currently in power. As political scientist V. O. Key famously once put it, the electorate plays the role of "rational god of vengeance and reward."[6] Does Key's idea help explain how people voted in the 2008 election?

A. Test this hypothesis: In a comparison of individuals, those whose personal finances improved during the year preceding the 2008 election were more likely to vote for the candidate of the incumbent party, John McCain, than were individuals whose personal finances worsened. Use these two variables from NES2008:

who08 (dependent variable) and r_econ_past3 (independent variable). Obtain a cross-tabulation of the relationship. Record the percentages voting for McCain and Obama in the table that follows:

| Pres vote 2008 | R better/same/worse off than 1 yr ago | | | |
|---|---|---|---|---|
| | Better | Same | Worse | Total |
| McCain | ? | ? | ? | ? |
| Obama | ? | ? | ? | ? |
| Total | 100.0 | 100.0 | 100.0 | 100.0 |

B.  What do you think? Are the data consistent with the hypothesis?

Yes     No

Write a paragraph explaining your reasoning. _____

_____

_____

_____

C.  *Loss aversion* is an interesting psychological phenomenon that can shape the choices people make.[7] One idea behind loss aversion is that losses loom larger than commensurate gains. According to this theory, for example, the psychological pain felt from losing $100 is greater than the pleasure felt from gaining $100. Applied to retrospective voting, loss aversion might suggest that the "vengeance" impulse is stronger than the "reward" impulse—that the anti-incumbent-party motivation among those whose financial fortunes worsened will be stronger than the pro-incumbent-party motivation among those whose financial fortunes improved.

With this idea in mind, examine the percentages in the table in part A. What do you think? Do the data suggest that Key's rational god of vengeance is stronger than his rational god of reward?

Yes     No

Write a few sentences explaining your reasoning. _____

_____

_____

_____

_____

_____

4.  (Dataset: NES2008. Variables: voter0408, age5, hispanic, race2, income_r3, contact_GOTV.) In 2008, 131 million Americans voted, 9 million more than in 2004. Who were these newly mobilized voters? The Obama campaign, in particular, staged get-out-the-vote (GOTV) drives among groups that are most likely to stay home on Election Day: young people, Hispanics, African Americans, and the poor. It stands to reason, then, that newly mobilized voters in 2008 will resemble nonvoters in most ways, except one: the newly mobilized

were more likely than nonvoters to be contacted by GOTV campaigns. Furthermore, both nonvoters and new voters will be demographically different from established voters, that is, people who voted in 2004 and 2008. Consider the following hypothetical expectations:

(i) Newly mobilized voters will be similar to nonvoters in age, Hispanic composition, racial composition, and income.

(ii) Newly mobilized voters and nonvoters will be different from established voters in age, Hispanic composition, racial composition, and income.

(iii) Newly mobilized voters will be more likely than nonvoters to report being contacted during the campaign by GOTV efforts.

NES2008 contains voter0408, which classifies each respondent as "Nonvoter" (did not vote in 2004 or 2008), "New voter" (voted in 2008 but not in 2004), or "Voter" (voted in both elections).[8] This is the independent variable.

A. Run Crosstabs to analyze the relationship between voter0408 and age5, hispanic, race2, income_r3, and contact_GOTV. (*Note*: You created income_r3 in Chapter 3.) When using SPSS to obtain a series of cross-tabulations having the same independent variable but different dependent variables, only one Crosstabs run is required. In the Crosstabs window, click voter0408 into the Column(s) panel. Click age5, hispanic, race2, income_r3, and contact_GOTV into the Row(s) panel. (Make sure to request column percentages in the Crosstabs Cell Display.)

B. Browse the cross-tabulation results. Write the percentage of each voter type having the attribute listed in the left-hand column. Along the "Percent 18-30 yrs of age" row, for example, write the percentage of nonvoters who are 18–30 years old, the percentage of new voters who are 18–30, and the percentage of voters who are 18–30.

| Attribute | Nonvoters | New voters | Voters |
|---|---|---|---|
| Percent 18-30 yrs of age | ? | ? | ? |
| Percent Hispanic | ? | ? | ? |
| Percent black | ? | ? | ? |
| Percent low income | ? | ? | ? |
| Percent "Yes" contacted | ? | ? | ? |

C. Examine the table in part B. Think about the hypothetical expectations regarding age, Hispanic composition, racial composition, income, and GOTV contact. In what ways are your findings consistent with the expectations? In what ways do your findings depart from expectations? _____

_____

_____

_____

_____

_____

5. (Dataset: GSS2008. Variables: polviews, fem_role.) Why do some people hold more traditional views about the role of women in society, whereas others take a less traditional stance? General ideological orientations, liberalism versus conservatism, may play an important role in shaping individuals' opinions on this cultural question. Thus it seems plausible to suggest that ideology (independent variable) will affect opinions about appropriate female roles (dependent variable). The hypothesis: In a comparison of individuals, liberals will be more likely than conservatives to approve of nontraditional female roles.

GSS2008 contains fem_role, a scale that measures opinions about the appropriate role of women. You analyzed this variable in Chapter 2. Recall that fem_role ranges from 0 (women "domestic") to 9 (women in "work"). That is, higher scores denote less traditional beliefs. This is the dependent variable. GSS2008 also has polviews, a 7-point ordinal scale measuring ideology. Scores on polviews can range from 1 ("Extremely liberal") to 7 ("Extremely conservative"). This is the independent variable.

A. According to the hypothesis, as the values of polviews increase, from 1 through 7, mean values of fem_role should (circle one)

decrease.    neither decrease nor increase.    increase.

B. Test the hypothesis using Compare Means → Means. Write the results in the table that follows:

| Female role: home, work | | |
|---|---|---|
| Ideological self ID | Mean | N |
| Extremely liberal | ? | ? |
| Liberal | ? | ? |
| Slightly liberal | ? | ? |
| Moderate | ? | ? |
| Slightly conservative | ? | ? |
| Conservative | ? | ? |
| Extremely conservative | ? | ? |
| Total | ? | ? |

C. Do the results support the hypothesis?

Yes    No

Write a few sentences explaining your reasoning. _____

_____

_____

_____

_____

D. Obtain a line chart of this relationship. Print the line chart you created.

6. (Dataset: GSS2008. Variables: educ4, muslim_tol.) Are people who have more education more likely than the less educated to be tolerant of radical Muslim clerics? The hypothesis: In a comparison of individuals, those having lower levels of education will be less tolerant of radical Muslims than will those having higher levels of education.

   A. GSS2008 contains educ4, which measures years of educational attainment in four categories: 0–11 years, 12 years, 13–15 years, and 16 years or more. The variable muslim_tol, which you created in Chapter 3, gauges tolerance toward Muslim clerics with three codes: 0 (low tolerance), 1 (middle), and 2 (high tolerance). Use Crosstabs to test the hypothesis. Print the cross-tabulation table.

   B. Create a bar chart of the relationship. *Hint*: You might decide to graph, for each value of the independent variable, the percentage of respondents falling into the least tolerant category of the dependent variable. Alternatively, you could graph the percentage of respondents falling into the most tolerant category of the dependent variable. In the Graph Editor, give the vertical axis a more descriptive label. Edit the chart for appearance. Print the chart.

   C. Consider the tabular and graphic evidence. Does your analysis support the hypothesis that those having lower levels of education are less tolerant of radical Muslims than those having higher levels of education? Explain. _____

   _____

   _____

   _____

   _____

7. (Dataset: GSS2008. Variables: intrace1, affrmact2, natrace.) Untruthful answers by survey respondents can create big headaches for public opinion researchers. Why might a respondent not tell the truth to an interviewer? Certain types of questions, combined with particular characteristics of the interviewer, can trigger a phenomenon called preference falsification: "the act of misrepresenting one's genuine wants under perceived social pressures."[9] For example, consider the difficulty in gauging opinions on affirmative action, hiring policies aimed at giving preference to black applicants. One might reasonably expect that people questioned by an African American interviewer would express greater support for such programs than would those questioned by a white interviewer. An affirmative action opponent, not wanting to appear racially insensitive to a black questioner, might instead offer a false pro–affirmative action opinion.[10]

   GSS2008 contains intrace1, coded 1 for respondents questioned by a white interviewer and coded 2 for those questioned by a black interviewer. This is the independent variable that will allow you to test two preference falsification hypotheses:

   Hypothesis 1: In a comparison of individuals, those questioned by a black interviewer will be more likely to express support for affirmative action than will those questioned by a white interviewer. (The dependent variable is affrmact2, coded 1 for "support" and 2 for "oppose.")

   Hypothesis 2: In a comparison of individuals, those questioned by a black interviewer will be more likely to say that we are spending too little to improve the condition of blacks than will those questioned by a white interviewer. (The dependent variable is natrace, which is coded 1 for respondents saying "too little," 2 for those saying "about the right amount," and 3 for "too much.")

   A. Run the Crosstabs analysis. In the table that follows, record the percentages that support affirmative action and the percentages that say we are spending too little to improve the condition of blacks:

|  | Interviewer's race | |
|---|---|---|
|  | White | Black |
| Percent "support" affirmative action | ? | ? |
| Percent spending "too little" to improve condition of blacks | ? | ? |

B. These findings (circle one)

               support Hypothesis 1.       do not support Hypothesis 1.

Briefly explain your reasoning. _____

_____

_____

_____

C. These findings (circle one)

               support Hypothesis 2.       do not support Hypothesis 2.

Briefly explain your reasoning. _____

_____

_____

_____

D. Produce a bar chart depicting the relationship between support for affirmative action and interviewer race. Give the vertical axis a more descriptive title and edit the bar color. Make the chart more readable. Print the chart.

8. (Dataset: GSS2008. Variables: egalit_scale, income06_5, degree.) It was among the most memorable sound bites of the 2008 presidential campaign. In response to a question from "Joe the Plumber," Barack Obama said that, to achieve better economic opportunity, we should "spread the wealth around."[11] *Economic egalitarianism*, the idea that wealth is unfairly concentrated and should be distributed more equitably throughout society, is a controversial ideology. Who supports it and who opposes it? Two differences between people in socioeconomic status, income and education, could help account for differences in egalitarian beliefs. The income hypothesis: In a comparison of individuals, people who have lower incomes will be stronger supporters of economic egalitarianism than will people having higher incomes.

A. Suppose someone proposes that, because income and education gauge social status in the same way, the education-egalitarianism relationship will be similar to the income-egalitarianism relationship. The education hypothesis: In a comparison of individuals, those having lower levels of education (complete the sentence) _____

_____

_____.

B. GSS2008 has egalit_scale, an interval-level variable that ranges from 0 (low egalitarianism) to 12 (high egalitarianism). The dataset also contains income06_5, which measures income in five categories, and degree, each respondent's highest academic degree. To test the income and education hypotheses, run the appropriate analyses. Print the tabular output.

C. Obtain line charts for the income-egalitarianism and education-egalitarianism relationships. Print the charts.

D. Consider the income-egalitarianism analysis. Does the analysis support the income hypothesis?

Yes    No

Explain your reasoning: _____

_____

_____

_____

E. Consider the education-egalitarianism analysis. In what ways is the education-egalitarianism relationship similar to the income-egalitarianism relationship? _____

_____

_____

F. In what ways is the education-egalitarianism relationship different from the income-egalitarianism relationship? _____

_____

_____

_____

9. (Dataset: States. Variables: union07, region.) Where are unions stronger? Where are they weaker? Consider the following two claims:

Claim 1: States in the northeastern United States are more likely to have unionized workforces than are states in the South.

Claim 2: States in the Midwest and West have levels of unionization more similar to those in the South than to those in the Northeast.

A. The States dataset contains union07, the percentage of each state's workforce who are union members. Another variable, region, is a four-category census classification of the states. Run a mean comparison analysis. Record your results in the table that follows:

| Percentage of workers who are union members | | |
|---|---|---|
| Census region | Mean | N |
| Northeast | ? | ? |
| Midwest | ? | ? |
| South | ? | ? |
| West | ? | ? |
| Total | ? | 50 |

B. Based on your analysis, would you say that Claim 1 is correct or incorrect? How do you know? Check the appropriate box and complete the sentence that follows.

❑ Claim 1 is correct, because _____

_____

_____ .

❑ Claim 1 is incorrect, because _____

_____

_____ .

C. Based on your analysis, would you say that Claim 2 is correct or incorrect? How do you know? Check the appropriate box and complete the sentence that follows.

❑ Claim 2 is correct, because _____

_____

_____ .

❑ Claim 2 is incorrect, because _____

_____

_____ .

D. Produce a bar chart of the relationship you just analyzed. Edit the chart for readability and appearance. Print the chart.

10. (Dataset: States. Variables: cigarettes, cig_tax_3.) Two policy researchers are arguing about whether higher taxes on cigarettes reduce cigarette consumption.

Policy researcher 1: "The demand for cigarettes is highly inelastic—smokers need to consume cigarettes, and they will buy them without regard to the cost. Raising taxes on a pack of cigarettes will have no effect on the level of cigarette consumption."

Policy researcher 2: "Look, any behavior that's taxed is discouraged. If state governments want to discourage smoking, then raising cigarette taxes will certainly have the desired effect. Higher taxes mean lower consumption."

Imagine a line chart of the relationship between cigarette taxes and cigarette consumption. The category axis measures state cigarette taxes in three categories, from lower taxes on the left to higher taxes on the right. The vertical axis records per-capita cigarette consumption. Now consider line charts X, Y, and Z, which follow:

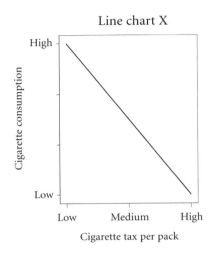

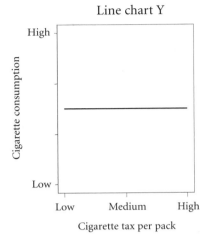

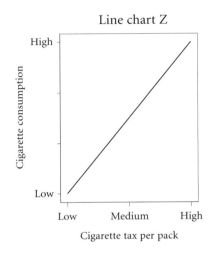

A. If policy researcher 1 were correct, which line chart would most accurately depict the relationship between states' cigarette taxes and cigarette consumption? (circle one)

Line chart X       Line chart Y       Line chart Z

B. If policy researcher 2 were correct, which line chart would most accurately depict the relationship between states' cigarette taxes and cigarette consumption? (circle one)

Line chart X       Line chart Y       Line chart Z

C. The States dataset contains the variables cigarettes and cig_tax_3. Run a mean comparison analysis, using cigarettes as the dependent variable and cig_tax_3 as the independent variable. Record your results in the table that follows:

| cigarettes Packs bimonthly per adult pop | | |
|---|---|---|
| cig_tax_3 Cigarette tax per pack: 3 categories | Mean | N |
| 1 $.07–$.64 | ? | ? |
| 2 $.695–$1.36 | ? | ? |
| 3 $1.41–$2.58 | ? | ? |
| Total | ? | 50 |

D. Create and print a line chart of the relationship.

E. Examine the mean comparison table and the line chart. Which policy researcher is more correct? (check one)

❑ Policy researcher 1 is more correct.

❑ Policy researcher 2 is more correct.

11. (Dataset: World. Variables: enpp3_democ, district_size3, frac_eth3.) Two scholars of comparative politics are discussing possible reasons why some democracies have many political parties and other democracies have only a few.

Scholar 1: "It all has to do with the rules of the election game. Some countries, such as the United Kingdom, have single-member electoral districts. Voters in each district elect only one representative. This militates in favor of fewer and larger parties, since small parties have less chance of winning enough votes to gain the seat. Other countries, like Switzerland, have multi-member districts. Because voters choose more than one representative per district, a larger number of smaller parties have a chance to win representation. It doesn't surprise me in the least, then, that the U.K. has fewer political parties than Switzerland."

Scholar 2: "I notice that your explanation fails to mention the single most important determinant of the number of political parties: social structural heterogeneity. Homogeneous societies, those with few linguistic or religious differences, have fewer conflicts and thus fewer parties. Heterogeneous polities, by the same logic, are more contentious and will produce more parties. By the way, the examples you picked to support your case also support mine: the U.K. is relatively homogeneous and Switzerland relatively heterogeneous. It doesn't surprise me in the least, then, that the U.K. has fewer political parties than Switzerland."

A.  Scholar 1's hypothesis: In a comparison of democracies, those having single-member districts will have (circle one)

    fewer political parties        more political parties

    than democracies electing multiple members from each district.

B.  State Scholar 2's hypothesis: _____

    _____

    _____

    _____

C.  The World dataset variable enpp3_democ measures, for each democracy, the number of effective parliamentary parties: 1–3 parties (coded 1), 4–5 parties (coded 2), or 6–11 parties (coded 3). Use enpp3_democ as the dependent variable to test each hypothesis. For independent variables, test Scholar 1's hypothesis using district_size3, which measures the number of seats per district: countries with single-member districts are coded 1, countries that average more than one but fewer than six members are coded 2, and countries with six or more members per district are coded 3. Test Scholar 2's hypothesis using frac_eth3, which classifies each country's level of ethnic/linguistic fractionalization as low (coded 1), medium (coded 2), or high (coded 3). Countries with higher codes on frac_eth3 have a higher level of ethnic conflict. Run Crosstabs to test the hypotheses. In the table that follows, record the percentages of cases falling into the lowest code of the dependent variable, 1–3 parties:

|  | Average # of members per district | | |
| --- | --- | --- | --- |
|  | Single-member | > 1 to 5 members | 6 or more members |
| Percentage having 1–3 parties | ? | ? | ? |
|  | Level of ethnic fractionalization | | |
|  | Low | Medium | High |
| Percentage having 1–3 parties | ? | ? | ? |

D.  Which of the following statements best summarizes your findings? (check one)

❑  Scholar 1's hypothesis is supported by the analysis, but Scholar 2's hypothesis is not supported by the analysis.

❑  Scholar 2's hypothesis is supported by the analysis, but Scholar 1's hypothesis is not supported by the analysis.

❑  Both hypotheses are supported by the analysis.

❑  Neither hypothesis is supported by the analysis.

E.  Making specific reference to your findings, write a paragraph explaining your choice in part D. _____

_____

_____

_____

_____

_____

_____

12. (Dataset: World. Variables: durable, regime_type3.) The two comparative politics scholars are still arguing, only now they're trying to figure out what sort of institutional arrangement produces the longest-lasting, most stable political system. Also, a third scholar joins the interchange of ideas.

Scholar 1: "Presidential democracies, like the United States, are going to be more stable than are any other type of system. In presidential democracies, the executive and the legislature have separate electoral constituencies and separate but overlapping domains of responsibility. The people's political interests are represented both by the president's national constituency and by legislators' or parliament members' more localized constituencies. If one branch does something that's unpopular, it can be blocked by the other branch. The result: political stability."

Scholar 2: "Parliamentary democracies are by far more stable than presidential democracies. In presidential systems, the executive and legislature can be controlled by different political parties, a situation that produces deadlock. Since the leaders of the legislature can't remove the president and install a more compliant or agreeable executive, they are liable to resort to a coup, toppling the whole system. Parliamentary democracies avoid these pitfalls. In parliamentary democracies, all legitimacy and accountability resides with the legislature. The parliament organizes the government and chooses the executive (the prime minister) from among its own leaders. The prime minister and members of parliament have strong incentives to cooperate and keep things running smoothly and efficiently. The result: political stability."

Scholar 3: "You two have made such compelling—if incorrect—arguments that I almost hesitate to point this out: Democracies of any species, presidential or parliamentary, are inherently unstable. Any system that permits the clamor of competing parties or dissident viewpoints is surely bound to fail. If it's stability that you value above all else, then dictatorships will deliver. Strong executives, feckless or nonexistent legislatures, powerful armies, social control. The result: political stability."

The World dataset contains the variable durable, which measures the number of years since the last regime transition. The more years that have passed since the system last failed (higher values on durable), the more stable a country's political system. The variable regime_type3 captures system type: dictatorship, parliamentary democracy, or presidential democracy.

A. Run Compare Means and examine the output. Which of the following inferences are supported by your analysis? (check all that apply)

❑ Parliamentary democracy is the most stable form of government.

❑ On average, parliamentary democracies last about 20 years longer than dictatorships.

❑ Presidential democracies are more stable than dictatorships.

❑ Of the 141 countries included in the analysis, dictatorships are the most prevalent.

❑ Among the 141 countries included in the analysis, the typical regime lasts about 24 years.

B. Create and print a bar chart depicting the relationship between regime durability and regime type.

13. (Dataset: World. Variables: fhrate08_rev, dem_other5.) Why do some countries develop democratic systems whereas others do not? Certainly the transition toward democracy—or away from it—is a complex process. Some scholars emphasize factors internal to countries, such as educational attainment or economic development. Others look to external factors, such as patterns of governance that are prevalent in a country's region of the world. Perhaps governmental systems are like infectious diseases: Similar systems diffuse among countries in close geographic proximity. According to the democratic diffusion hypothesis, countries in regions having fewer democracies are themselves less likely to be democratic than are countries in regions having more democracies.[12]

A. Suppose you had two variables for a large number of countries: a dependent variable that measured democracy along an interval-level scale, with higher scores denoting higher levels of democracy, and an independent variable measuring the number of democracies in each country's region, from fewer democracies to more democracies. According to the democratic diffusion hypothesis, if you were to compare mean values of the dependent variable for countries having different values on the independent variable, you should find (check one)

❑ a lower mean of the dependent variable for countries in regions having fewer democracies than for countries in regions having more democracies.

❑ a higher mean of the dependent variable for countries in regions having fewer democracies than for countries in regions having more democracies.

❑ no difference between the means of the dependent variable for countries in regions having fewer democracies and countries in regions having more democracies.

B. The World dataset contains fhrate08_rev, an interval-level measure of democracy based on the Freedom House rating system. Scores range from 0 (least democratic) to 12 (most democratic).[13] This is the dependent variable. World also has dem_other5, a 5-category ordinal measure that divides countries into five groups, based on the percentage of democracies in each country's geographic region: 10 percent, approximately 40 percent, approximately 60 percent, approximately 90 percent, or 100 percent. This is the independent variable. Run the appropriate mean comparison analysis. Label and record the results in the table that follows:

| FH 1–7 scale reversed, rescaled 0–12 | | |
|---|---|---|
| dem_other5 * Percentage of other democracies in region: 5 cats | Mean | N |
| 1 10% | ? | ? |
| 2 Approx 40% | ? | ? |
| 3 Approx 60% | ? | ? |
| 4 Approx 90% | ? | ? |
| 5 100% | ? | ? |
| Total | ? | 188 |

C. Create and print a line chart of the relationship you just analyzed.

D. Examine the mean comparison table and the line chart. Which of the following statements are supported by your analysis? (check all that apply)

❑ Countries in regions having fewer democracies are more likely to be democratic than are countries in regions having more democracies.

❑ The relationship between the independent and dependent variables is positive.

❑ The democratic diffusion hypothesis is incorrect.

❑ Countries in regions having fewer democracies are less likely to be democratic than are countries in regions having more democracies.

❑ The relationship between the independent and dependent variables is negative.

14. (Dataset: World. Variables: decentralization4, effectiveness, confidence.) Are decentralized governments more effective than centralized governments? In decentralized systems, local officials have politically autonomous authority to raise public money and administer government programs. This leads to more effective governance and, the argument goes, inspires confidence among citizens. Centralized systems, remote from local problems and burdened by red tape, may be both less effective and less likely to be viewed positively by citizens. Given these putative benefits, it is little wonder that "policy makers and politicians have frequently pushed for decentralisation as a panacea for the ills of poor governance."[14]

The World dataset variable decentralization4 measures, for each country, the level of decentralization by four ordinal categories, from less decentralization (coded 1) to more decentralization (coded 4). The assessments of a panel of expert observers were used to create the variable effectiveness, which measures government effectiveness on a 100-point scale (higher scores denote greater effectiveness). Confidence, also on a 100-point scale, gauges the degree to which a country's citizens have "a great deal" or "quite a lot" of confidence in state institutions (higher scores denote higher confidence).

A. The World variables will permit you to test two hypotheses about the effects of decentralization on effectiveness and confidence. State the two hypotheses.

Effectiveness hypothesis: _____

_____

_____

Confidence hypothesis: _____

_____

_____

B. Test the hypotheses with Compare Means. Write the results in the table that follows:

| Level of decentralization | | Government effectiveness scale | Confidence in institutions scale |
|---|---|---|---|
| Low | Mean | ? | ? |
| | N | ? | ? |
| 2 | Mean | ? | ? |
| | N | ? | ? |
| 3 | Mean | ? | ? |
| | N | ? | ? |
| High | Mean | ? | ? |
| | N | ? | ? |
| Total | Mean | ? | ? |
| | N | ? | ? |

C. The effectiveness hypothesis (circle one)

　　　　　　　　　　is supported　　　　is not supported

by the analysis.

Explain your reasoning. _____

_____

_____

D. The confidence hypothesis (circle one)

　　　　　　　　　　is supported　　　　is not supported

by the analysis.

Explain your reasoning. _____

_____

_____

_____

That concludes the exercises for this chapter. Before exiting SPSS, be sure to save your output file.

## NOTES

1. Because we are graphing one relationship, we want a single line. And because we are comparing groups of partisans, we want SPSS to display a summary measure, the mean, for each group.

2. Unless we modify the Line Represents panel to suit our analysis, SPSS will produce a line chart for the number of cases (N of cases) in each category of partyid7.

3. Of course, you will encounter situations in which you do not want mean values. Later in this chapter we review the procedure for Change Statistic.

4. The Change Statistic button will not be available unless the variable in the Variable box is highlighted. A variable is highlighted automatically when you click it into the Variable box. If you are experimenting and lose the highlighting, simply click directly on the variable in the Variable box. This restores the highlighting.

5. The same result can be achieved by clicking the Percentage below radio button and typing a "3" in the Value box. Because code 1 is the only value of camp_int lower than 3, SPSS will return the percentage of respondents having code 1 on camp_int.

6. V. O. Key, *Politics, Parties, and Pressure Groups,* 5th ed. (New York: Crowell, 1964), 568.

7. George A. Quattrone and Amos Tversky, "Contrasting Rational and Psychological Analyses of Political Choice," *American Political Science Review* 82 (1988): 719–736.

8. A fourth category, citizens who voted in 2004 but not in 2008, contains only 86 respondents. This category was set to missing.

9. Timur Kuran, *Private Truths, Public Lies: The Social Consequences of Preference Falsification* (Cambridge: Harvard University Press, 1995), 3.

10. It may have occurred to you that this effect might be greater for white respondents than for black respondents, with white subjects more likely to hide their true preferences in the presence of a black interviewer. An exercise in Chapter 5 will give you a chance to investigate this possibility.

11. Obama made the remark on October 12, 2008, while campaigning in Toledo, Ohio.

12. See Jeffrey S. Kopstein and David A. Reilly, "Geographic Diffusion and the Transformation of the Postcommunist World," *World Politics* 53 (2000): 1–37. Noting that "[a]ll of the big winners of postcommunism share the trait of being geographically close to the former border of the noncommunist world" (p. 1), Kopstein and Reilly use geographic proximity to the West as the independent variable in testing the democratic diffusion hypothesis. By and large, the authors find "that the farther away a country is from the West, the less likely it is to be democratic" (p. 10).

13. The original Freedom House scale ranges from 1 (most democratic) to 7 (least democratic). The author reversed the coding and rescaled fhrate08_rev so that it ranges from 0 (least democratic) to 12 (most democratic).

14. Conor O'Dwyer and Daniel Ziblatt, "Does Decentralisation Make Government More Efficient and Effective?" *Commonwealth & Comparative Politics* 44 (November 2006): 1–18. This quote is from page 2. O'Dwyer and Ziblatt use sophisticated multivariate techniques to test the hypothesis that decentralized systems are more effective. Interestingly, they find that decentralization produces a higher quality of governance in richer countries but a lower quality of governance in poorer countries.

# 5

# Making Controlled Comparisons

Procedures Covered

Analyze → Descriptive Statistics → Crosstabs (with Layers)

Analyze → Compare Means → Means (with Layers)

Graphs → Legacy Dialogs → Multiple Line

Political analysis often begins by making simple comparisons using cross-tabulation analysis or mean comparison analysis. Simple comparisons allow the researcher to examine the relationship between an independent variable, X, and a dependent variable, Y. However, there is always the possibility that alternative causes—rival explanations—are at work, affecting the observed relationship between X and Y. An alternative cause is symbolized by the letter Z. If the researcher does not control for Z, then he or she may misinterpret the relationship between X and Y.

What can happen to the relationship between an independent variable and a dependent variable, controlling for an alternative cause? One possibility is that the relationship between the independent variable and the dependent variable is spurious. In a spurious relationship, once the researcher controls for a rival causal factor, the original relationship becomes very weak, perhaps disappearing altogether. The control variable does all the explanatory work. In another possibility, the researcher observes an additive relationship between the independent variable, the dependent variable, and the control variable. In an additive relationship, two sets of meaningful relationships exist. The independent variable maintains a relationship with the dependent variable, and the control variable helps to explain the dependent variable. A third possibility, interaction, is somewhat more complex. If interaction is occurring, then the effect of the independent variable on the dependent variable depends on the value of the control variable. The strength or tendency of the relationship is different for one value of the control variable than for another value of the control variable.

These situations—a spurious relationship, an additive relationship, and interaction—are logical possibilities. Of course, SPSS cannot interpret a set of controlled comparisons for you. But it can produce tabular analysis and graphics that will give you the raw material you need to evaluate controlled comparisons.

In this chapter you will learn to use Crosstabs to analyze relationships when all three variables—the independent variable, the dependent variable, and the control variable—are nominal or ordinal. In this chapter you will also learn to use Compare Means to analyze relationships in which the dependent variable is interval level and the independent and control variables are nominal or ordinal level. Because graphic displays are especially valuable tools for evaluating complex relationships, we will demonstrate how to obtain and edit multiple line charts. These skills are natural extensions of the procedures you learned in Chapter 4.

## CROSS-TABULATION ANALYSIS WITH A CONTROL VARIABLE

To demonstrate how to use SPSS Crosstabs to obtain control tables, we will work through an example with GSS2008. This guided example uses one of the variables you created in Chapter 3, polview3.

We begin with this hypothesis: In a comparison of individuals, liberals will be more likely to favor the legalization of marijuana than will conservatives. In this hypothesis, polview3, which categorizes respondents as liberal, moderate, or conservative, is the independent variable. GSS2008 contains the variable grass, which records respondents' opinions on the legalization of marijuana. (Code 1 is "Legal," and code 2 is "Not legal.") To stay acquainted with cross-tabulation analysis, we will start by looking at the uncontrolled relationship between polview3 and grass. In addition to considering whether the hypothesis has merit, we will note the tendency of the relationship, and we will apply a nonstatistical measure of the relationship's strength. By determining tendency and gauging strength, you are better able to interpret relationships involving control variables.

Open GSS2008. Go through the following steps, as covered in Chapter 4: Click Analyze → Descriptive Statistics → Crosstabs. Find the dependent variable, grass, in the left-hand variable list and click it into the Row(s) panel. Find the independent variable, polview3, and click it into the Column(s) panel. Click the Cells button and select the box next to "Column" in the Percentages panel. Click Continue, and then click OK. SPSS reports the results:

grass  Should marijuana be made legal  * polview3 Ideology: 3 categories Crosstabulation

| | | | polview3 Ideology: 3 categories | | | |
|---|---|---|---|---|---|---|
| | | Statistics | 1 Liberal | 2 Moderate | 3 Conservative | Total |
| grass Should marijuana be made legal | 1 Legal | Count | 164 | 185 | 113 | 462 |
| | | % within polview3 Ideology: 3 categories | 54.7% | 41.0% | 25.6% | 38.7% |
| | 2 Not legal | Count | 136 | 266 | 329 | 731 |
| | | % within polview3 Ideology: 3 categories | 45.3% | 59.0% | 74.4% | 61.3% |
| Total | | Count | 300 | 451 | 442 | 1193 |
| | | % within polview3 Ideology: 3 categories | 100.0% | 100.0% | 100.0% | 100.0% |

Clearly, the hypothesis has merit. If no relationship existed between marijuana opinions and ideology, then roughly equal percentages of liberals, moderates, and conservatives would favor legalization. Obviously, the percentages are not equal. Of the liberals, 54.7 percent favor legalization, compared with 41.0 percent of moderates and 25.6 percent of conservatives. And note that, given the way polview3 is coded—increasing values denote increasing conservatism—a negative relationship exists between ideology and the percentage favoring legalization. As conservatism increases, the percentage favoring legalization declines. (If you interpret the cross-tab by examining the "Not legal" row, then the tendency is positive. As conservatism increases, the percentage opposing legalization increases.) How strong is the relationship? You can arrive at a quick and easy measure of strength by figuring out the percentage-point change in the dependent variable across the full range of the independent variable. At one pole, 54.7 percent of liberals favor legalization. At the other pole, 25.6 percent of conservatives are in favor. Therefore, the percentage favoring legalization drops by 54.7 − 25.6 = 29.1, or about 29 percentage points. By this rudimentary measure, the relationship's strength is 29. (In Chapter 7, we consider statistical measures of strength.)

What other factors, besides ideology, might account for differing opinions on marijuana legalization? A plausible answer: whether the respondent has children. Regardless of ideology, people with children may be less inclined to endorse the legalization of an illegal drug than are people who do not have children. And here is an interesting (if complicating) fact: Conservatives are more likely to have children than are liberals.[1] Thus, when we compare the marijuana opinions of liberals and conservatives, as we have just done, we are also comparing people who are less likely to have children (liberals) with people who are more likely to have children (conservatives). It could be that liberals are more inclined to favor legalization, not because they are liberal per se, but because they are less likely to have children. By the same token, conservatives might oppose legalization for reasons unrelated to their ideology: They're more likely to have children. The only way to isolate the effect of ideology on marijuana opinions is to compare liberals who do not have children with conservatives who do not have children, and to compare liberals who have children with conservatives who have children. In other words, we need to control for the effect of having children by holding it constant. Crosstabs with layers will perform the controlled comparison we are after.

GSS2008 contains the variable kids, which classifies respondents into one of two categories: those with children (coded 1 and labeled "Yes" on kids) or those without (coded 0 and labeled "No" on kids). Let's run the analysis again, this time adding kids as a control variable.

Again click Analyze → Descriptive Statistics → Crosstabs, returning to the Crosstabs window. You will find the dependent variable, grass, and the independent variable, polview3, just where you left them. To obtain a controlled comparison—the relationship between grass and polview3, controlling for kids—scroll down the variable list until you find kids and click it into the box labeled "Layer 1 of 1," as shown in Figure 5-1. SPSS will run a separate cross-tabulation analysis for each value of the variable that appears in the Layer box. And that is precisely what we want: a cross-tabulation of grass and polview3 for respondents without children and a separate analysis for those with children. Click OK. SPSS returns its version of a control table:

**grass Should marijuana be made legal * polview3 Ideology: 3 categories * kids Does R have children? Crosstabulation**

| kids Does R have children? | | | Statistics | polview3 Ideology: 3 categories | | | Total |
|---|---|---|---|---|---|---|---|
| | | | | 1 Liberal | 2 Moderate | 3 Conservative | |
| 0 No | grass Should marijuana be made legal | 1 Legal | Count | 59 | 53 | 32 | 144 |
| | | | % within polview3 Ideology: 3 categories | 59.0% | 49.5% | 33.3% | 47.5% |
| | | 2 Not legal | Count | 41 | 54 | 64 | 159 |
| | | | % within polview3 Ideology: 3 categories | 41.0% | 50.5% | 66.7% | 52.5% |
| | Total | | Count | 100 | 107 | 96 | 303 |
| | | | % within polview3 Ideology: 3 categories | 100.0% | 100.0% | 100.0% | 100.0% |
| 1 Yes | grass Should marijuana be made legal | 1 Legal | Count | 105 | 132 | 82 | 319 |
| | | | % within polview3 Ideology: 3 categories | 52.2% | 38.5% | 23.6% | 35.8% |
| | | 2 Not legal | Count | 96 | 211 | 265 | 572 |
| | | | % within polview3 Ideology: 3 categories | 47.8% | 61.5% | 76.4% | 64.2% |
| | Total | | Count | 201 | 343 | 347 | 891 |
| | | | % within polview3 Ideology: 3 categories | 100.0% | 100.0% | 100.0% | 100.0% |
| Total | grass Should marijuana be made legal | 1 Legal | Count | 164 | 185 | 114 | 463 |
| | | | % within polview3 Ideology: 3 categories | 54.5% | 41.1% | 25.7% | 38.8% |
| | | 2 Not legal | Count | 137 | 265 | 329 | 731 |
| | | | % within polview3 Ideology: 3 categories | 45.5% | 58.9% | 74.3% | 61.2% |
| | Total | | Count | 301 | 450 | 443 | 1194 |
| | | | % within polview3 Ideology: 3 categories | 100.0% | 100.0% | 100.0% | 100.0% |

SPSS output using Crosstabs with layers can be a bit confusing at first, so let's consider closely what SPSS has produced. There are two cross-tabulations, appearing as one table. To the left-hand side of the table you will see the label of the control variable, kids: "Does R have children?" The first value of kids, "No," appears beneath that label. So the top cross-tabulation shows the grass-polview3 relationship for people who do not have children. The bottom cross-tabulation shows the relationship for respondents with children, respondents with the value "Yes" on the control variable.

**Figure 5-1** Crosstabs with Layers

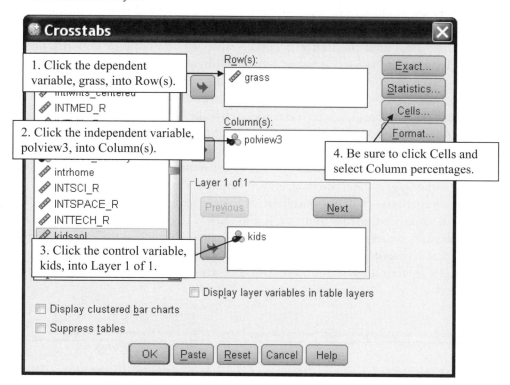

First assess the tendency and strength of the relationship between ideology and support for marijuana legalization among respondents who do not have children. Then assess tendency and strength among respondents who have children. Among people without children, the tendency is negative. As the values of polview3 increase from liberal to conservative, support for legalization declines: 59.0 percent of the liberals favor legalization, compared with 49.5 percent of the moderates and 33.3 percent of the conservatives. How large is the drop? Across the full range of ideological self-identification, the percentage favoring legalization declines from 59.0 among liberals to 33.3 among conservatives—an "ideology effect" of about 26 percentage points. Turn your attention to respondents with children. Note that the tendency, once again, is negative: 52.2 percent of the liberals favor legalization, compared with 38.5 percent of moderates and 23.6 percent of conservatives. Indeed, the strength of the relationship is about the same for people who have children as for people who do not have children. Among respondents with kids, the percentage who favor legalization drops from 52.2 among liberals to 23.6 among conservatives—an "ideology effect" of about 29 points.

To help you make correct interpretations of controlled comparisons, it is a good idea to evaluate the relationship between the control variable and the dependent variable, controlling for the independent variable. In the current example, we would determine the tendency and strength of the relationship between the control variable, kids, and marijuana attitudes, controlling for ideology. This is accomplished by jumping between the top cross-tabulation and the bottom cross-tabulation, comparing marijuana opinions of people who share the same ideology but who differ on the control variable, kids. Consider liberals. You can see that liberals without kids are more likely to favor legalization than are liberals with kids. When the control variable switches from "No" (top cross-tab) to "Yes" (bottom cross-tab), the percentage of marijuana supporters drops. Among liberals without children, 59.0 percent favor legalization, compared with 52.2 percent for liberals with children—a "kid effect" of about 7 percentage points. How about moderates? For moderates, as with liberals, the kid effect is negative, although the effect is a slightly stronger 11 points: 49.5 percent compared with 38.5 percent. Conservatives add to the repetitiveness of the pattern. Among conservatives without children, 33.3 percent favor legalization, a number that declines to 23.6 percent among conservatives, about a 10-point drop.

How would you characterize this set of relationships? Does a spurious relationship exist between grass and polview3? Or are these additive relationships, with polview3 helping to explain legalization opinions and kids adding to the explanation? Or is interaction going on? If the grass-polview3 relationship were spurious, then the relationship would weaken or disappear after controlling for kids. Among respondents without children, liberals, moderates, and conservatives would all hold the same opinion about marijuana legalization. Ditto for people with children: Ideology would not play a role in explaining the dependent variable. Because the relationship persists after controlling for kids, we can rule out spuriousness. Now, it is sometimes difficult to distinguish between additive relationships and interaction relationships, so let's dwell on this question. In additive relationships, the effect of the independent variable on the dependent variable is the same or quite similar for each value of the control variable. In interaction relationships, by contrast, the effect of the independent variable on the dependent variable varies in tendency or strength for different values of the control variable.

According to the analysis, the grass-polview3 relationship has the same tendency for people with and without children: For both values of the control, liberals are more pro-legalization than are conservatives. Tellingly, the grass-polview3 relationships are quite similar in strength for people who do not have children and for people who do have children. For respondents without kids, the ideology effect is 26. For those with kids, the effect is 29. Notice, too, that the kid effect has the same tendency and roughly the same strength for liberals, moderates, and conservatives. To be sure, the grass-polview3-kids relationships are not paragons of symmetrical perfection—real-world relationships rarely are—but the pattern more closely approximates an additive pattern than an interactive pattern.

Additive relationships always take the same form: the relationship between the independent and dependent variables has the same tendency and the same or very similar strength at all values of the control variable. Interaction relationships are more complex—and they are probably more common. So that you can become comfortable recognizing interaction in cross-tabulations, we will present another example related to the line of analysis we have been pursuing. How do liberals, moderates, and conservatives differ in the behavioral traits they consider important in children, such as obedience to authority? It seems plausible to suggest that liberals are less likely than conservatives to stress obedience. Furthermore, regardless of ideology, people with children may place more importance on obedience than will people without children. Let's run the analysis and see whether these ideas are borne out by the data.

GSS2008 contains obey2, coded 1 for respondents who think that obedience is among a child's "more important" traits, and coded 2 for respondents who think obedience is "less important." Return to the previous Crosstabs window. Click grass back into the Variable list. Find obey2 and click it into the Row(s) panel. Click OK. An SPSS cross-tabulation control table is perhaps more familiar this time around.

obey2 Obedience important in children? * polview3 Ideology: 3 categories * kids Does R have children? Crosstabulation

| kids Does R have children? | | Statistics | polview3 Ideology: 3 categories | | | Total |
|---|---|---|---|---|---|---|
| | | | 1 Liberal | 2 Moderate | 3 Conservative | |
| 0 No | obey2 Obedience important in children? 1 More important | Count | 19 | 26 | 33 | 78 |
| | | % within polview3 Ideology: 3 categories | 17.3% | 21.7% | 32.0% | 23.4% |
| | 2 Less important | Count | 91 | 94 | 70 | 255 |
| | | % within polview3 Ideology: 3 categories | 82.7% | 78.3% | 68.0% | 76.6% |
| | Total | Count | 110 | 120 | 103 | 333 |
| | | % within polview3 Ideology: 3 categories | 100.0% | 100.0% | 100.0% | 100.0% |
| 1 Yes | obey2 Obedience important in children? 1 More important | Count | 53 | 97 | 107 | 257 |
| | | % within polview3 Ideology: 3 categories | 25.2% | 25.9% | 29.3% | 27.1% |
| | 2 Less important | Count | 157 | 277 | 258 | 692 |
| | | % within polview3 Ideology: 3 categories | 74.8% | 74.1% | 70.7% | 72.9% |
| | Total | Count | 210 | 374 | 365 | 949 |
| | | % within polview3 Ideology: 3 categories | 100.0% | 100.0% | 100.0% | 100.0% |
| Total | obey2 Obedience important in children? 1 More important | Count | 72 | 123 | 140 | 335 |
| | | % within polview3 Ideology: 3 categories | 22.5% | 24.9% | 29.9% | 26.1% |
| | 2 Less important | Count | 248 | 371 | 328 | 947 |
| | | % within polview3 Ideology: 3 categories | 77.5% | 75.1% | 70.1% | 73.9% |
| | Total | Count | 320 | 494 | 468 | 1282 |
| | | % within polview3 Ideology: 3 categories | 100.0% | 100.0% | 100.0% | 100.0% |

To ensure consistency with the analysis of marijuana opinions, in which we tracked the more permissive response ("legal"), let's focus here on the percentage of respondents saying "less important." Clearly enough, ideology has a big effect on the dependent variable for people without children. As the independent variable changes from liberal to moderate to conservative, the percentage of respondents who think obedience is "less important" declines by nearly 15 points: from 82.7 percent of liberals, to 78.3 percent of moderates, to 68.0 percent among conservatives. So the relationship has a negative tendency, as viewed along the "less important" row, and a strength equal to about 15. Do we see the same tendency and similar strength among respondents who have kids? Not at all. For respondents with kids, ideology has virtually no effect on beliefs about obedience. As the independent variable changes from liberal to moderate, the percentage of respondents who think obedience is "less important" does not budge: 74.8 percent of liberals compared with 74.1 percent of moderates. Conservatives with kids, 70.7 percent of whom say "less important," show only slightly less agreement with their liberal and moderate counterparts. Gauged end-to-end, the percentages drop from 74.8 to 70.7, a paltry 4-point decline. To confirm an interaction interpretation, evaluate the kid effect separately for liberals, moderates, and conservatives. For liberals the effect is negative. Among liberals without children, 82.7 percent say "less important," compared with 74.8 percent among liberals with children, about an 8-point difference. For moderates, the negative effect is a more modest 4 points (78.3 percent compared with 74.1). Among conservatives, the kid effect shows a weakly positive tendency: 68.0 percent compared with 70.7 percent.

Although a relationship exists between ideology and childrearing beliefs at one value of the control—among people with no childrearing experience—ideological predispositions have very little effect at the other value of the control—among people who have childrearing experience. Moreover, the relationship between obey2 and the control variable, kids, varies in tendency and strength across the values of the independent variable, from a noticeable 8-point negative tendency among liberals to a slight 3-point positive tendency among conservatives. This sort of asymmetrical pattern, in which a relationship exists at one value of the control but is nonexistent at another value of the control, is a common form of interaction. It bears emphasizing, however, that interaction may take other forms. One of these alternative forms is discussed later in the chapter.

## GRAPHING RELATIONSHIPS WITH A CONTROL VARIABLE

In Chapter 4 you learned how to obtain a bar chart or line chart depicting the relationship between an independent variable and a dependent variable. SPSS also produces two types of graphs for controlled comparisons, clustered bar charts and multiple line charts. For clarifying controlled comparisons, multiple line charts are preferred. Compared with bar charts, line charts are simpler and more elegant, and they have a more favorable data-ink ratio, defined as "the proportion of a graphic's ink devoted to the nonredundant display of data-information."[2] In other words, if one were to "add up" all the ink used in a graph, line charts tend to devote a larger proportion of the total ink to the essential communication of the data.[3] In the following guided example, we produce a multiple line chart for the additive relationship analyzed at the beginning of the chapter, the relationship between grass and polview3, controlling for kids.

Click Graphs → Legacy Dialogs → Line. When the Line Charts window opens, click Multiple and make sure that the Summaries for Groups of Cases radio button is selected (the default). Click Define. The Define Multiple Line: Summaries for Groups of Cases window appears (Figure 5-2). What do we want the line chart to depict? We want to see the percentage of respondents who think marijuana should be legal (code 1 on grass) for each value of the independent variable (polview3). Furthermore, we want to see the grass-polview3 relationship separately for each value of the control variable, kids. In all SPSS charts, the values of the independent variable appear along the axis labeled "Category Axis." Because polview3 is the independent variable, click polview3 into the Category Axis box, as shown in Figure 5-3. For each value of polview3, we want to see the relationship separately for different values of the control variable, kids. In a multiple line chart, the values of the control variable "define" the lines. The variable kids is the control variable, so click kids into the Define Lines by box.

Now we need to make sure that the lines will represent the percentages of respondents saying "legal." In the Lines Represent panel, select the Other statistic radio button, which activates the Variable box. Find grass in the variable list and then click it into the Variable box. By default, SPSS will display the mean value of grass, "MEAN(grass)," which does not suit our purpose (see Figure 5-3). Click Change Statistic. In the Statistic

**Figure 5-2**  Opening the Define Multiple Line Window

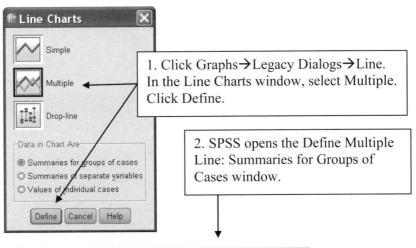

1. Click Graphs→Legacy Dialogs→Line. In the Line Charts window, select Multiple. Click Define.

2. SPSS opens the Define Multiple Line: Summaries for Groups of Cases window.

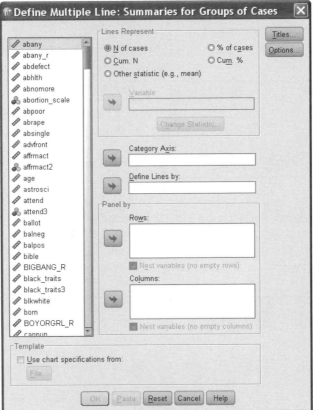

window, click the Percentage inside radio button. Type "1" in the Low box and "1" in the High box. As in Figure 5-3, these instructions tell SPSS to display the percentage of respondents in one category of the dependent variable, the percentage coded 1 on grass. Click Continue, returning to the Define Multiple Line window. The Variable box should now read "PIN(1 1)(grass)," meaning "The bars will display the percentages of respondents inside the value of 1 on grass at the low end and the value of 1 on grass at the high end." Click OK.

The multiple line chart, constructed to our specifications, appears in the Viewer (Figure 5-4). This graphic greatly facilitates interpretation of the relationship. The upper line shows the relationship between grass and polview3 for people without children, and the lower line depicts the relationship for people with children. Trace the effect of the independent variable by moving from left to right along each line, across the values of polview3. As we learned from the cross-tabulation, the lines drop by about the same amount, 26 points for the "No kids" line and 29 points for the "Yes kids" line. Note the effect of the control variable by jumping between the lines, at each value of polview3. As we saw earlier, the kid effect is quite similar at each

**Figure 5-3** Obtaining a Multiple Line Chart

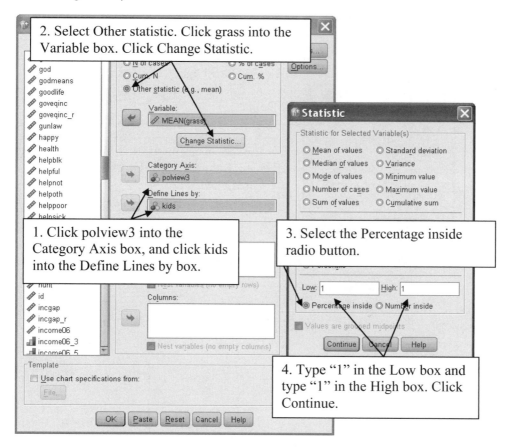

**Figure 5-4** Multiple Line Chart of Additive Relationships with Negative Tendency

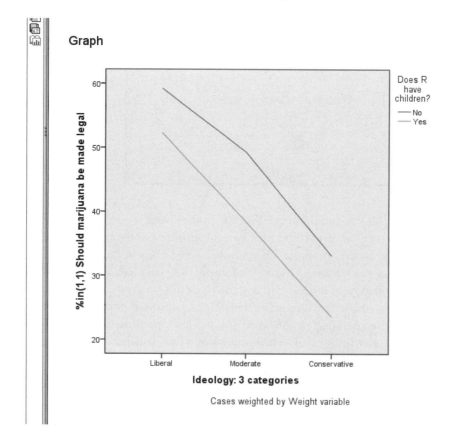

value of the independent variable: 7 points for liberals, 11 points for moderates, and 10 points for conservatives. This is a beautiful line chart. But let's spruce it up using the Chart Editor.

We will make three changes to the chart: First, we will change the title on the y-axis. Second, we will make the lines thicker. Finally, we will change the style of one of the lines, so that the legend clearly communicates the categories of the control variable, kids. (If you print graphics in black and white, as we do in this book, it is sometimes difficult to distinguish subtle differences in the colors of the lines.)

Place the cursor anywhere on the chart and double-click. This invokes the Chart Editor. To change the y-axis title (Figure 5-5), first select it with a single-click. Single-click again to edit it. Replace the current title with this new title: "Percent saying marijuana should be legal." (Clicking anywhere else on the chart returns the axis title to its proper position.) Next let's make the lines thicker. Double-click on one of the lines, as shown in Figure 5-6. The Chart Editor selects both lines and opens the Properties window. In the Lines panel of the Lines tab, click the Weight drop-down and select a heavier weight, such as 2. Click Apply. The Editor makes both lines thicker. To edit the properties of a single line, single-click on the line. The Editor will select only that line. (See Figure 5-7.) To edit the line's style, click the Style drop-down and choose from

**Figure 5-5**  Changing the Y-axis Title

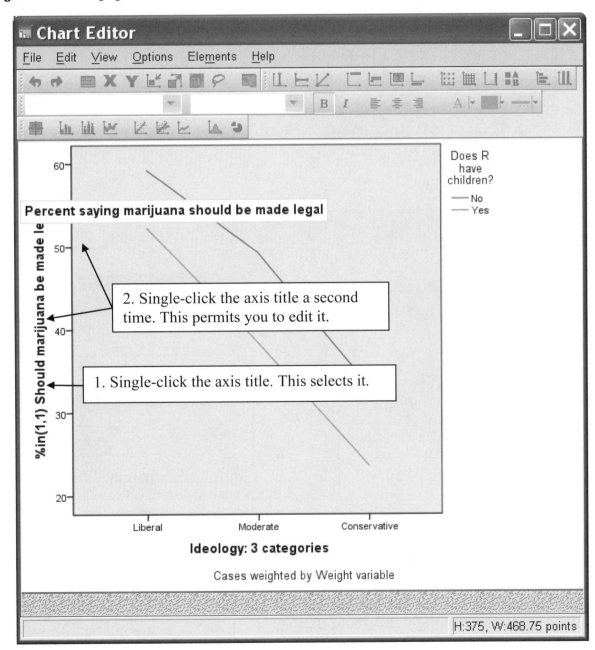

**Figure 5-6**   Changing Line Weights

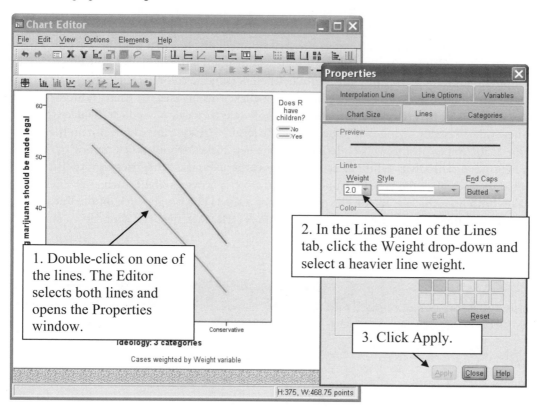

**Figure 5-7**   Changing Line Style

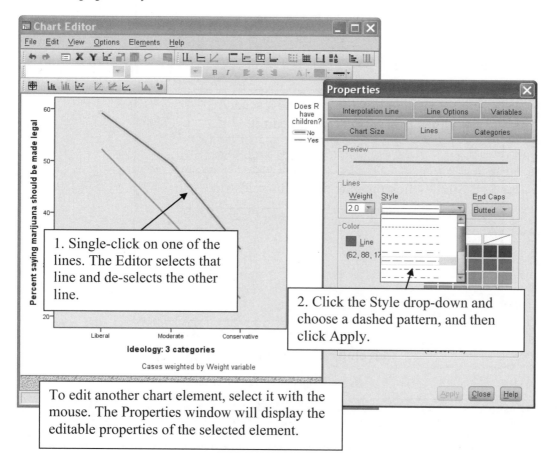

**Figure 5-8**    Multiple Line Chart of Additive Relationships with Negative Tendency (edited)

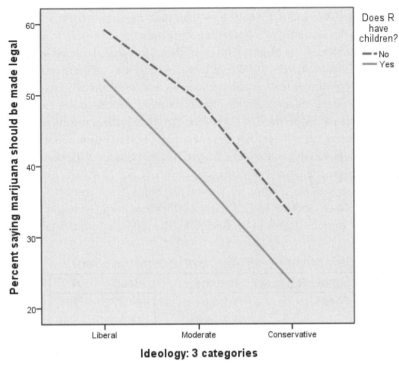

Cases weighted by Weight variable

among the many dashed patterns. When you click Apply, the Editor modifies the line's style and makes a corresponding change in the legend. Close the Properties window and exit the Chart Editor. A newly edited multiple line chart appears in the Viewer (Figure 5-8).

## MEAN COMPARISON ANALYSIS WITH A CONTROL VARIABLE

Mean comparison analysis is used when the dependent variable is interval level and the independent variable and the control variable are nominal or ordinal level. In most ways, the procedure for using Compare Means with layers to obtain controlled comparisons is similar to that for using Crosstabs. However, the two procedures differ in one important way. We will work through two guided examples using NES2008. The first example shows an interesting pattern of interaction. The second example gives you a chance to identify a set of additive relationships. Open NES2008 and let's begin the first guided example.

### Example of an Interaction Relationship

One of the most durable fixtures in U.S. politics is the relationship between partisanship and income. For many years, people who made less money were attracted to the Democratic Party more strongly than were more affluent individuals. Yet newer partisan divisions, such as that based on gender, may be altering the relationship between partisanship and income. How can this be? Think about this question for a moment. Suppose that, among men, traditional partisan differences persist, with lower-income males much more pro-Democratic than higher-income males. But suppose that, among women, these income-based partisan differences are weaker, with lower-income females only somewhat more pro-Democratic than higher-income females. This idea suggests that the relationship between income and partisanship is weaker for females than for males. If this idea is correct, then we should find a set of interaction relationships between income, partisanship, and gender. Let's investigate.

As you know, NES2008 contains a number of feeling thermometer variables, which record respondents' ratings of different political groups and personalities on a scale from 0 (cold or negative) to 100 (warm or positive). One of these variables, dem_therm, which gauges feelings toward the Democratic Party, will be the dependent variable in the current example. The independent variable is income_r3, the ordinal measure of income that you created in Chapter 3. Recall that income_r3 classifies respondents into three income levels: "low" (coded 1), "middle" (coded 2), and "high" (coded 3). The control variable is gender, coded 1 for males and coded 2 for females. We will use Analyze → Compare Means → Means to produce mean values of dem_therm for each value of income_r3, controlling for gender.

Click Analyze → Compare Means → Means. Find dem_therm in the variable list and click it into the Dependent List box. Now we want SPSS to proceed as follows. First, we want it to separate respondents into two groups, men and women, on the basis of the control variable, gender. Second, we want SPSS to calculate mean values of dem_therm for each category of the independent variable, income_r3. SPSS handles mean comparisons by first separating cases on the variable named in the first Layer box. It then calculates means of the dependent variable for variables named in subsequent Layer boxes. For this reason, it is best to put the control variable in the first layer and to put the independent variable in the second layer. Because gender is the control variable, locate gender in the variable list and click it into the Layer 1 of 1 box (Figure 5-9). Click Next. The next Layer box, labeled "Layer 2 of 2," opens. The independent variable, income_r3, goes in this box. Click income_r3 into the Layer 2 of 2 box (Figure 5-10). One last thing: Click Options. In Cell Statistics, click Standard Deviation back into the left-hand Statistics box, and then click Continue. Ready to go. Click OK. The following control table appears in the Viewer:

dem_therm Feeling Thermometer: Democratic Party

| gender R gender | income_r... | Mean | N |
|---|---|---|---|
| 1 Male | 1 Low | 61.27 | 251 |
| | 2 Middle | 58.29 | 314 |
| | 3 High | 47.21 | 413 |
| | Total | 54.38 | 978 |
| 2 Female | 1 Low | 60.95 | 518 |
| | 2 Middle | 61.16 | 378 |
| | 3 High | 55.15 | 263 |
| | Total | 59.70 | 1159 |
| Total | 1 Low | 61.06 | 770 |
| | 2 Middle | 59.86 | 692 |
| | 3 High | 50.30 | 676 |
| | Total | 57.27 | 2137 |

**Figure 5-9** Means Window with Dependent Variable and Control Variable

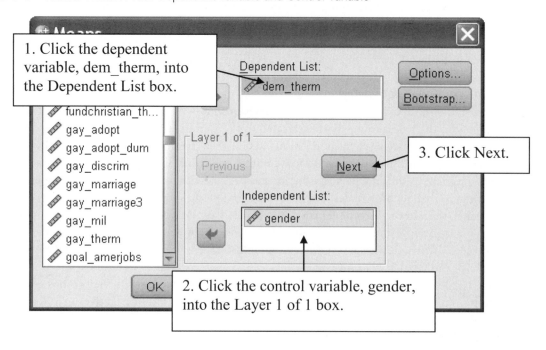

**Figure 5-10**   Means Window with Independent Variable in Layer Box

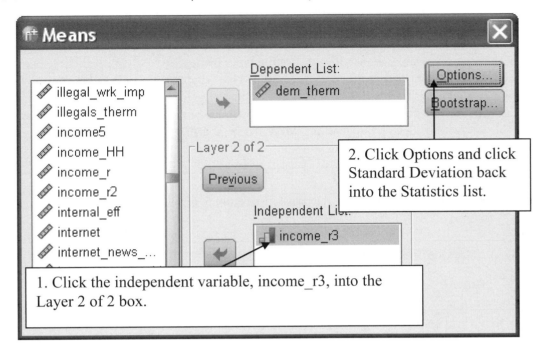

This is a highly readable table. The values of the control variable, gender, appear along the left-hand side of the table. The topmost set of mean comparisons shows the mean Democratic Party thermometer ratings of males, the next set is for females, and the bottom set (labeled "Total") shows the uncontrolled relationship between dem_therm and income_r3—for males and females combined. What is the relationship for men? Overall, the relationship is negative: As income increases, mean ratings of the Democratic Party decline. But notice the mild difference between low- and middle-income males, and the much larger decline between middle- and high-income males. Males in the lowest income category average 61.27, only about 3 "degrees" warmer than males in the middle-income category, who average 58.29. When we move to high-income men, feelings toward the Democratic Party turn frosty indeed. High-income males rate the Democrats at 47.21 degrees, 11 degrees cooler than middle-income males and 14 degrees cooler than low-income males. So, the central feature of the relationship is the stark difference between high-income males and everybody else.

Now consider the dem_therm-income_r3 relationship for women. Here we find a greatly toned down version of the male pattern: similar Democratic ratings for low- and middle-income females, and a some-what less positive reception among high-income females. Between the low- and middle-income categories, Democratic ratings stick stubbornly at a quite-positive 61 degrees (60.95 and 61.16, respectively). Moving to higher-income women, feelings toward the Democrats turn slightly chillier. High-income females rate the Democrats at 55.15, about 5 degrees below the other two groups. Clearly, the relationship has the same tendency for women as for men. However, for men the relationship is substantially stronger. A situation such as this—same tendency, different strength—is a common form of interaction.

Notice, too, that the effect of gender varies by income. Return to the mean comparison table. Evaluate the relationship between Democratic ratings and the control variable, gender, by comparing men and women at the same income level. No gender gap exists for low-income individuals. Males (61.27) and females (60.95) report the same mean ratings. Middle-income people are quite similar as well: 58.29 for men compared with 61.16 for women, a 3-degree gap. However, among high-income people we find an explicit parting of the partisan ways. A comparison of high-income males (47.21) with high-income females (55.15) reveals an 8-degree gender difference.

A line chart will illuminate the interaction relationships we have been discussing. The steps for obtain-ing a multiple line chart for an interval-level dependent variable are the same steps you learned earlier, with one work-saving exception.

1. Click Graphs → Legacy Dialogs → Line. This opens the Line Charts window.
2. In the Line Charts window, click Multiple, and then click Define. This opens the Define Multiple Line window.
3. Click the independent variable, income_r3, into the Category Axis box.
4. Click the control variable, gender, into the Define Lines by box.
5. In the Lines Represent panel of the Define Multiple Line window, select the Other statistic radio button. This activates the Variable box.
6. Click the dependent variable, dem_therm into the Variable box. SPSS moves dem_therm into the Variable box with its default designation, "MEAN(dem_therm)." For an interval-level dependent variable, this default is precisely what you want.
7. Click OK.

We now have a tailor-made line chart of the relationships (Figure 5-11).

**Figure 5-11**  Multiple Line Chart of Interaction Relationships

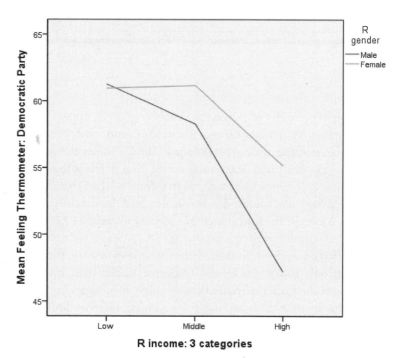

You can see why line charts are essential for correctly interpreting controlled comparisons. By tracing along each line, from lower income to higher income, you can see the effect of income on thermometer ratings. Among males, the line drops mildly between low and middle, and declines sharply between middle and high. The female line is a muted version of the same pattern. Notice, too, the relationship between gender and the dependent variable at different levels of income. For the lowest-income group, men and women give the same ratings to the Democratic Party—no gender gap. The gap opens up slightly among middle-income people, and then becomes a gender chasm at the highest income level. This pattern suggests some general questions that you may wish to investigate in your own research. Do male-female differences on other political and social issues become most pronounced among high-income people? Do some issues produce a larger gender gap than others?

Thus far in this chapter we found two instances of interaction: the obey2-polview3-kids relationships and the dem_therm-income_r3-gender relationships. In the former, we saw a prominent relationship for one value of the control (people without children), but no relationship for the other value (people with children). In the latter case, the relationship between the independent and dependent variables had the same tendency for both values of the control variable, but the relationship was stronger for one value (males) than for the other value (females). That's what makes identifying interaction such a challenge—it can take on several different forms. Interaction has two field marks, though, that will give it away. First, when you examine the relationship between the independent variable and the dependent variable at different values of the control variable, you may find that the relationship varies in tendency, perhaps positive for one value of the control variable, zero or negative for other control values. Second, the relationship may have the same tendency for all control values but differ in strength, negative-weak versus negative-strong or positive-weak versus positive-strong. In identifying interaction, practice makes perfect. And, believe it or not, statistics can help (see Chapter 9).

### Example of an Additive Relationship

Compared with the protean complexity of interaction, additive relationships are the soul of simplicity. In a set of additive relationships, both the independent and the control variables help to explain the dependent variable. More than this, the effect of the independent variable is the same or very similar—same tendency, same strength—for all values of the control variable. Interaction relationships assume several forms. Additive relationships assume only one.

A final example will illustrate this point. Consider this dependent variable: the NES feeling thermometer toward gays (gay_therm). Why might some people have negative feelings about homosexuals while others have more positive feelings? Two sorts of variables, one political and one personal, may have effects here. On the political side, we can plausibly expect that liberals will give gays higher average ratings than will conservatives. NES2008 has liberalism3, coded 1 for "more conservative," 2 for "middle," and 3 for "more liberal." This is the independent variable. On the personal side, it stands to reason that people who have family members or friends who are gay will score higher on the rating scale than will people whose personal associates do not include homosexuals. For the control variable, we will use sexl_orient_famfrnds, which is coded 1 for respondents who have—and coded 5 for respondents who do not have—associates who are gay, lesbian, or bisexual (GLB). Let's run the analysis and find out what's going on.

Click Analyze → Compare Means → Means. Everything is still in place from our dem_therm-income_r3-gender analysis. Click Reset. Click gay_therm into the Dependent list. Click the control variable, sexl_orient_famfrnds, into the first Layer box. Click Next. Click the independent variable, liberalism3, into the second Layer box. Click Options and remove the Standard Deviation from Cell Statistics. Click OK. Another compact and readable table is at hand:

gay_therm Feeling therm: Homosexuals

| sexl_orient_famfrnds ... | liberalism3 How ... | Mean | N |
|---|---|---|---|
| 1 1. Yes | 1 More conservative | 41.53 | 98 |
| | 2 Middle | 55.67 | 736 |
| | 3 More liberal | 77.65 | 199 |
| | Total | 58.55 | 1033 |
| 5 5. No | 1 More conservative | 23.67 | 175 |
| | 2 Middle | 42.78 | 662 |
| | 3 More liberal | 56.88 | 86 |
| | Total | 40.48 | 923 |
| Total | 1 More conservative | 30.11 | 273 |
| | 2 Middle | 49.56 | 1398 |
| | 3 More liberal | 71.39 | 285 |
| | Total | 50.02 | 1956 |

Liberalism has a remarkably strong effect on the dependent variable among people with, and without, GLB family and friends. Among "GLB-Yes" respondents, mean ratings of gays rise more than 36 degrees, from 41.53 among conservatives to 77.65 among liberals. This positive tendency has practically the same strength, 33 degrees, among "GLB-No" respondents (23.67 for conservatives compared with 56.88 for liberals). So, the liberalism effect is 30-plus degrees at both values of the control. How about the GLB effect, controlling for liberalism? At the ideological poles, the effects are quite similar. The GLB effect is almost 18 for conservatives (41.53 for GLB-Yes conservatives, compared with 23.67 for GLB-No conservatives) and 21 for liberals (77.65 compared with 56.88). For respondents who are ideologically in the middle, the effect is less pronounced, about 13 degrees (55.67 compared with 42.78). This minor hiccup aside, the relationship between gay thermometer ratings and liberalism, controlling for the sexual orientation of family and friends, closely approximates an additive pattern. The effect of the independent variable is pretty much the same at all values of the control variable, and the effect of the control variable is pretty much the same at all values of the independent variable.

By now, obtaining a multiple line chart of the gay_therm-income_r3-gender relationships is a straightforward exercise.

1. Click Graphs → Legacy Dialogs → Line.
2. In the Line Charts window, click Multiple, and then click Define. (Click Reset to clear the panels.)
3. Click the independent variable, liberalism3, into the Category Axis box.
4. Click the control variable, sexl_orient_famfrnds, into the Define Lines by box.
5. In the Lines Represent panel of the Define Multiple Line window, select the Other statistic radio button.
6. Click the dependent variable, gay_therm, into the Variable box.
7. Click OK.

You can see how this line chart (Figure 5-12) communicates the additive relationship. Moving from left to right, from more conservative to more liberal, each line rises by more than 30 degrees. That's the liberalism effect. The GLB effect is conveyed by the distance between the lines. Despite a slight narrowing at the middle category of the liberalism measure, the GLB effect is fairly consistent. Now, you might encounter

**Figure 5-12**   Multiple Line Chart of Additive Relationships with Positive Tendency

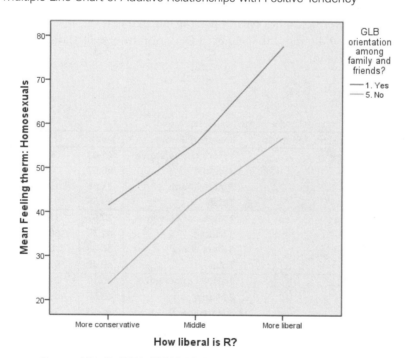

Cases weighted by WT.1. CROSS-SECTION SAMPLE WEIGHT - PRE-ELECTION: centered

**Figure 5-13**  Multiple Line Chart of Additive Relationships with Positive Tendency (edited)

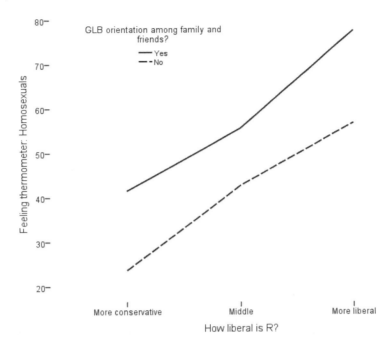

additive relationships in which the lines slope downward, imparting a negative relationship between the independent and dependent variables. And the lines might "float" closer together, suggesting a consistent but weaker effect of the control variable on the dependent variable, controlling for the independent variable. But you will always see symmetry in the relationships. The effect of the independent variable on the dependent variable will be the same or very similar for all values of the control variable, and the effect of the control variable on the dependent variable will be the same or very similar for all values of the independent variable.

Invoke the Chart Editor and make some improvements to this multiple line chart. Experiment with the Chart Editor. Ask the Properties window to do new things. Figure 5-13 may serve as inspiration. (Advanced chart editing is covered in Chapter 8.)

## EXERCISES

1.  (Dataset: World. Variables: democ_regime, frac_eth3, gdp_cap2.) Some countries have democratic regimes, and other countries do not. What factors help to explain this difference? One idea is that the type of government is shaped by the ethnic and religious diversity in a country's population. Countries that are relatively homogeneous, with most people sharing the same language and religious beliefs, are more likely to develop democratic systems than are countries having more linguistic conflicts and religious differences. Consider the ethnic heterogeneity hypothesis: Countries with lower levels of ethnic heterogeneity will be more likely to be democracies than will countries with higher levels of ethnic heterogeneity.

    A.  According to the ethnic heterogeneity hypothesis, if you were to compare countries having lower heterogeneity with countries having higher heterogeneity, you should find (check one)

        ❑  a lower percentage of democracies among countries having lower heterogeneity.

        ❑  a higher percentage of democracies among countries having lower heterogeneity.

        ❑  no difference between the percentage of democracies among countries having lower heterogeneity and the percentage of democracies among countries with higher heterogeneity.

    B.  World contains the variable democ_regime, which classifies each country as a democracy (coded 1) or a dictatorship (coded 0). This is the dependent variable. World also contains frac_eth3, which classifies countries according to their level of ethnic heterogeneity: low (coded 1), moderate (coded 2), or high (coded 3). This is the independent variable. Run Crosstabs, testing the ethnic heterogeneity hypothesis. Fill in the percentages of democracies:

| | Ethnic heterogeneity | | |
|---|---|---|---|
| | Low | Moderate | High |
| Percentage of democracies | ? | ? | ? |

C. Based on these results, you could say that (check one)

❑ as ethnic heterogeneity increases, the percentage of democracies increases.

❑ as ethnic heterogeneity increases, the percentage of democracies decreases.

❑ as ethnic heterogeneity increases, there is little change in the percentage of democracies.

D. A country's level of economic development also might be linked to its type of government. According to this perspective, countries with higher levels of economic development are more likely to be democracies than are countries with lower levels. The World dataset contains the variable gdp_cap2. This variable, based on gross domestic product per capita, is an indicator of economic development. Countries are classified as low (coded 1) or high (coded 2). Obtain a cross-tabulation analysis of the democ_regime-frac_eth3 relationship, controlling for gdp_cap2. Fill in the percentages of democracies:

| | Ethnic heterogeneity | | |
|---|---|---|---|
| | Low | Moderate | High |
| Low GDP per capita Percentage of democracies | ? | ? | ? |
| High GDP per capita Percentage of democracies | ? | ? | ? |

E. Consider the democ_regime-frac_eth3 relationship for low-GDP countries. Examine the difference between the percentage of democracies for "low" heterogeneity and the percentage of democracies for "high" heterogeneity. This difference shows that the percentage of democracies among low-heterogeneity countries is (fill in the blank) _____ percentage points (circle one)

lower than      higher than

the percentage of democracies among high-heterogeneity countries.

Now consider the democ_regime-frac_eth3 relationship for high-GDP countries. Examine the difference between the percentage of democracies for low heterogeneity and the percentage of democracies for high heterogeneity. This difference shows that the percentage of democracies among low-heterogeneity countries is (fill in the blank) _____ percentage points (circle one)

lower than      higher than

the percentage of democracies among high-heterogeneity countries.

F.  Think about the set of relationships you just analyzed. How would you describe the relationship between ethnic heterogeneity and democracy, controlling for GDP per capita? (circle one)

Spurious     Additive     Interaction

Explain your reasoning. _____

_____

_____

_____

_____

G.  Obtain a multiple line chart depicting the percentage of democracies for each value of frac_eth3, controlling for gdp_cap2. (Remember that democracies are coded 1 on democ_regime.) In the Chart Editor, give the scale axis this new title: "Percentage of democracies." Edit the line weights. Change the style of one of the lines. Make other desired changes to enhance appearance and readability. Print the chart you created.

2.  (Dataset: World, Variables: women09, pr_sys, womyear2). In Chapter 2 you analyzed the distribution of the variable women09, the percentage of women in the lower house of the legislatures in a number of countries. In this exercise you will analyze the relationship between women09 and two variables that could have an impact on the number of women serving in national legislatures.

First, consider the role of the type of electoral system. Many democracies have proportional representation (PR) systems. PR systems foster multiple parties having diverse ideological positions—and, perhaps, having diverse demographic compositions as well. Non-PR systems, like the system used in U.S. elections, militate in favor of fewer and more homogeneous parties. Thus you might expect that non-PR countries will have fewer women in their national legislatures than will countries with PR-based electoral systems.

Now consider the role of history and tradition. In some countries, women have had a long history of political empowerment. New Zealand, for example, gave women the right to vote in 1893. In other countries, such as Switzerland (where women were not enfranchised until 1971), women have had less experience in the electoral arena. Thus it seems reasonable to hypothesize that countries with longer histories of women's suffrage will have higher percentages of women in their national legislatures than will countries in which women's suffrage is a more recent development. In this exercise you will isolate the effect of the type of electoral system on the percentage of women in parliament, controlling for the timing of women's suffrage. However, before running any analyses, you will graphically depict different possible scenarios for the relationships you might discover.

Parts A, B, and C contain graphic shells showing the percentage of women in parliament along the vertical axis and the type of electoral system along the horizontal axis. Countries without PR systems are represented by the tick mark on the left and countries with PR systems by the tick mark on the right. For each shell, you will draw two lines within the graphic space, a solid line depicting the relationship for countries having a longer history of women's suffrage and a dashed line depicting the relationship for countries having a shorter history of women's suffrage.

A. Draw an additive relationship fitting this description: Countries with PR systems have higher percentages of women in parliament than do countries with non-PR systems, and countries with a longer history of women's suffrage have higher percentages of women in parliament than do countries with a shorter history of women's suffrage. (*Hint*: In additive relationships, the strength and tendency of the relationship is the same or very similar for all values of the control variable.) Remember to use a solid line to depict the relationship for countries having a longer history of women's suffrage and a dashed line to depict the relationship for countries having a shorter history of women's suffrage.

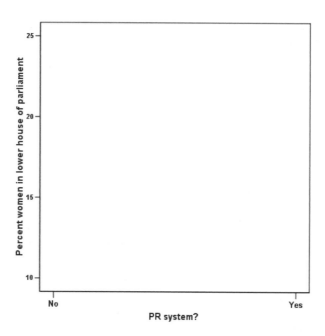

B. Draw a spurious relationship: Type of electoral system has no effect on the percentage of women in parliament; timing of women's suffrage has a big effect on the percentage of women in parliament.

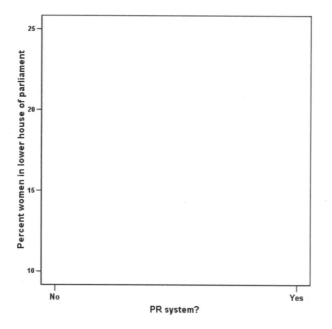

C. Draw a set of interaction relationships fitting this description: For countries with a longer history of women's suffrage, those with PR systems have higher percentages of women in parliament than do countries with non-PR systems. For countries with a shorter history of women's suffrage, the type of electoral system has no effect on the percentage of women in parliament.

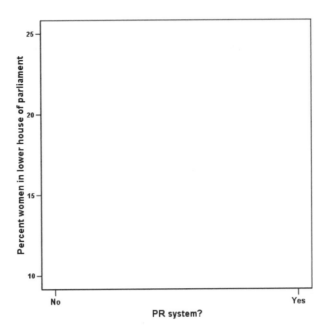

D. In addition to the dependent variable, women09, world contains pr_sys, coded 0 for countries with non-PR systems and coded 1 for those having PR systems. Use pr_sys as the independent variable. World also contains womyear2, which measures the timing of women's suffrage by two values: 1944 or before (coded 0) and after 1944 (coded 1). Use womyear2 as the control variable.

Run Compare Means to determine the mean of women09 for each value of pr_sys, controlling for womyear2. (Remember to put the control variable, womyear2, in the first Layer box.) Record the mean values in the table that follows:

| Women's suffrage | PR system? | Mean |
|---|---|---|
| 1944 or before | No | ? |
|  | Yes | ? |
| After 1944 | No | ? |
|  | Yes | ? |

E. Obtain a line chart of the relationship between women09 and pr_sys, controlling for womyear2. Open the Chart Editor. Edit the line weights, and change the style of one of the lines. Make other desired changes to enhance appearance and readability. Print the chart you created.

F. Examine the table (part D) and the chart (part E). Consider the women09-pr_sys relationship for countries that enfranchised women in 1944 or before. Examine the difference between the means for non-PR countries and PR countries. This difference shows that the mean for PR countries is (fill in the blank) _____ points (circle one)

lower than     higher than

the mean for non-PR countries.

Now consider the women09-pr_sys relationship for countries that enfranchised women after 1944. Examine the difference between the means for non-PR countries and PR countries. This difference shows that the mean for PR countries is (fill in the blank) _____ points (circle one)

lower than     higher than

the mean for non-PR countries.

G.  Which of the following statements best characterizes the women09-pr_sys relationship, controlling for womyear2? (check one)

❑  The women09-pr_sys relationships have the same tendency and very similar strengths at both values of womyear2.

❑  The women09-pr_sys relationships have the same tendency but very different strengths at each value of womyear2.

❑  The women09-pr_sys relationships have different tendencies at each value of womyear2.

H.  Review your artistic work in parts A–C. Examine the table (part D) and the line chart (part E). Consider your conclusions in parts F and G. Which possible scenario—the line chart you drew in A, B, or C—most closely resembles the pattern shown in the data? (circle one)

The line chart in A     The line chart in B     The line chart in C

3.  (Dataset: NES2008. Variables: voted, memnum2, educ_r3.) Two political analysts are discussing factors that affect turnout in U.S. elections.

Political analyst 1: "I think that the role of voluntary association memberships has been overlooked as an important causal factor. People who belong to community associations, interest groups, or labor unions will interact with other individuals, discuss politics, and become aware of important issues. As a result, they will be more likely to vote than will people who do not belong to any groups."

Political analyst 2: "Your idea is interesting . . . but flawed. We know that education is strongly linked to voter turnout. People with more education are more likely to vote. Plus—and here is the relationship you are overlooking—people with more education are also more likely to join voluntary organizations. So as education goes up, people will be more likely to join groups *and* to vote. But there is no causal connection between belonging to groups and voting."

A.  According to political analyst 1, if you were to compare the voter turnout of people who are not members of voluntary groups with that of people who are members, you would find (check one)

❑  people who are members of voluntary groups are more likely to vote than are people who are not members of voluntary groups.

❑  people who are members of voluntary groups are less likely to vote than are people who are not members of voluntary groups.

❑  no difference in turnout between people who are members of voluntary groups and those who are not members of voluntary groups.

B.  According to political analyst 2, if you were to compare the group memberships of people who have less education with the group memberships of people who have more education, you would find (check one)

❑ people with more education are more likely to be members of voluntary groups than are people with less education.

❑ people with more education are less likely to be members of voluntary groups than are people with less education.

❑ no difference in group membership between people with more education and people with less education.

C. Think about an analysis showing three cross-tabulations: the relationship between turnout (dependent variable) and group membership (independent variable) for people with low education, medium education, and high education. According to political analyst 2, such an analysis would reveal (check all that apply)

❑ that at each level of education, people who are members of voluntary groups are no more likely to vote than are people who are not members of voluntary groups.

❑ that regardless of group memberships, people with more education are more likely to vote than are people with less education.

❑ that the relationship between voter turnout and group membership is spurious.

D. NES2008 contains voted (coded 1 for voters and 0 for nonvoters). This is the dependent variable. The independent variable, memnum2, is coded 0 for respondents having no voluntary group memberships and coded 1 for respondents having at least one membership. The control variable, educ_r3, categorizes respondents into three education levels: 0–11 years (coded 1), 12 years (coded 2), and 13 or more years (coded 3). Use Crosstabs to analyze the relationship between voted and memnum2, controlling for educ_r3. In the table that follows, record the *percentages who voted*:

| | Percentages who voted: | |
| --- | --- | --- |
| | Belong to group? | |
| Education level: | No | Yes |
| 0–11 years | ? | ? |
| 12 years | ? | ? |
| 13-more years | ? | ? |

E. Examine the tabular evidence (Part D). Consider the set of relationships you just analyzed. How would you describe the relationship between the independent variable, memnum2, and turnout, controlling for education? (circle one)

Spurious     Additive     Interaction

Explain your reasoning. _____

_____

_____

_____

_____

F.  Which political analyst, political analyst 1 or political analyst 2, is more correct? (circle one)

Political analyst 1        Political analyst 2

Explain your reasoning. _____

_____

_____

_____

_____

_____

4.  (Dataset: NES2008. Variables: voted_04, contact_GOTV, voted.) The two political analysts are at it again. This time they are discussing the effectiveness of get-out-the-vote (GOTV) campaigns.

Political analyst 1: "GOTV campaigns are a big waste of time and money."

Political analyst 2: "Your dyspepsia is showing. Care to share your reasoning?"

Political analyst 1: "Suppose you were in charge of a GOTV campaign. How would you locate people to contact? Simple: You'd go through voter registration records and identify people who are registered to vote or who voted in the previous election. Well, people who voted in the past election are very likely to vote in the current election, GOTV contact or no GOTV contact. A big waste of time and money. Now, suppose you managed to contact a few habitual nonvoters. People with no history of voting are unlikely to vote in the current election. A friendly 'Don't forget to vote' reminder will have no effect on them. Another big waste of time and money."

Political analyst 2: "Look, GOTV campaigns are a lot more sophisticated than you seem to think. My guess is that individuals' voting histories make little or no difference in whether they are contacted by GOTV mobilizers. Plus, you are way off base on the effect of GOTV contact. People who are contacted are much more likely to vote than are people who are not contacted, past voters and past nonvoters alike."

A.  Imagine performing two analyses. The first analysis: the relationship between GOTV contact (dependent variable) and past voting turnout (independent variable). The second analysis: the relationship between current voting turnout (dependent variable) and GOTV contact (independent variable), controlling for past voting behavior (control variable).

(i)  If political analyst 1 is correct, these analyses will show (check three)

❑  Past voters are much more likely to be contacted than are past nonvoters.

❑  Past voters and past nonvoters are about equally likely to be contacted.

❑  Among past voters, those who are contacted are much more likely to vote than are those who are not contacted.

❑  Among past voters, those who are contacted and those who are not contacted are about equally likely to vote.

❑ Among past nonvoters, those who are contacted are much more likely to vote than are those who are not contacted.

❑ Among past nonvoters, those who are contacted and those who are not contacted are about equally likely to vote.

(ii) If political analyst 2 is correct, these analyses will show (check three)

❑ Past voters are much more likely to be contacted than are past nonvoters.

❑ Past voters and past nonvoters are about equally likely to be contacted.

❑ Among past voters, those who are contacted are much more likely to vote than are those who are not contacted.

❑ Among past voters, those who are contacted and those who are not contacted are about equally likely to vote.

❑ Among past nonvoters, those who are contacted are much more likely to vote than are those who are not contacted.

❑ Among past nonvoters, those who are contacted and those who are not contacted are about equally likely to vote.

NES2008 contains three variables: voted_04, coded 0 for respondents who did not vote in the previous presidential election (2004) and coded 1 for respondents who did vote; contact_GOTV, coded 0 for respondents who were not contacted in 2008, and coded 1 for those who were contacted; and voted, coded 0 for those who did not vote in 2008 and coded 1 for those who voted in 2008.

B. Analyze the contact_GOTV-voted_04 relationship to find out whether past voters are more likely to be contacted than are past nonvoters. According to your analysis, _____ percent of past nonvoters were contacted, compared with _____ of past voters.

Based on this analysis, you can conclude (check one)

❑ Past voters are much more likely to be contacted than are past nonvoters.

❑ Past voters and past nonvoters are about equally likely to be contacted.

C. Analyze the voted-contact_GOTV relationship, controlling for voted_04. In the table that follows, record the *percentages who voted*:

| | Anyone talk to R about registering or getting out to vote? (contact_GOTV) | |
|---|---|---|
| Did R vote in 2004? (voted_04) | No | Yes |
| No | ? | ? |
| Yes | ? | ? |

D. Examine the tabular results (part C). Based on this analysis, you can conclude (check two)

❑ Among past voters, those who are contacted are much more likely to vote than are those who are not contacted.

❑ Among past voters, those who are contacted and those who are not contacted are about equally likely to vote.

❑ Among past nonvoters, those who are contacted are much more likely to vote than are those who are not contacted.

❑ Among past nonvoters, those who are contacted and those who are not contacted are about equally likely to vote.

E. Consider your findings and conclusions in parts B–D. Which political analyst is more correct, political analyst 1 or political analyst 2? (circle one)

Political analyst 1     Political analyst 2

Explain your reasoning. _____

_____

_____

_____

_____

F. Think for a few minutes about the set of relationships between voted and contact_GOTV, controlling for voted_04. How would you describe this set of relationships? (circle one)

Spurious     Additive     Interaction

Explain your reasoning. _____

_____

_____

_____

_____

5. (Dataset: NES2008. Variables: environ_therm, enviro_jobs_r3, enviro_jobs_imp2.) Who has positive feelings toward environmentalists? Who views this group with greater suspicion? It seems plausible to suggest that ratings of environmentalists are shaped by positions on important environmental issues. For example, we could hypothesize that individuals who favor environmental protection over jobs will give environmentalists higher ratings than will individuals who favor jobs over environmental protection. Yet this relationship may not be the same for everyone. Imagine a group of people for whom the environment-jobs issue is not very important. To be sure, these individuals have positions on the tradeoff—some favor the environment, some take a middle position, and some favor jobs—but the issue, to them, has *low salience.* Contrast these low-salience individuals to a group for whom the environment-jobs tradeoff is very important, for whom the issue has *high salience.* We might expect that the relationship between issue position (environment vs. jobs) and ratings of environmentalists will be stronger for high-salience individuals than for low-salience individuals. In this exercise you will test two hypothetical claims:

Claim x: Controlling for issue salience, individuals who favor the environment over jobs will give environ-mentalists higher ratings than will individuals who favor jobs over the environment.

Claim y: The relationship between environmental-issue position (environment vs. jobs) and ratings of environmentalists will be stronger for high-salience individuals than for low-salience individuals.

NES2008 contains environ_therm, a feeling thermometer of environmentalists. Ratings can range from 0 (negative ratings) to 100 (positive ratings). This is the dependent variable. For the independent variable, use enviro_jobs_r3, coded 1 ("Protect the environment, even if it costs jobs and standard of living"), 2 (middle position), or 3 ("Jobs and standard of living are more important than the environment"). For the control variable, issue salience, use enviro_jobs_imp2, which is coded 0 (the issue is "not very" important) or 1 (the issue is "very/extremely" important).

A. Run Compare Means. Record the mean values of environ_therm in the table that follows:

| How imp envir/jobs issue? (enviro_jobs_imp2) | Protect environment or jobs? (enviro_jobs_r3) | Mean |
|---|---|---|
| Not very | 1 Environ | ? |
| | 2 Middle | ? |
| | 3 Jobs | ? |
| Very/extrmly | 1 Environ | ? |
| | 2 Middle | ? |
| | 3 Jobs | ? |

B. Does your analysis support claim x? (circle one)

Yes, the analysis supports claim x.     No, the analysis does not support claim x.

Explain your reasoning. _____

_____

_____

_____

C. Does your analysis support claim y? (circle one)

Yes, the analysis supports claim y.     No, the analysis does not support claim y.

Explain your reasoning. _____

_____

_____

_____

D. Produce an edited, presentable multiple line chart of the environ_therm-enviro_jobs_r3 relationship, controlling for enviro_jobs_imp2. Use the Chart Editor to make necessary text edits and appearance enhancements. Print the chart.

E. Consider how issue salience affects the relationship between environmental issues and ratings of environmentalists. How would you describe this set of relationships? (circle one)

Spurious     Additive     Interaction

Explain your reasoning. _____

_____

_____

_____

_____

6. (Dataset: GSS2008. Variables: closeblk_3, closewht_3, intrace_2, race2.) For an exercise in Chapter 4, you tested for the presence of preference falsification, the tendency for respondents to offer false opinions that they nonetheless believe to be socially desirable under the circumstances. You evaluated the hypothesis that respondents are more likely to express support for government policies aimed at helping blacks when questioned by a black interviewer than when questioned by a white interviewer. But you did not control for the respondent's race. That is, you did not look to see whether whites are more (or less) likely than are blacks to misrepresent their racial feelings, depending on the race of the interviewer.[4]

In what sorts of survey situations are respondents more prone to falsify their true racial opinions? The raceof-interviewer effect should be weaker when respondents are asked opinions about their own race. Imagine white respondents being asked how close they feel to whites, or blacks being asked how close they feel to blacks. Even when posed by a different-race interviewer, respondents will feel little pressure to falsify socially acceptable, positive responses to questions about their own race.

More potent race-of-interviewer effects should occur when whites are asked by blacks how close they feel to blacks, or when blacks are asked by whites how close they feel to whites. The social pressure is clear—offer an opinion that is almost certainly the same as the interviewer's. It is an open question whether white respondents and black respondents are equally susceptible to falsification pressures.

In addition to variables measuring interviewer race (intrace_2) and respondent race (race2), GSS2008 contains closeblk_3 and closewht_3. Each of these records responses to the questions: "In general, how close do you feel to blacks?" "And in general, how close do you feel to whites?" Reponses are coded –1 ("not close"), 0 ("neutral"), or 1 ("close"). Perform two cross-tabulation analyses, the first using closeblk_3 as the dependent variable and the second using closewht_3 as the dependent variable. In both analyses, use intrace_2 as the independent variable and race2 as the control variable. Test the following hypotheses (and proposition):

Hypothesis 1: In a comparison of individuals, those who are questioned by an interviewer of a different race will be no more likely to express positive feelings toward members of their own race than will those who are questioned by an interviewer of the same race.

Hypothesis 2: In a comparison of individuals, those who are questioned by an interviewer of a different race will be more likely to express positive feelings toward members of the interviewer's race than will those who are questioned by an interviewer of the same race.

Proposition 1: The race-of-interviewer effects described in hypothesis 1 and hypothesis 2 will be the same for white respondents and black respondents.

A.  In the table that follows, record the percentages of respondents saying "close" in each cross-tabulation analysis. For each analysis, record the race-of-interviewer effect by obtaining the absolute value (ignoring the sign) of the difference between the two percentages. For example, if 50.0 percent of respondents said "close" when questioned by a white and 70.0 percent said "close" when questioned by a black, then the race-of-interview effect would be: $|50.0 - 70.0| = 20.0$. The table entries are arranged so that the top two sets of comparisons will help you test hypothesis 1, and the bottom two sets will help you test hypothesis 2.

| Race of respondent | Percent saying | Race of interviewer | | |
|---|---|---|---|---|
| | | White | Black | Race-of-interviewer effect \|White % − Black%\| |
| White | "close" to whites | ? | ? | ? |
| Black | "close" to blacks | ? | ? | ? |
| | | | | |
| White | "close" to blacks | ? | ? | ? |
| Black | "close" to whites | ? | ? | ? |

B.  Hypothesis 1 is (circle one)

                    not supported.       supported.

Explain your reasoning. _____

_____

_____

_____

_____

_____

C.  Hypothesis 2 is (circle one)

                    not supported.       supported.

Explain your reasoning. _____

_____

_____

_____

_____

_____

D.  Proposition 1 is (circle one)

not supported.     supported.

Explain your reasoning. _____

_____

_____

_____

_____

_____

That concludes the exercises for this chapter. Before exiting SPSS, be sure to save your output file.

## NOTES

1.  A cross-tabulation analysis reveals that 65.1 percent of liberals have children, compared with 79.0 percent of conservatives.
2.  Edward R. Tufte, *The Visual Display of Quantitative Information,* 2nd ed. (Cheshire, Conn.: Graphics Press, 2001), 93.
3.  The Graphs → Legacy Dialogs → Bar (Clustered) interface is identical to the multiple line chart interface in every detail. If you prefer clustered bar charts, you can directly apply the skills you will learn in this chapter.
4.  See Darren W. Davis and Brian D. Silver, "Stereotype Threat and Race of Interviewer Effects in a Survey of Political Knowledge," *American Journal of Political Science* 47 (January 2003): 33–45.

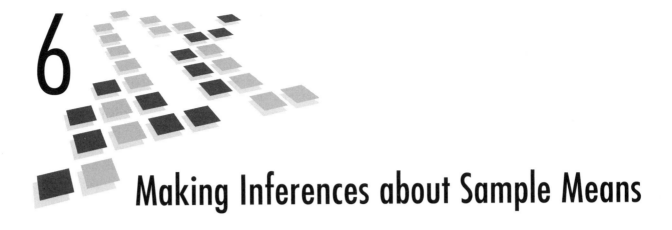

# 6

# Making Inferences about Sample Means

Procedures Covered

Analyze → Descriptive Statistics → Descriptives

Analyze → Compare Means → One-Sample T Test

Analyze → Compare Means → Independent-Samples → T Test

Political research has much to do with observing patterns, creating explanations, framing hypotheses, and analyzing relationships. In interpreting their findings, however, researchers often operate in an environment of uncertainty. This uncertainty arises, in large measure, from the complexity of the political world. As we have seen, when we infer a causal connection between an independent variable and a dependent variable, it is hard to know for sure whether the independent variable is causing the dependent variable. Other, uncontrolled variables might be affecting the relationship, too. Yet uncertainty arises, as well, from the simple fact that research findings are often based on random samples. In an ideal world, we could observe and measure the characteristics of every element in the population of interest—every voting-age adult, every student enrolled at a university, every bill introduced in every state legislature, and so on. In such an ideal situation, we would enjoy a high degree of certainty that the variables we have described and the relationships we have analyzed mirror what is really going on in the population. But of course we often do not have access to every member of a population. Instead we rely on a sample, a subset drawn at random from the population. By taking a random sample, we introduce random sampling error. In using a sample to draw inferences about a population, therefore, we never use the word *certainty*. Rather, we talk about *confidence* or *probability*. We know that the measurements we make on the sample will reflect the characteristics of the population, within the boundaries of random sampling error.

What are those boundaries? If we calculate the mean income of a random sample of adults, for example, how confident can we be that the mean income we observe in our sample is the same as the mean income in the population? The answer depends on the standard error of the sample mean, the extent to which the mean income of the sample departs by chance from the mean income of the population. If we use a sample to calculate a mean income for women and a mean income for men, how confident can we be that the difference between these two sample means reflects the true income difference between women and men in the population? Again, the answer depends on the standard error—in this case, the standard error of the *difference* between the sample means, the extent to which the difference in the sample departs from the difference in the population.

In this chapter you will use three procedures to explore and apply inferential statistics. First, you will learn to use Descriptives to obtain basic information about interval-level variables. Second, using the One-Sample T Test procedure, you will obtain confidence intervals for a sample mean. The 95 percent confidence interval will tell you the boundaries within which there is a .95 probability that the true population mean falls. You will also find the 90 percent confidence interval, which is applied in testing hypotheses at the .05 level of significance. Third, using the Independent-Samples T Test procedure, you will test for statistically significant differences between two sample means.

## DESCRIPTIVES AND ONE-SAMPLE T TEST

To gain insight into the properties and application of inferential statistics, we will work through an example using NES2008. We begin by looking at the Descriptives procedure, which yields basic information about interval-level variables. We then demonstrate the fundamentals of inference using the One-Sample T Test procedure.

NES2008 contains spend10, a measure of individuals' opinions about federal spending. Respondents were asked by how much federal spending should be increased or decreased on each of 10 programs. Spend10 was created by adding up the number of times the respondent said "increased a great deal" or "increased a moderate amount."[1] We can use Descriptives to obtain summary information about spend10. Open NES2008. Click Analyze → Descriptive Statistics → Descriptives. In the main Descriptives window, scroll down the left-hand variable list until you find spend10. Click spend10 into the Variable(s) list (Figure 6-1). Click the Options button. Now you can specify which descriptive statistics you would like SPSS to produce. These defaults should already be checked: mean, standard deviation, minimum, and maximum. That's fine. Also check the box beside "S.E. mean," which stands for "standard error of the mean," as shown in Figure 6-1. Click Continue, and then click OK.

**Figure 6-1** Descriptives Window and Descriptives: Options Window (modified)

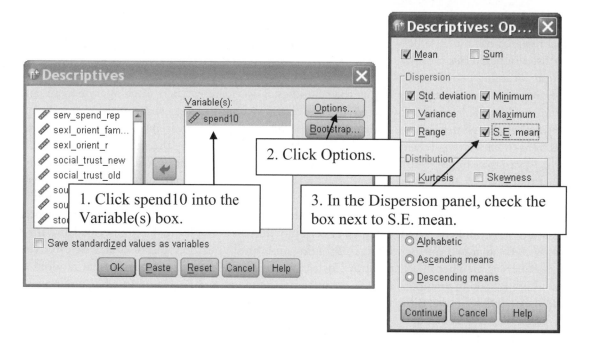

**Descriptive Statistics**

|  | N | Minimum | Maximum | Mean | | Std. Deviation |
|---|---|---|---|---|---|---|
|  | Statistic | Statistic | Statistic | Statistic | Std. Error | Statistic |
| spend10 Incr federal $:10 programs | 2323 | 0 | 10 | 4.56 | .052 | 2.497 |
| Valid N (listwise) | 2323 | | | | | |

SPSS has reported the requested statistics for spend10: number of cases analyzed (N), minimum and maximum observed values for spend10, mean value of spend10, standard error of the mean, and standard deviation. Among the 2,323 respondents, scores on spend10 range from 0 (increase spending on none of the programs) to 10 (increase spending on all the programs). The mean value of spend10 is 4.56, with a standard deviation of 2.497 (which rounds to 2.50).[2] How closely does the mean of 4.56 reflect the true mean in the population from which this sample was drawn? If we had measured spend10 for every U.S. citizen of voting age and calculated a population mean, how far off the mark would our sample estimate of 4.56 be?

The answer depends on the standard error of the sample mean. The standard error of a sample mean is based on the standard deviation and the size of the sample. SPSS determines the standard error just as you would—by dividing the standard deviation by the square root of the sample size. For spend10, the standard error is the standard deviation, 2.50, divided by the square root of 2,323. Performed with a hand calculator: 2.50 / sqrt(2323) = 2.50 / 48.20 ≈ .05.

This number, .05, tells us the extent to which the sample mean of 4.56 departs by chance from the population mean. The standard error is the essential ingredient for making inferences about the population mean. But let's get SPSS to help us make these inferences. Specifically, we will use the One-Sample T Test procedure to do three things: find the 95 percent confidence interval of the mean, calculate the 90 percent confidence interval of the mean, and test a hypothetical claim about the population mean using the .05 level of significance.

First we will use One-Sample T Test to obtain the 95 percent confidence interval of spend10. Click Analyze → Compare Means → One-Sample T Test, causing the One-Sample T Test window to open (Figure 6-2). The user supplies SPSS with information in two places: the Test Variable(s) panel and the Test Value box, which currently contains the default value of 0. Now, One-Sample T Test is not naturally designed to report the 95 percent confidence interval for a mean. Rather, it is set up to compare the mean of a variable in the Test Variable(s) panel with a hypothetical mean (provided by the user in the Test Value box) and to see if random error could account for the difference. (We will discuss this calculation below.) However, if you run One-Sample T Test on its defaults, it will provide the 95 percent confidence interval. To obtain the 95 percent confidence interval for the mean of a variable, do three things. First, click the variable into the Test Variable(s) panel, as shown in Figure 6-2. Second, make sure that the Test Value box contains the default value of 0. Third, click Options and ensure that the confidence interval is set at 95 percent. To follow these steps for spend10, we would click it into the Test Variable(s) panel and leave the Test Value box as it is. After clicking Options to make sure that the confidence interval is set at 95 percent, we would click Continue, and then click OK to run the analysis.

**Figure 6-2**  One-Sample T Test Window: Setting the Confidence Interval

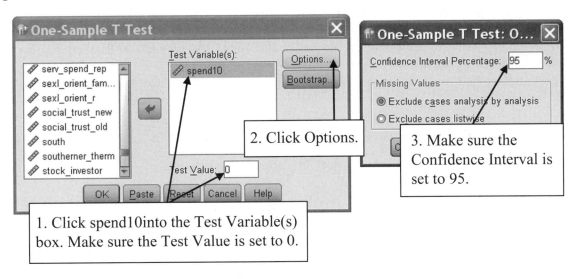

One-Sample Statistics

|  | N | Mean | Std. Deviation | Std. Error Mean |
|---|---|---|---|---|
| spend10 Incr federal $:10 programs | 2323 | 4.56 | 2.497 | .052 |

One-Sample Test

|  | Test Value = 0 | | | | | |
|---|---|---|---|---|---|---|
|  |  |  |  |  | 95% Confidence Interval of the Difference | |
|  | t | df | Sig. (2-tailed) | Mean Difference | Lower | Upper |
| spend10 Incr federal $:10 programs | 88.014 | 2322 | .000 | 4.561 | 4.46 | 4.66 |

SPSS output for One-Sample T Test includes two tables. In the One-Sample Statistics table, SPSS reports summary information about spend10. This information is similar to the Descriptives output discussed earlier. Again, we can see that spend10 has a mean of 4.56, a standard deviation of 2.50, and a standard error of .05 (which, reassuringly, is the same number we calculated by hand). We are interested mainly in the second table, the One-Sample Test table. In fact, when using One-Sample T Test to obtain confidence intervals, you may safely ignore all the information in the One-Sample Test table except for the rightmost cells. The values appearing under the label "95% Confidence Interval of the Difference," 4.46 and 4.66, define the lower and upper boundaries of the 95 percent confidence interval.

Now, we are trying to determine how much confidence to invest in our sample mean of 4.56. Is the true population mean right around 4.56? The 95 percent confidence interval provides a probabilistic—not a definitive—answer. There is a high probability, a 95 percent probability, that the true population mean lies in the region between 4.46 at the low end and 4.66 at the high end. If we were to take a very large number of random samples from the population and calculate, for each sample, the mean of spend10, 95 percent of those calculated means would fall in the interval between 4.46 and 4.66. To be sure, there is random "noise" in each random sample. Yet 95 percent of the time, that noise will give us a sample mean within the bandwidth of 4.46 to 4.66. On uncommon occasions—5 percent of the time—we would obtain a sample mean that falls outside those boundaries, below 4.46 or above 4.66. Therefore, we can infer that there is a 95 percent chance that the true population mean is in the 4.46–4.66 range and a 5 percent probability that the population mean lies outside these boundaries—a 2.5 percent chance that the population mean is less than 4.46 and a 2.5 percent chance that it is greater than 4.66. You might be wondering where the "2.5 percent chance" came from. Aren't we discussing the 95 percent confidence interval? Shouldn't the statement instead say "5 percent chance"? The "2.5 percent" terminology is correct. This is an understandable confusion, so let's pause and clear things up.

Figure 6-3 shows a bell-shaped curve. The 2008 NES sample mean, 4.56, is in the center of the distribution, bisecting the curve, with half the distribution of possible sample means falling in the negative tail below 4.56 and half falling in the positive part of the tail, above 4.56. Now consider the two shorter lines, one in the negative tail, drawn at 4.46, and one in the positive tail, at 4.66. These numbers, as we have seen, define the 95 percent confidence interval. Because the 95 percent confidence interval brackets 95 percent of all possible sample means, the remaining 5 percent must fall below 4.46 *or* above 4.66. Furthermore, because the curve is symmetrical, one-half of the remaining 5 percent (2.5 percent) will fall in the negative tail, below 4.46, and one-half (the other 2.5 percent) will fall in the positive tail, above 4.66.

**Figure 6-3**  Bell-shaped Curve Showing 95 Percent Confidence Interval of the NES Sample Mean

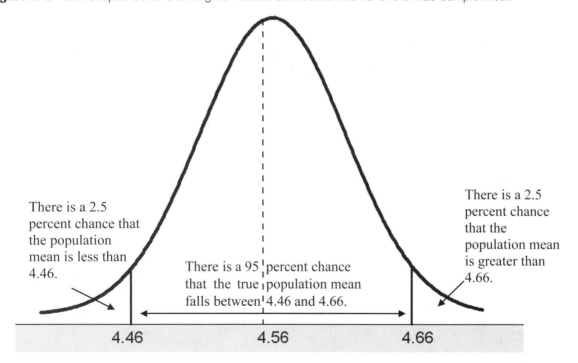

Suppose someone made the (weak) hypothetical claim that the true population mean was *different from* 4.56. This person didn't say whether he or she thought the true mean was greater than 4.56 or less than 4.56. Just different from 4.56. Such a claim—a claim that the real population mean is different from the observed sample mean—requires a two-tailed test of statistical significance. Because we don't know the direction of the hypothetical difference, we would apply the 95 percent confidence interval and use a .05 two-tailed test. We would simply say that there is a 95 percent chance that the population mean falls between 4.46 and 4.66 and a 5 percent chance that it falls above or below these numbers.

Most hypothesis-testing in political science uses a one-tailed test, not a two-tailed test. Why? Because any properly constructed hypothesis communicates the direction of a relationship. A gender-gap hypothesis, for example, would not suggest ambiguously that women and men will have different means on spend10. The hypothesis will tell you that the female mean on spend10 should be found to be higher (or lower) than the male mean on spend10. Because hypothetical relationships imply direction, only one-half of the bell-shaped curve (either the half with the negative tail or the half with the positive tail) comes into play in testing the hypothesis. For this reason, the 95 percent confidence interval provides a stringent .025 one-tailed test of statistical significance for hypothetical claims about the population mean.

Let's apply this one-tailed logic. Suppose someone were to suggest that spend10's population mean is 4.70 and that we obtained a sample mean of 4.56 by chance when we drew the sample. Notice that this hypothetical assertion, 4.70, locates the population mean above the observed sample mean of 4.56, out in the positive region of the curve. For the purposes of testing this claim, then, we ignore the negative tail and use the positive tail. Return your attention to Figure 6-3, focusing only on the upper half, the half with the line drawn at 4.66, the upper boundary of the 95 percent confidence interval. Ask yourself this question: If the unseen population mean is equal to the observed sample mean of 4.56, what is the probability that a random sample would produce a mean of 4.70? As Figure 6-3 illustrates, we know that if we were to draw a large number of random samples from a population in which the true mean is 4.56, only 2.5 percent of those samples would yield sample means of 4.66 *or higher*. Using the .025 standard, then, we would reject the claim that the population mean is 4.70. If the population mean is equal to the observed sample mean of 4.56, then random processes would yield a mean of 4.70 less than 2.5 percent of the time. Naturally, the same logic applies to any hypothetical claim that locates the population mean below the observed sample mean, down in the negative tail of the distribution.

A .025 one-tailed test of statistical significance is stringent and conservative—it makes it tougher to reject hypothetical challenges to relationships observed in a sample. Researchers tend to err on the side of caution, and you will never be criticized for using the .025 criterion. However, a somewhat less stringent test, the .05 one-tailed test of statistical significance, is perhaps a more widely applied standard. There are two ways to apply this standard—the confidence interval approach and the P-value approach. In the confidence interval approach, the researcher finds the 90 percent confidence interval of the mean. Why the 90 percent confidence interval? For the same reasons that the 95 percent confidence interval provides a .025 one-tailed test, the 90 percent confidence interval sets the limits of random sampling error in applying the .05 standard. There is a 90 percent probability that the population mean falls between the lower value of the 90 percent confidence interval and the higher value of the 90 percent confidence interval. There is a 10 percent chance that the population mean falls outside these limits—5 percent *below* the lower boundary and 5 percent *above* the upper boundary.

In the P-value approach the researcher determines the exact probability associated with a hypothetical claim about the population mean. First, let's find the 90 percent confidence interval for spend10. Then we will use the P-value approach to evaluate a hypothetical claim about the mean of spend10.

Click Analyze → Compare Means → One-Sample T Test. Make sure that spend10 is still in the Test Variable(s) panel and that the default value of 0 appears in the Test Value box. Click Options. This time, type "90" in the confidence interval box. Click Continue, and then click OK.

One-Sample Test

| | Test Value = 0 | | | | | |
|---|---|---|---|---|---|---|
| | | | | | 90% Confidence Interval of the Difference | |
| | t | df | Sig. (2-tailed) | Mean Difference | Lower | Upper |
| spend10 lncr federal $:10 programs | 88.014 | 2322 | .000 | 4.561 | 4.48 | 4.65 |

You can be confident, 90 percent confident anyway, that the population mean of spend10 lies between 4.48 at the low end and 4.65 at the high end. There is a probability of .10 that the population mean lies beyond these limits—a .05 probability that it is less than 4.48 and a .05 probability that it is greater than 4.65.

We can apply this knowledge to the task of testing a hypothetical claim. Suppose you hypothesized that political science majors will be more likely to favor government spending than will most people. To test this idea, you ask a group of political science majors a series of 10 questions about federal programs—the same set of questions that appears in the 2008 National Election Study. Whereas the 2008 NES reports a mean value of 4.56 on spend10, you find a higher mean value, 4.63, among the respondents in your study. Thus it would appear that your respondents are, on average, more supportive of government spending than are the individuals in the NES random sample of U.S. adults. But is this difference, 4.56 versus 4.63, *statistically* significant at the .05 level? No, it is not. Why can we say this? Because the political science majors' mean, 4.63, does not exceed the NES sample's upper confidence boundary, 4.65.

Think about it this way. Imagine a population of U.S. adults in which spend10's true mean is equal to 4.56. Now suppose you were to draw a random sample from this population and calculate the mean of spend10. The upper confidence boundary tells you that such a sample would yield a mean of greater than 4.65 *less frequently* than 5 times out of 100. The upper confidence boundary also says that such a sample would produce a mean of less than 4.65 *more frequently* than 5 times out of 100. Because 4.63 is less than 4.65, you must conclude that the political science majors' mean is not significantly higher than the NES mean. Put another way, there is a probability of greater than .05 that your sample of political science majors and the NES sample of adults were both drawn from the same population—a population in which spend10 has a mean equal to 4.56.

Confidence interval approaches to statistical significance work fine. Find the 90 percent confidence interval and compare the hypothetical mean to the appropriate interval boundary. If the hypothetical mean falls above the upper boundary (or below the lower boundary), then conclude that the two numbers are significantly different at the .05 level. If the hypothetical mean falls below the upper boundary (or above the lower boundary), then conclude that the two numbers are not significantly different at the .05 level. Thus the confidence interval approach tells you that a random sample of U.S. adults would produce a sample mean of 4.63 more frequently than 5 percent of the time. However, the P-value approach to statistical significance is more precise. The P-value will tell you *exactly* how frequently a sample mean of 4.63 would occur.

Let's run One-Sample T Test on spend10 one more time and obtain the information we need to determine the P-value associated with the political science majors' mean of 4.63. Click Analyze → Compare Means → One-Sample T Test. Spend10 should still be in the Test Variable(s) panel. Now click in the Test Value box and type "4.63" (Figure 6-4). SPSS will calculate the difference between the mean of the test

**Figure 6-4**   Testing a Hypothetical Claim about a Sample Mean

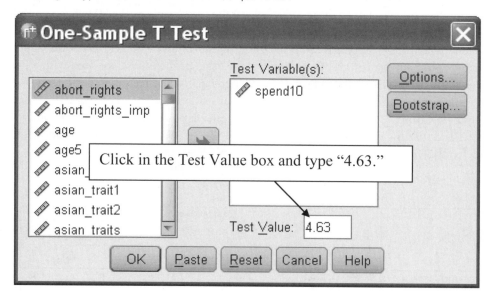

**Figure 6-5**  Testing for Statistical Significance

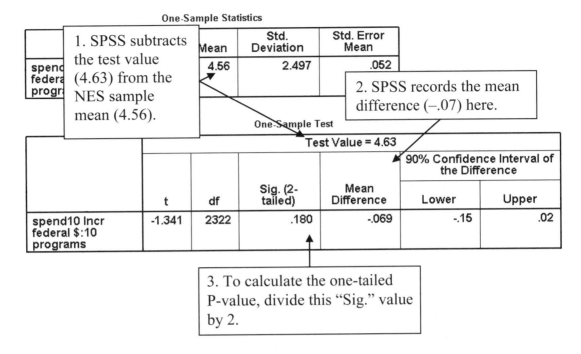

variable, spend10, and the test value, 4.63. SPSS will then report the probability that the test value, 4.63, came from the same population as did the mean of the test variable, spend10. Click OK. Again, we have One-Sample T Test output (Figure 6-5).

Consider the One-Sample Test table. What do these numbers mean? SPSS follows this protocol in comparing the mean of a test variable with a test value. First, it finds the difference between them: test variable mean minus test value. So SPSS calculates $4.56 - 4.63$ and reports the result, $-.07$, in the "Mean Difference" column of the One-Sample Test table.

Obviously, if the political science majors' mean were the same as the NES mean of 4.56, then SPSS would report a mean difference of 0. But the calculated difference, $-.07$, is not 0. In fact, given the way that SPSS evaluates test values, the difference between the political science majors' mean and the NES mean falls in the hypothesized direction. Consider another bell-shaped curve, shown in Figure 6-6. This time the curve is centered at 0, the difference we would observe if the political science majors' mean were the same as the NES mean. The curve tapers off into a skinnier and skinnier tail above 0. This is the region of the curve you would use to test the hypothesis that the majors' mean is less than the NES mean. Because our hypothesis stipulated that the majors' mean would be greater than the NES mean, we can ignore the positive tail of the curve. The curve also tapers off into a skinnier tail below 0. This region, the half of the curve below 0, is the region we must focus on to test the hypothesis. The inferential question: If the true difference between the NES mean and the test value is equal to 0, how often would random processes produce an observed difference of $-.07$? If the answer is "less than 5 percent of the time," then we can infer that the two means are significantly different from one another. If the answer is "more than 5 percent of the time," then we must conclude that the two means are not significantly different.

Unhelpfully, SPSS does not report the one-tailed P-value that would permit us to settle the issue. As Figure 6-6 illustrates, SPSS does not ignore the positive half of the bell-shaped curve, the half that, in the present example, is irrelevant to testing the hypothesis. Instead, SPSS answers this inferential question: If in the population there is really no difference between the test variable mean and the test value, how often would one obtain a difference of at least .07 in *either direction*? How often would one observe a difference of less than $-.07$ (in the negative region) *or* more than $+.07$ (in the positive region)? To answer this inferential question, SPSS calculates a Student's t-test statistic, or t-ratio, the mean difference divided by spend10's standard error.[3] This value of t, $-1.34$, appears in the leftmost cell of the One-Sample Test table. Thus spend10's mean falls 1.34 standard errors below the political science majors' mean. Turn your attention to the cell

**Figure 6-6** SPSS Reports Two-Tailed P-values

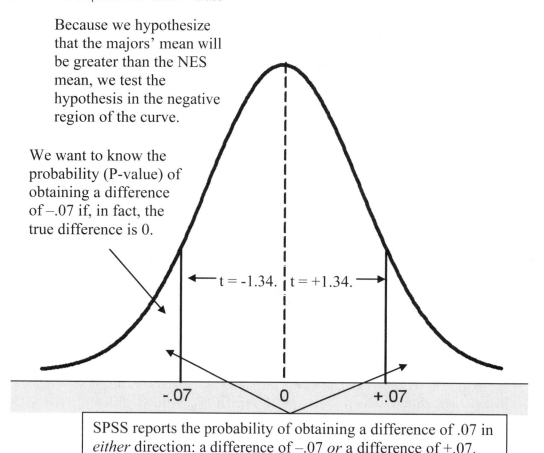

Because we hypothesize that the majors' mean will be greater than the NES mean, we test the hypothesis in the negative region of the curve.

We want to know the probability (P-value) of obtaining a difference of –.07 if, in fact, the true difference is 0.

t = -1.34.  t = +1.34.

-.07          0          +.07

SPSS reports the probability of obtaining a difference of .07 in *either* direction: a difference of –.07 *or* a difference of +.07.

labeled "Sig. (2-tailed)," which contains the number .180. This two-tailed value tells you the proportion of the curve that lies below t = –1.34 *and* above t = +1.34. Thus, despite SPSS's willingness to report a t-test statistic whose negative sign communicates the region of the curve we are interested in, it disappoints by reporting the proportion of both tails—the proportion below t = –1.34 plus the proportion above t = +1.34.

Again, we are not testing the hypothesis that the political science majors' mean is *different from* the mean of the adult population. We are not asking SPSS to tell us if the majors' mean is less than or greater than the population mean. Rather, we are testing the hypothesis that the majors' mean is *greater than* the population mean. In testing this hypothesis, we do not want to know the proportion of the curve that lies above t = +1.34. We only want to know the P-value associated with t = –1.34, the test statistic for spend10's mean of 4.56 minus the political science majors' mean of 4.63. Because the Student's t-distribution is perfectly symmetrical, one-half of .180, or .09, falls below t = –1.34. From the confidence interval approach, we already know that there is a probability of greater than .05 that the political science majors' mean and the NES sample mean came from the same population, that the true difference between them is equal to 0. We can now put a finer point on this probability. We can say that, if the true difference between the majors' mean and the NES mean is equal to 0, then the observed difference between them (–.07) would occur, by chance, .09 of the time. Because .09 exceeds the .05 threshold, we would conclude again that the test subjects' mean is not significantly higher than the population mean.

Before proceeding, let's pause to review the steps for obtaining a P-value from the One-Sample T Test procedure. This is also the appropriate place to introduce a useful template for writing an interpretation of your results.

1. Click the variable of interest into the Test Variable(s) panel.
2. Type the hypothetical value in the Test Value box and click OK.
3. Obtain a P-value by dividing the number in the "Sig. (2-tailed)" cell by 2.
4. Use the P-value to fill in the blank in the following template:

If in the population there is no difference between the mean of [the test variable] and [the test value], then the observed difference of [the mean difference] would occur _____ of the time by chance.

Of course, you can embellish the template to make it fit the hypothesis you are testing. For the spend10 example, you could complete the sentence this way: "If in the population there is no difference between the NES mean of spend10 and the political science majors' mean of 4.63, then the observed difference of –.07 would occur .09 of the time by chance." It is also acceptable to express the P-value as a percentage, as in: "would occur 9 percent of the time by chance." The .05 benchmark is the standard for testing your hypothesis. If the P-value is less than or equal to .05, then you can infer that the test value is significantly greater than (or less than) the mean of the test variable. If the P-value is greater than .05, then you can infer that the test value is not significantly greater than (or less than) the mean of the test variable.

## INDEPENDENT-SAMPLES T TEST

We now turn to a common hypothesis-testing situation: comparing the sample means of a dependent variable for two groups that differ on an independent variable. Someone investigating the gender gap, for example, might test a series of hypotheses about the political differences between men and women. In the next guided example, we test two gender gap hypotheses:

Hypothesis 1: In a comparison of individuals, men will give gays lower feeling thermometer ratings than will women.

Hypothesis 2: In a comparison of individuals, men will place more importance on international policy goals than will women.

The first hypothesis suggests that when we divide the sample on the basis of the independent variable, gender, and compare mean values of the gay feeling thermometer, the male mean will be lower than the female mean. The second hypothesis suggests that when we compare men and women on a 100-point scale that gauges the importance of international policy goals, the male mean will be higher than the female mean.

The researcher always tests his or her hypotheses against a skeptical foil, the null hypothesis. The null hypothesis claims that, regardless of any group differences that a researcher observes in a random sample, no group differences exist in the population from which the sample was drawn. How does the null hypothesis explain apparently systematic patterns that might turn up in a sample, such as a mean difference between women and men on the gay feeling thermometer? It points to random sampling error. In essence the null hypothesis says, "You observed such and such a difference between two groups in your random sample. But, in reality, no difference exists in the population. When you took the sample, you introduced random sampling error. Thus random sampling error accounts for the difference you observed." For both hypotheses 1 and 2 above, the null hypothesis says that there are no real differences between men and women in the population, that men do not give gays lower ratings or place greater importance on international policies. The null hypothesis asserts further that any observed differences in the sample can be accounted for by random sampling error.

The null hypothesis is so central to the methodology of statistical inference that we always begin by assuming it to be correct. We then set a fairly high standard for rejecting it. The researcher's hypotheses—such as the gay thermometer hypothesis and the international policy hypothesis—are considered alternative hypotheses. The Independent-Samples T Test procedure permits us to test each alternative hypothesis against the null hypothesis and to decide whether the observed differences between males and females are too large to have occurred by random chance when the sample was drawn. For each mean comparison, the Independent-Samples T Test procedure will give us a P-value: the probability of obtaining the sample difference under the working assumption that the null hypothesis is true.

Click Analyze → Compare Means → Independent-Samples T Test. The Independent-Samples T Test window appears (Figure 6-7). SPSS wants to know two things: the name or names of the test variable(s) and the name of the grouping variable. SPSS will calculate the mean values of the variables named in the Test Variable(s) panel for each category of the variable named in the Grouping Variable box. It will then test to see if the differences between the means are significantly different from 0.

**Figure 6-7** Independent-Samples T Test Window

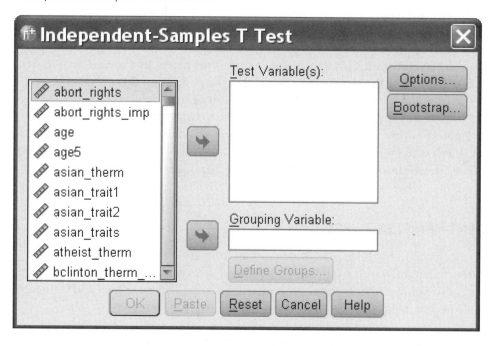

**Figure 6-8** Specifying Test Variables and Setting the Confidence Interval

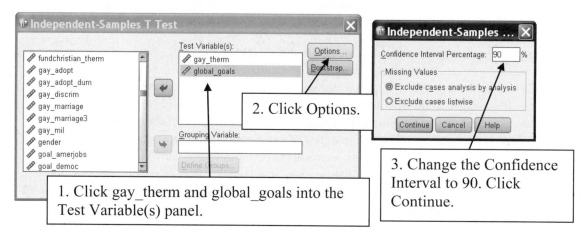

We want to compare the means for men and women on the gay feeling thermometer (gay_therm) and on the international policy scale (global_goals). Find gay_therm and global_goals in the variable list and click both of them into the Test Variable(s) panel. Click Options and change the confidence interval to 90 percent and click Continue (Figure 6-8). Because you want the means of these variables to be calculated separately for each sex, gender is the grouping variable. When you click gender into the Grouping Variable box, SPSS moves it into the box with the designation "gender(? ?)." The Define Groups button is activated (Figure 6-9). SPSS needs more information. It needs to know the codes of the two groups you wish to compare. Men are coded 1 on gender and women are coded 2. (Recall that by right-clicking on a variable you can reacquaint yourself with that variable's codes.) Click Define Groups (Figure 6-9). There are two ways to define the groups you want to compare: Use specified values (the default) and Cut point. The choice depends on the situation. If you opt for Use specified values, then SPSS will divide the cases into two groups based on the codes you supply for Group 1 and Group 2. If the grouping variable has more than two categories, then you may wish to use Cut point. SPSS will divide the cases into two groups based on the code entered in the Cut point box—one group for all cases having codes equal to or greater than the Cut point code and one group having codes less than the Cut point code. (You will use Cut point in one of the

**Figure 6-9** Defining the Grouping Variable

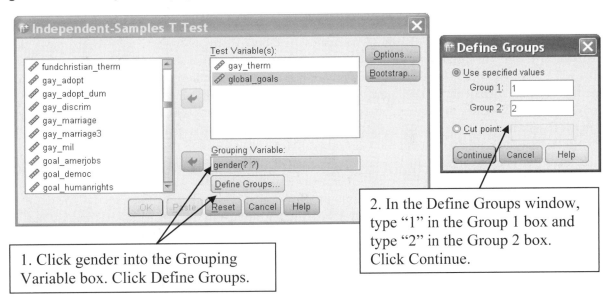

1. Click gender into the Grouping Variable box. Click Define Groups.

2. In the Define Groups window, type "1" in the Group 1 box and type "2" in the Group 2 box. Click Continue.

exercises at the end of this chapter.) Because gender has two codes—1 for males and 2 for females—we will go with the Use specified values option in this example. Click in the Group 1 box and type "1," and then click in the Group 2 box and type "2." Click Continue, as shown in Figure 6-9. Notice that SPSS has replaced the question marks next to gender with the codes for males and females. All set. Click OK.

SPSS runs both mean comparisons and reports the results in the Viewer (Figure 6-10). The top table, labeled "Group Statistics," shows descriptive information about the means of gay_therm and global_goals. The bottom table, "Independent Samples Test," tests for statistically significant differences between men and women on each dependent variable. There is a lot of information to digest here, so let's take it one step at a time.

**Figure 6-10** Results of Independent Samples T Test

Group Statistics

| | R gender | N | Mean | Std. Deviation | Std. Err Mean |
|---|---|---|---|---|---|
| Feeling therm: Gay men and lesbians (homosexuals) | 1 Male | 923 | 43.16 | 26.855 | .8 |
| | 2 Female | 1142 | 54.44 | 27.340 | .8 |
| Imp intl policy: 10 goals | 1 Male | 939 | 55.06 | 26.635 | .869 |
| | 2 Female | 1148 | 53.54 | 25.227 | .745 |

1. SPSS subtracts the female mean from the male mean.

2. SPSS reports the mean difference here.

Independent Samples Test

| | | Levene's Test for Equality of Variances | | t-test for Equality of Means | | | | | 90% Confidence Interval of the Difference | |
|---|---|---|---|---|---|---|---|---|---|---|
| | | F | Sig. | t | df | Sig. (2-tailed) | Mean Difference | Std. Error Difference | Lower | Upper |
| Feeling therm: Gay men and lesbians (homosexuals) | Equal variances assumed | .395 | .530 | -9.393 | 2063 | .000 | -11.277 | 1.201 | -13.253 | -9.301 |
| | Equal variances not assumed | | | -9.411 | 1986.642 | .000 | -11.277 | 1.198 | -13.249 | -9.305 |
| Imp intl policy: 10 goals | Equal variances assumed | 3.918 | .048 | 1.341 | 2085 | .180 | 1.526 | 1.138 | -.347 | 3.399 |
| | Equal variances not assumed | | | 1.334 | 1957.666 | .182 | 1.526 | 1.144 | -.357 | 3.410 |

3. If this "Sig." value is greater than or equal to .05, then use the "Equal variances assumed" row. If it is less than .05, the use the "Equal variances not assumed" row.

We will evaluate the gender difference on the gay feeling thermometer first. From the Group Statistics table we can see that males, on average, rated gays at 43.16, whereas females had a higher mean, 54.44. It would appear that our alternative hypothesis has merit. The difference between these two sample means is 43.16 minus 54.44, or −11.277 (SPSS always calculates the difference by subtracting the Group 2 mean from the Group 1 mean. This value appears in the "Mean Difference" column of the Independent-Samples Test table.) The null hypothesis claims that this difference is the result of random sampling error and, therefore, that the true male-female difference in the population is 0. Using the information in the Independent-Samples Test table, we test the null hypothesis against the alternative hypothesis that the male mean is lower than the female mean.

Notice that there are two rows of numbers for each dependent variable. One row is labeled "Equal variances assumed" and the other "Equal variances not assumed." What is this all about? One of the statistical assumptions for comparing sample means is that the population variances of the dependent variable are the same for both groups—in this case, that the amount of variation in gay_therm among men in the population is equal to the amount of variation in gay_therm among women in the population. If this assumption holds up, then you would use the first row of numbers to test the hypothesis. If this assumption is incorrect, however, then you would use the second row of numbers. SPSS evaluates the assumption of equal variances using Levene's test, which tests the hypothesis that the two variances are equal.

Here is how to proceed. Look at the "Sig." value that appears under "Levene's Test for Equality of Variances." If this value is greater than or equal to .05, then use the "Equal variances assumed" row. If the "Sig." value under "Levene's Test for Equality of Variances" is less than .05, then use the "Equal variances not assumed" row. Because this "Sig." value for gay_therm is greater than .05, we would use the first row of numbers ("Equal variances assumed") to evaluate the mean differences between men and women.

Let's test the gay_therm hypothesis using the 90 percent confidence interval of the mean difference to apply the .05 standard. There are two boundaries: a lower boundary, −13.25, and an upper boundary, −9.30. Here is a foolproof method for testing a hypothesis using the confidence interval approach. First, make sure that the direction of the mean difference is consistent with the alternative hypothesis. If the mean difference runs in the direction opposite from the one you hypothesized, then the game is already up—the alternative hypothesis is incorrect. Second, determine whether the confidence interval of the difference—the interval between the lower boundary and the upper boundary—includes 0. If the confidence interval includes 0, then the mean difference is not significant at the .05 level. It would occur by chance more than 5 times out of 100. If the confidence interval does not include 0, then the mean difference is statistically significant at the .05 level. The difference would occur by chance fewer than 5 times out of 100.

In the gay_therm example, we have seen that the difference between men and women on gay_therm runs in the hypothesized direction: Males have a lower mean rating than do females. So the alternative hypothesis is still alive. Does the confidence interval, the interval between −13.25 and −9.30 include 0? No, it doesn't. Conclusion: If the null hypothesis is correct, the probability of observing a mean difference of −11.28 is less than .05. Reject the null hypothesis.

Just as with the One-Sample T Test, we can arrive at an exact probability, or P-value, by dividing the "Sig. (2-tailed)" value by 2. The value in the "Sig. (2-tailed)" cell is .000. Dividing by 2, we still have .000. More precise conclusion: If the null hypothesis is correct, then random sampling error would produce a mean difference of −11.28 virtually 0 percent of the time. The alternative hypothesis is on safe inferential ground. Reject the null hypothesis.[4]

All right, so men score significantly lower on gay_therm than do women. Are men more internationally minded, as hypothesis 2 suggests? Again, the information on global_goals in the Group Statistics table would appear consistent with the hypothesis. Men, with a mean of 55.06, score higher than do women, who averaged 53.54. SPSS reports the mean difference, 1.53, in the Independent-Samples Test table. Does this difference pass muster with the null hypothesis? Let's look at the Independent-Samples Test table and decide which row of numbers to use. In this case the "Sig." value under "Levene's Test for Equality of Variances" is less than .05, so we will use the "Equal variances not assumed" row. Now pan to the right and consider the 90 percent confidence interval. Does the 90 percent confidence interval, −.36 to 3.41, include 0, the null's talisman? Yes, the confidence interval brackets 0. Thus we know that the mean difference would occur by chance more frequently than 5 times out of 100. There is indeed a fairly high probability of observing a difference of 1.53, if in fact the null hypothesis is correct. Dividing .182 by 2 gives us a P-value of .091. If the null hypothesis is correct, then we would observe by chance a sample difference of 1.53 about 9 times out of 100. Accept the null hypothesis.

## EXERCISES

1. (Dataset: GSS2008. Variable: egalit_scale.) The 2008 General Social Survey asked people a series of questions designed to measure how egalitarian they are—that is, the extent to which they think economic opportunities and rewards should be distributed more equally in society. The GSS2008 variable egalit_scale ranges from 0 (low egalitarianism) to 12 (high egalitarianism). The 2008 GSS, of course, is a random sample of U.S. adults. In this exercise you will analyze egalitarianism using One-Sample T Test. You then will draw inferences about the population mean.

   A. Egalitarianism has a sample mean of (fill in the blank) _____.

   B. There is a probability of .95 that the true population mean falls between an egalitarianism score of (fill in the blank) _____ at the low end and a score of (fill in the blank) _____ at the high end.

   C. There is a probability of .90 that the true population mean falls between an egalitarianism score of _____ (fill in the blank) at the low end and a score of _____ (fill in the blank) at the high end.

   D. A student researcher hypothesizes that social work majors will score significantly higher on the egalitarianism scale than the typical adult. The student researcher also hypothesizes that business majors will score significantly lower on the egalitarianism scale than the average adult. After administering the scale to a number of social work majors and a group of business majors, the researcher obtains these results: Social work majors' mean, 7.10; business majors' mean, 6.85.

   Using the confidence interval approach to apply the .05 one-tailed test of significance, based on your answer in C, you can infer that (check one)

   ❑ social work majors probably are not more egalitarian than most adults.

   ❑ social work majors probably are more egalitarian than most adults.

   Using the confidence interval approach to apply the .05 one-tailed test of significance, based on your answer in C, you can infer that (check one)

   ❑ business majors probably are not less egalitarian than most adults.

   ❑ business majors probably are less egalitarian than most adults.

   E. Obtain a P-value from your analysis of the business majors' mean. (fill in the blanks)

   If in the population there is no difference between the mean of egalitarianism and the business majors' mean of 6.85, the observed difference of _____ would occur _____ of the time by chance.

2. (Dataset: GSS2008. Variables: int_info_scale, age.) Are older people interested in a wider variety of social, economic, political, and scientific issues than are younger people? Or do younger people and older people not differ significantly in the scope of their interests? In this exercise you will use Independent-Samples T Test to test this hypothesis: In a comparison of individuals, people who are 30 or older are interested in a wider range of current issues than are people who are younger than 30.

GSS2008 contains int_info_scale, which measures respondents' level of interest in ten different issue areas. Scores on int_info_scale range from 0 (low interest) to 20 (high interest). This is the dependent variable and will go in the Test Variable(s) panel. The independent variable, age, goes in the Grouping Variable box. You want SPSS to create two groups from age: a group of respondents who are 30 or older and a group of respondents who are younger than 30. You can tell SPSS to do so by using Cut point in the Define Groups window. When you open the Define Groups window, click the radio button next to "Cut point" and type "30" in the Cut point box. Click Continue. Don't forget to click Options and ask for the 90 percent confidence interval.

A.  Examine the results of your analysis. Fill in the table below:

| Statistics for int_info_scale Level of interest in current issues | |
|---|---|
| Mean for older group (>=30 years old) | ? |
| Mean for younger group (<30 years old) | ? |
| Mean difference | ? |
| Equal variances assumed? (Yes or No) | ? |
| Lower 90 percent confidence boundary of mean difference | ? |
| Upper 90 percent confidence boundary of mean difference | ? |
| Does confidence interval contain 0? (Yes or No) | ? |
| t-statistic | ? |
| Sig. (2-tailed) | ? |
| Sig. (2-tailed) divided by 2 | ? |

B.  According to the null hypothesis, in the population from which the sample was drawn, the difference between the mean for people 30 or older and the mean for people younger than 30 is equal to (fill in the blank) _____.

C.  Based on your tabular entries in part A, you can infer that (check one)

❏  the older age group and the younger age group do not differ significantly in their level of interest in current issues.

❏  the older age group scores significantly higher on the level of interest scale than does the younger age group.

❏  the older age group scores significantly lower on the level of interest scale than does the younger age group.

D.  Your inferential decision, therefore, is to (check one)

❏  accept the null hypothesis.

❏  reject the null hypothesis.

3.  (Dataset: GSS2008. Variables: childs, attend3.) The role of religion lies at the center of an interesting debate about the future of U.S. partisan politics. Republican presidential candidates do much better among people who frequently attend religious services than among people who are less observant. However, religious attendance has been waning. This growing secularization, according to some observers, portends a weakening of the Republican base and a growing opportunity for the Democratic Party.[5] But we also know that religious beliefs and affiliations (or the lack thereof) are strongly shaped by childhood socialization. Are less-religious people raising and socializing children at the same rate as the more religious? In this exercise you will test this

hypothesis: In a comparison of individuals, those with lower levels of religiosity will have fewer children than will those with higher levels of religiosity. This hypothesis says that as religious attendance goes up, so will the average number of children.

Dataset GSS2008 contains childs, the respondent's number of children. This is the dependent variable. The independent variable is attend3, which measures religious attendance by three ordinal categories: "low" (coded 1), "med" (coded 2), and "high" (coded 3).

A. Exercise a familiar skill you acquired in Chapter 4. Perform a mean comparison analysis, obtaining mean values of childs (and numbers of cases) for each value of attend3. Fill in the following table:

|  | Summary of Number of Children | |
| --- | --- | --- |
| Religious attendance | Mean | N |
| Low | ? | ? |
| Med | ? | ? |
| High | ? | ? |
| Total | ? | ? |

B. Do these findings support the childs-attend3 hypothesis? (circle one)

Yes     No

Briefly explain your reasoning. _____

_____

_____

_____

C. Focus your analysis on a comparison between respondents with low attendance (coded 1) and respondents with high attendance (coded 3). Run Independent-Samples T Test to find out if high attenders have significantly more children than do low attenders. Apply the .05 level of significance. (fill in the blanks)

"Low" attendance mean: _____

"High" attendance mean: _____

Mean difference: _____

90 percent confidence interval of the mean difference: between _____ and _____

D. Does the statistical evidence support the hypothesis that people who are more religious have significantly more children than do people who are less religious? (check one)

❑ Yes, the statistical evidence supports the hypothesis.

❑ No, the statistical evidence does not support the hypothesis.

4. (Dataset: GSS2008. Variables: sibs, relig, authoritarianism, sex.) Here are two bits of conventional wisdom, beliefs that are accepted widely as accurate descriptions of the world.

Conventional wisdom 1: Catholics have bigger families than do Protestants.

Conventional wisdom 2: Men have stronger authoritarian tendencies than do women.

In this exercise you will test these ideas and see how well they stand up to the statistical evidence. Test conventional wisdom 1 by comparing the average number of siblings (GSS2008 variable sibs) for Protestants and Catholics (relig). Test conventional wisdom 2 by comparing mean authoritarianism scale scores (authoritarianism) for males and females (sex). The authoritarianism scale ranges from 0 (low authoritarianism) to 7 (high authoritarianism). Run the analyses. Record the results in the following table:

|  | Conventional wisdom 1 | Conventional wisdom 2 |
|---|---|---|
| Mean difference | ? | ? |
| Equal variances assumed? (Yes or No) | ? | ? |
| Lower 90 percent confidence boundary of mean difference | ? | ? |
| Upper 90 percent confidence boundary of mean difference | ? | ? |
| Does confidence interval contain 0? (Yes or No) | ? | ? |
| t-statistic | ? | ? |
| Sig. (2-tailed) | ? | ? |
| Sig. (2-tailed) divided by 2 | ? | ? |
| Does statistical evidence support conventional wisdom? (Yes or No) | ? | ? |

That concludes the exercises for this chapter. Before exiting SPSS, be sure to save your output file.

## NOTES

1.  Respondents were asked whether federal spending should be increased a great deal, increased a moderate amount, increased a little, kept the same, decreased a little, decreased a moderate amount, or decreased a great deal on each of these programs or policies: child care, crime, environment, foreign aid, highways, aid to poor, public schools, science and technology, social security, and welfare. Spend10 records the number of "increased a great deal" or "increased a moderate amount" responses.
2.  To simplify the presentation of the material in this part of the chapter, values will be rounded to two decimal places.
3.  SPSS performs this calculation at 32-decimal precision. So if you were to check SPSS's math, using the mean difference and standard error that appear in the SPSS Viewer, you would arrive at a slightly different t-ratio than the value of t reported in the One-Sample Test table.
4.  A P-value cannot exactly equal 0. There is always some chance, however remote, that random error produced the difference observed in the sample. Although the editing of SPSS tabular output is not covered in this book, it is possible to double-click on a table (SPSS refers to tabular objects as "pivot tables") to make appearance-enhancing changes or to look at the mind-numbingly precise numbers that SPSS rounds to produce the digits on display in the table. Double-clicking on the Independent-Samples T Test output shown in Figure 6-10, and then double-clicking on the "Sig. (2-tailed)" value just discussed (the value .000), reveals this number: 1.493049096232529E-20. That's scientific notation for .00000000000000000001493049096232529. Of course, this is a two-tailed probability. To arrive at a one-tailed P-value, we would divide this vanishingly small number by 2.
5.  John B. Judis and Ruy Teixeira, *The Emerging Democratic Majority* (New York: Scribner, 2002).

# 7

# Chi-square and Measures of Association

In the preceding chapter you learned how to test for mean differences on an interval-level dependent variable. But what if you are not dealing with interval-level variables? What if you are doing cross-tabulation analysis and are trying to figure out whether an observed relationship between two nominal or ordinal variables mirrors the true relationship in the population? Just as with mean differences, the answer depends on the boundaries of random sampling error, the extent to which your observed results "happened by chance" when you took the sample. The Crosstabs procedure can provide the information needed to test the statistical significance of nominal or ordinal relationships, and it will yield appropriate measures of association.

You are familiar with the Crosstabs procedure. For analyzing datasets that contain a preponderance of categorical variables—variables measured by nominal or ordinal categories—cross-tabulation is by far the most common mode of analysis in political research. In this section we will revisit Crosstabs and use the Statistics subroutine to obtain the oldest and most widely applied test of statistical significance in cross-tabulation analysis, the chi-square test. With rare exceptions, chi-square can always be used to determine whether an observed cross-tab relationship departs significantly from the expectations of the null hypothesis. In the first guided example, you will be introduced to the logic behind chi-square, and you will learn how to interpret SPSS's chi-square output.

In this chapter you will also learn how to obtain measures of association for the relationships you are analyzing. SPSS is programmed to produce, at the user's discretion, a large array of such measures. In doing your own analysis, you will have to tell SPSS which measure or measures you are after. If one or both variables in the cross-tabulation are nominal level, then you need to request lambda. If both are ordinal-level variables, then SPSS offers several choices: gamma, Kendall's tau-b, Kendall's tau-c, and Somers' d. Gamma tends to overstate the strength of a relationship, and so it is not recommended. The Kendall's statistics are *symmetrical* measures. Each will yield the same value, regardless of whether the independent variable is used to predict the dependent variable or the dependent variable is used to predict the independent variable. Somers' d is an *asymmetrical* measure. It will report different measures of the strength of a relationship, depending on whether the independent variable is used to predict the dependent variable or the dependent variable is used to predict the independent variable. Lambda, too, is an asymmetrical measure. Asymmetrical measures of association generally are preferred over symmetrical measures.[1] Therefore, in this book we will cover lambda for nominal relationships and Somers' d for ordinal relationships.[2]

Somers' d is a directional measure that ranges from –1 to +1. A plus (+) sign says that increasing values of the independent variable are associated with increasing values of the dependent variable. A minus (–) sign says that increasing values of the independent variable are related to decreasing values of the dependent variable. Both lambda and Somers' d are proportional reduction in error (PRE) measures of strength. A PRE measure tells you the extent to which the values of the independent variable predict the values of the dependent variable. A value close to 0 says that the independent variable provides little predictive leverage; the relationship is weak. Values close to the poles—to –1 for negative associations or to +1 for positive relationships—tell you that the independent variable provides a lot of help in predicting the dependent variable; the relationship is strong.

Lambda's PRE status stands it in good stead with political researchers because PRE measures are generally preferred over measures that do not permit a PRE interpretation. Even so, lambda tends to underestimate the strength of a relationship, especially when one of the variables has low variation. Therefore, when you are analyzing a relationship in which one or both of the variables are nominal, it is a good practice to request Cramer's V as well as lambda. Cramer's V, one of a variety of chi-square-based measures, does not measure strength by the PRE criterion. However, it is bounded by 0 (no relationship) and 1 (a perfect relationship).

## ANALYZING AN ORDINAL-LEVEL RELATIONSHIP

We will begin by using NES2008 to analyze an ordinal-level relationship. (If you are running Student Version, you will need to open NES2008B_Student. For the remainder of the book, when a guided example or exercise calls for NES2008, you will use NES2008B_Student.) Consider this hypothesis: In a comparison of individuals, people with higher incomes will be more likely to support abortion rights than will people with lower incomes. NES2008 contains abort_rights3, an ordinal measure of abortion opinions. Respondents who think abortion should "never" be allowed are coded 1, those who think it "depends" are coded 2, and those who say it should "always" be permitted are coded 3. Abort_rights3 is the dependent variable. For the independent variable, we will use income_r3, a three-category measure of income that you created in Chapter 3. On income_r3, respondents are classified as "low" (coded 1), "middle" (coded 2), or "high" (coded 3).

Let's first test this hypothesis the old-fashioned way—by running the Crosstabs analysis and comparing column percentages. Click Analyze → Descriptive Statistics → Crosstabs. Remember to put the dependent variable, abort_rights3, on the rows and the independent variable, income_r3, on the columns. Request column percentages. Run the analysis and consider the output.

abort_rights3 When should abortion be permitted? * income_r3 R income: 3 categories Crosstabulation

|  |  | Statistics | income_r3 R income: 3 categories | | | Total |
|  |  |  | 1 Low | 2 Middle | 3 High |  |
|---|---|---|---|---|---|---|
| abort_rights3 When should abortion be permitted? | 1 Never | Count | 64 | 42 | 35 | 141 |
|  |  | % within income_r3 R income: 3 categories | 18.2% | 11.7% | 12.3% | 14.2% |
|  | 2 Depends | Count | 154 | 173 | 128 | 455 |
|  |  | % within income_r3 R income: 3 categories | 43.9% | 48.2% | 44.9% | 45.7% |
|  | 3 Always | Count | 133 | 144 | 122 | 399 |
|  |  | % within income_r3 R income: 3 categories | 37.9% | 40.1% | 42.8% | 40.1% |
| Total |  | Count | 351 | 359 | 285 | 995 |
|  |  | % within income_r3 R income: 3 categories | 100.0% | 100.0% | 100.0% | 100.0% |

How would you evaluate the abort_rights3-income_r3 hypothesis in light of this analysis? Focus on the column percentages in the "Always" row. According to the hypothesis, as we move along this row, from lower income to higher income, the percentages of respondents supporting abortion rights should increase. Is this what happens? Sort of. The percentages increase from 37.9 percent among low-income respondents to 42.8 percent among high-income respondents. So the difference between low-income and high-income individuals is something on the order of 5 percentage points—not a terribly robust relationship between the independent and dependent variables. Using your own interpretive skills, you would probably say that the findings are too weak to support the abort_rights3-income_r3 hypothesis.

Now let's reconsider the abort_rights3-income_r3 cross-tabulation from the standpoint of inferential statistics, the way the chi-square test of statistical significance would approach it. Chi-square begins by looking at the "Total" column, which contains the distribution of the entire sample across the values of the dependent variable, abort_rights3. Thus 14.2 percent of the sample opposes abortion under all circumstances, 45.7 percent think it depends, and 40.1 percent say that abortion should always be allowed. Chi-square then frames the null hypothesis, which claims that, in the population, abort_rights3 and income_r3 are not related to each other, that individuals' incomes are unrelated to their opinions about abortion. If the null hypothesis is correct, then a random sample of low-income people would produce the same distribution of abortion opinions as the total distribution. By the same token, a random sample of middle-income people would yield a distribution that looks just like the total distribution, as would a random sample of high-income individuals. If the null hypothesis is correct, then the distribution of cases down each column of the table will be the same as in the "Total" column. Of course, the null hypothesis asserts that any observed departures from this monotonous pattern resulted from random sampling error.

Now reexamine the table and make a considered judgment. Would you say that the observed distribution of cases within each category of income_r3 conforms to the expectations of the null hypothesis? For the low-income category, the distribution is pretty close to the total distribution, with some departures—for example, a somewhat higher percentage in the "Never" category than would be expected according to the null hypothesis and a slightly lower percentage in the "Always" category. In fact, for each value of income_r3, there is fairly close conformity to what we would expect to find if the null hypothesis were true. The small departures from these expectations, furthermore, might easily be explained by random sampling error, the null's explanation for everything.

Let's rerun the analysis and find out if our considered judgment is borne out by the chi-square test. We will also obtain a measure of association for the relationship. Return to the Crosstabs window. Click Statistics. The Crosstabs: Statistics window pops up (Figure 7-1). There are many choices here, but we know what we want: We would like SPSS to perform a chi-square test on the table. Check the Chi-square box. We also know which measure of association to request. Because both abort_rights3 and income_r3 are ordinal-level variables, we will request Somers' d. Check the box next to "Somers' d." Click Continue, and then click OK.

**Figure 7-1** Requesting Statistics (ordinal-level relationship)

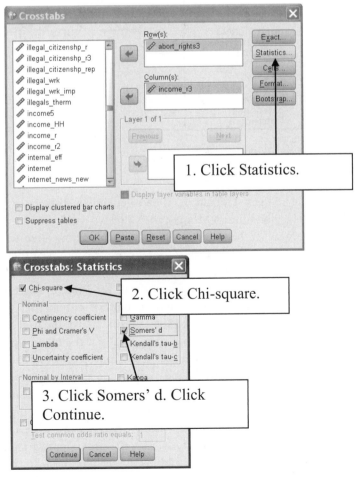

SPSS runs the cross-tabulation analysis again, and this time it has produced two additional tables of statistics: Chi-Square Tests and Directional Measures (Figure 7-2). Given the parsimony of our requests, SPSS has been rather generous in its statistical output. In the Chi-Square Tests table, focus exclusively on the row labeled "Pearson Chi-Square." The first column, labeled "Value," provides the chi-square test statistic. A test statistic is a number that SPSS calculates from the observed data. Generally speaking, the larger the magnitude of a test statistic, the less likely that the observed data can be explained by random sampling error. The smaller the test statistic, the more likely that the null's favorite process—random chance—accounts for the observed data. So, if the observed data perfectly fit the expectations of the null hypothesis, then the chi-square test statistic would be 0. As the observed data depart from the null's expectations, this value grows in size, allowing the researcher to begin entertaining the idea of rejecting the null hypothesis.

**Figure 7-2**   Chi-square Tests and Directional Measures Results (ordinal-level relationship)

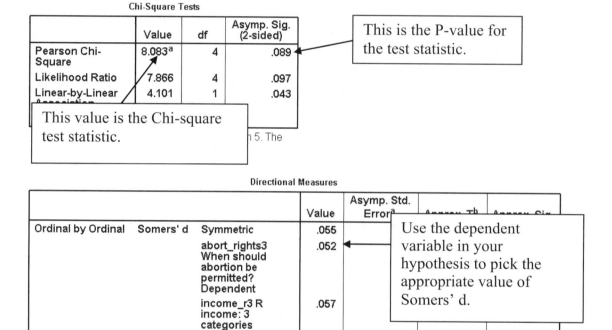

a. Not assuming the null hypothesis.
b. Using the asymptotic standard error assuming the null hypothesis.

For the abort_rights3-income_r3 cross-tabulation, SPSS calculated a chi-square test statistic equal to 8.083. Is this number, 8.083, statistically different from 0, the value we would expect to obtain if the null hypothesis were true? Put another way: If the null hypothesis is correct, how often will we obtain a test statistic of 8.083 by chance? The answer is contained in the rightmost column of the Chi-Square Tests table, under the label "Asymp. Sig. (2-sided)." For the chi-square test of significance, this value *is* the P-value.[3] In our example SPSS reports a P-value of .089. If the null hypothesis is correct in its assertion that no relationship exists between the independent and dependent variables, then we will obtain a test statistic of 8.083, by chance, about 9 percent of the time. Because .089 exceeds the .05 standard, the null hypothesis is on safe ground. From our initial comparison of percentages, we suspected that the relationship might not trump the null hypothesis. The chi-square test has confirmed that suspicion. Accept the null hypothesis.

Turn your attention to the Directional Measures table, which reports the requested measure of Somers' d. In fact, three Somers' d statistics are displayed. Because SPSS doesn't know (or care) how we framed the hypothesis, it provided values of Somers' d for every scenario: symmetric (no hypothetical expectations about the relationship), abort_rights3 dependent (abort_rights3 is the dependent variable and income_r3 is the independent variable), and income_r3 dependent (income_r3 is the dependent variable and abort_rights3 is the independent variable). Always use the dependent variable in your hypothesis to choose the correct value of Somers' d. Because abort_rights3 is our dependent variable, we would report the Somers's d value, .052.

What does this value, .052, tell us about the relationship? Because the statistic is positive, it tells us that increasing codes on income_r3 are associated with increasing codes on abort_rights3. We can discern a faint pattern: Low-income respondents (coded 1 on income_r3) are slightly more likely to oppose abortion (coded 1 on abort_rights3), and high-income respondents (coded 3 on income_r3) are slightly more likely to support abortion rights (coded 3 on abort_rights3). Because the statistic is puny—it registers .052 on a scale that runs from 0.000 to 1.000—it tells us that the relationship is weak. How weak? Because Somers' d is a PRE measure of association, we can say this: Compared to how well we can predict abortion opinions without knowing respondents' incomes, we can improve our prediction by 5.2 percent by knowing respondents' incomes.

Not much going on there. Obviously, we need to frame another hypothesis, using different variables, and see if our luck changes. But before moving on, let's quickly review the interpretation of the chi-square statistic and Somers's d. This is also a good place to introduce templates that will help you describe your findings.

### Summary

SPSS reports a chi-square test statistic, labeled "Pearson Chi-Square." This test statistic is calculated from the observed tabular data. Values close to 0 are within the domain of the null hypothesis. As chi-square increases in magnitude (the chi-square statistic cannot assume negative values), the null's explanation for the observed data—"it all happened by chance"—becomes increasingly implausible.

The chi-square statistic is accompanied by a P-value, which appears beneath the label "Asymp. Sig. (2-sided)." Here is a template for writing an interpretation of the P-value:

If the null hypothesis is correct that, in the population from which the sample was drawn, there is no relationship between [independent variable] and [dependent variable], then random sampling error will produce the observed data [P-value] of the time.

For our example: "If the null hypothesis is correct that, in the population from which the sample was drawn, there is no relationship between income and opinions about abortion, then random sampling error will produce the observed data .089 of the time." (If you prefer percentages, you can make this substitution: ". . . will produce the observed data 8.9 percent of the time.") Use the .05 benchmark. If the P-value is less than or equal to .05, then reject the null hypothesis. If the P-value is greater than .05, accept the null hypothesis.

For ordinal-by-ordinal relationships, request Somers' d. Somers' d is a directional measure, ranging from −1 to +1. Somers' d has a PRE interpretation. Here is a template for writing an interpretation of Somers' d or, for that matter, any PRE measure:

Compared to how well we can predict [dependent variable] by not knowing [independent variable], we can improve our prediction by [value of PRE measure] by knowing [independent variable].

Our example: "Compared to how well we can predict opinions on abortion by not knowing respondents' incomes, we can improve our prediction by .052 by knowing respondents' incomes." (Actually, percentages may sound better here: ". . . we can improve our prediction by 5.2 percent by knowing respondents' incomes.") Note that a negative sign on a PRE measure imparts the direction of the relationship, but it does not affect the PRE interpretation. Thus, if Somers' d were −.052, we would still conclude that the independent variable increases our predictive power by .052 or 5.2 percent.

## ANALYZING AN ORDINAL-LEVEL RELATIONSHIP WITH A CONTROL VARIABLE

The abort_rights3-income_r3 hypothesis didn't fare too well against the null hypothesis. Here is a hypothesis that sounds more promising: In a comparison of individuals, younger people will be more likely than older people to think that gay marriage should be allowed. For the dependent variable we will use gay_marriage3, which measures support for same-sex marriage by three ordinal response categories, "yes, allow" (coded 1), "no, but permit civil unions" (coded 2), and "no, do not allow" (coded 3). For the independent variable we will use yob3, which you created in Chapter 3. Recall that yob3 measures year of birth in three ordinal categories: before 1950 (coded 1), 1950–1965 (coded 2), and after 1965 (coded 3). When testing hypotheses about political opinions in which age is an independent variable, it is often a good idea to control for education. Older people tend to have lower levels of educational attainment than do younger people. And because education might itself be related to opinions about gay marriage—people with more education may be more willing to allow the practice—we don't want our analysis to confuse the effects of age with the effects of education. Thus we will include a control variable, educ2, coded 1 for respondents with a high school education or less and coded 2 for respondents having more than a high school education.

By this point in the book, cross-tabulation analysis has become routine. Return to the Crosstabs window. Click abort_rights3 back into the variable list and click gay_marriage3 into the Row(s) panel. Click income_r3 back into the variable list and click yob3 into the Column(s) panel. We're running a controlled comparison this time, so click the control variable, educ2, into the Layer box. The other necessary choices—column percentages in Cells, Chi-square and Somers's d in Statistics—should still be in place from the previous analysis. Click OK.

gay_marriage3 R position on gay marriage * yob3 Three generations * educ_r2 RECODE of educ_r (Highest grade of school or year of college R completed) Crosstabulation

| educ_r2 RECODE of educ_r (Highest grade of school or year of college R completed) | | | Statistics | yob3 Three generations | | | Total |
|---|---|---|---|---|---|---|---|
| | | | | 1 Before 1950 | 2 1950-1965 | 3 After 1965 | |
| 1 HS or less | gay_marriage3 R position on gay marriage | 1 Yes | Count | 51 | 76 | 202 | 329 |
| | | | % within yob3 Three generations | 17.6% | 29.5% | 52.5% | 35.3% |
| | | 2 No/Civ unions yes | Count | 69 | 59 | 79 | 207 |
| | | | % within yob3 Three generations | 23.8% | 22.9% | 20.5% | 22.2% |
| | | 3 No | Count | 170 | 123 | 104 | 397 |
| | | | % within yob3 Three generations | 58.6% | 47.7% | 27.0% | 42.6% |
| | Total | | Count | 290 | 258 | 385 | 933 |
| | | | % within yob3 Three generations | 100.0% | 100.0% | 100.0% | 100.0% |
| 2 >HS | gay_marriage3 R position on gay marriage | 1 Yes | Count | 71 | 142 | 319 | 532 |
| | | | % within yob3 Three generations | 25.6% | 34.2% | 56.5% | 42.3% |
| | | 2 No/Civ unions yes | Count | 101 | 129 | 129 | 359 |
| | | | % within yob3 Three generations | 36.5% | 31.1% | 22.8% | 28.6% |
| | | 3 No | Count | 105 | 144 | 117 | 366 |
| | | | % within yob3 Three generations | 37.9% | 34.7% | 20.7% | 29.1% |
| | Total | | Count | 277 | 415 | 565 | 1257 |
| | | | % within yob3 Three generations | 100.0% | 100.0% | 100.0% | 100.0% |
| Total | gay_marriage3 R position on gay marriage | 1 Yes | Count | 122 | 218 | 521 | 861 |
| | | | % within yob3 Three generations | 21.5% | 32.4% | 54.8% | 39.3% |
| | | 2 No/Civ unions yes | Count | 170 | 188 | 208 | 566 |
| | | | % within yob3 Three generations | 30.0% | 27.9% | 21.9% | 25.8% |
| | | 3 No | Count | 275 | 267 | 221 | 763 |
| | | | % within yob3 Three generations | 48.5% | 39.7% | 23.3% | 34.8% |
| | Total | | Count | 567 | 673 | 950 | 2190 |
| | | | % within yob3 Three generations | 100.0% | 100.0% | 100.0% | 100.0% |

Where people stand on gay marriage would appear to have something to do with when they were born. Consider the gay_marriage3-yob3 relationship among the less educated. There are large increases in the percentages saying "yes" as we move across the columns: from 17.6 percent among the oldest generation to 29.5 among the baby boomers to 52.5 for the post-boom cohorts. Thus, from oldest to youngest, there is about a 35-point increase in the percentage of respondents who would allow gay marriage. In fact, we see a similar pattern among the more highly educated group. The percentage who approve of gay marriage rises from 25.6 percent among the oldest cohort to 56.5 percent among the youngest—about a 31-point increase. Education, too, has fairly consistent (if surprisingly weak) effects, controlling for age. For the oldest group, only 17.6 percent of the less educated fall into the "yes" category, compared with 25.6 percent of their better educated cohorts, an 8-point difference. Somewhat weaker effects occur for the baby boomers (about a 5-point difference between the less and more educated) and for the post-1965 generation (a 4-point difference). Overall, it would appear that this controlled cross-tabulation—gay_marriage-yob3, controlling for educ2—depicts a set of additive relationships. The effect of the independent variable on the dependent variable has the same tendency and about the same strength for both values of the control variable.

Reading tables and discussing patterns are familiar tasks. But do the statistics support our interpretation? First let's check for statistical significance by examining the Chi-Square Tests.

**Chi-Square Tests**

| educ_r2 RECODE of educ_r (Highest grade of school or ye... | | Value | df | Asymp. Sig. (2-sided) |
|---|---|---|---|---|
| 1 HS or less | Pearson Chi-Square | 102.390[a] | 4 | .000 |
| | Likelihood Ratio | 105.501 | 4 | .000 |
| | Linear-by-Linear Association | 97.414 | 1 | .000 |
| | N of Valid Cases | 933 | | |
| 2 >HS | Pearson Chi-Square | 90.342[b] | 4 | .000 |
| | Likelihood Ratio | 91.879 | 4 | .000 |
| | Linear-by-Linear Association | 72.404 | 1 | .000 |
| | N of Valid Cases | 1257 | | |
| Total | Pearson Chi-Square | 194.202[c] | 4 | .000 |
| | Likelihood Ratio | 198.824 | 4 | .000 |
| | Linear-by-Linear Association | 177.075 | 1 | .000 |
| | N of Valid Cases | 2190 | | |

Judging from the chi-square tests at both values of the control—for the less educated (chi-square = 102.390, P-value = .000) and for the more educated (chi-square = 90.342, P-value = .000)—it is extremely unlikely that the observed patterns were produced by random sampling error. So the chi-square statistics invite us to reject the null hypothesis. How strong is the relationship between gay_marriage3 and yob3 for each education group? Refer to the Directional Measures table.

**Directional Measures**

| educ_r2 RECODE of educ_r (Highest grade of school or year of college R completed) | | | | Value | Asymp. Std. Error[a] | Approx. T[b] | Approx. Sig. |
|---|---|---|---|---|---|---|---|
| 1 HS or less | Ordinal by Ordinal | Somers' d | Symmetric | -.290 | .027 | -10.732 | .000 |
| | | | gay_marriage3 R position on gay marriage Dependent | -.287 | .027 | -10.732 | .000 |
| | | | yob3 Three generations Dependent | -.292 | .027 | -10.732 | .000 |
| 2 >HS | Ordinal by Ordinal | Somers' d | Symmetric | -.224 | .024 | -9.334 | .000 |
| | | | gay_marriage3 R position on gay marriage Dependent | -.226 | .024 | -9.334 | .000 |
| | | | yob3 Three generations Dependent | -.221 | .024 | -9.334 | .000 |
| Total | Ordinal by Ordinal | Somers' d | Symmetric | -.258 | .018 | -14.397 | .000 |
| | | | gay_marriage3 R position on gay marriage Dependent | -.259 | .018 | -14.397 | .000 |
| | | | yob3 Three generations Dependent | -.257 | .018 | -14.397 | .000 |

a. Not assuming the null hypothesis.
b. Using the asymptotic standard error assuming the null hypothesis.

For less educated respondents we would say that, compared to how well we can predict their gay marriage opinions without knowing when they were born, we can improve our prediction by 28.7 percent by knowing the independent variable. Similarly, for the more educated group, knowledge of the independent variable increases our predictive accuracy by 22.6 percent. Thus, Somers' d supports our interpretation of a set of additive relationships. For both values of the control variable, the relationship is negative—as yob3 shades from lower codes (older cohorts) to higher codes (younger cohorts), respondents tend to shift from higher codes of gay_marriage3 (the "no" responses) to the lower code of gay_marriage3 ("yes")—and the relationships are of similar strength at both levels of education.

## ANALYZING A NOMINAL-LEVEL RELATIONSHIP WITH A CONTROL VARIABLE

All the variables analyzed thus far have been ordinal level. Many social and political characteristics, however, are measured by nominal categories—gender, race, region, or religious denomination, to name a few. In this example we will use gender to help frame the following hypothesis: In a comparison of individuals, women are more likely than men to favor gun control. To make things more interesting, we will control for another variable that might affect attitudes toward gun control, whether the respondent resides in the South. Would the gender gap on gun opinions be the same for southerners and non-southerners? Or might the gender gap be weaker in the South than in the non-South? Let's investigate.

NES2008 contains gunlaw2, coded 1 for respondents who think it should be "more difficult" to buy a gun and coded 2 for those who think that laws should stay the same or make it easier to buy a gun. Gunlaw2 is the dependent variable. The independent variable is gender. For the control variable we will use south, coded 0 for non-southerners and 1 for southerners. Click Analyze → Descriptive Statistics → Crosstabs. Click Reset to clear the panels. Now set up our new analysis: gunlaw2 on the rows, gender on the columns, and south in the Layer panel (Figure 7-3). Click Cells and request column percentages. Click Statistics. Again we want chi-square, so make sure that the Chi-square box is checked. Because gender is a nominal-level variable, we can't use any of the ordinal statistics. But we can use lambda and Cramer's V. In the Nominal panel, check the box next to "Phi and Cramer's V" and check the box next to "Lambda." Click Continue, and then click OK.

**Figure 7-3**   Requesting Chi-square, Lambda, and Cramer's V (nominal-level relationship)

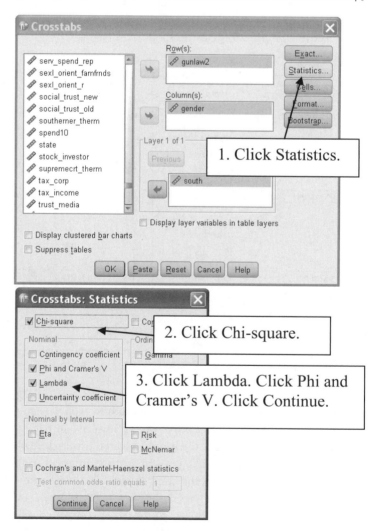

gunlaw2 Fed gov make it more difficult to buy a gun? * gender R gender * south R resides in South Crosstabulation

| south R resides in South | | | Statistics | gender R gender | | Total |
|---|---|---|---|---|---|---|
| | | | | 1 Male | 2 Female | |
| 0 Nonsouth | gunlaw2 Fed gov make it more difficult to buy a gun? | 1 More difficult | Count | 220 | 431 | 651 |
| | | | % within gender R gender | 35.9% | 61.3% | 49.5% |
| | | 2 Same/easier | Count | 392 | 272 | 664 |
| | | | % within gender R gender | 64.1% | 38.7% | 50.5% |
| | Total | | Count | 612 | 703 | 1315 |
| | | | % within gender R gender | 100.0% | 100.0% | 100.0% |
| 1 South | gunlaw2 Fed gov make it more difficult to buy a gun? | 1 More difficult | Count | 154 | 288 | 442 |
| | | | % within gender R gender | 36.2% | 51.9% | 45.1% |
| | | 2 Same/easier | Count | 272 | 267 | 539 |
| | | | % within gender R gender | 63.8% | 48.1% | 54.9% |
| | Total | | Count | 426 | 555 | 981 |
| | | | % within gender R gender | 100.0% | 100.0% | 100.0% |
| Total | gunlaw2 Fed gov make it more difficult to buy a gun? | 1 More difficult | Count | 374 | 719 | 1093 |
| | | | % within gender R gender | 36.0% | 57.2% | 47.6% |
| | | 2 Same/easier | Count | 664 | 539 | 1203 |
| | | | % within gender R gender | 64.0% | 42.8% | 52.4% |
| | Total | | Count | 1038 | 1258 | 2296 |
| | | | % within gender R gender | 100.0% | 100.0% | 100.0% |

Before examining the statistics, consider the substantive relationships depicted in the cross-tabulations. Among non-southerners, 35.9 percent of the males favor stronger gun control, compared with 61.3 percent of the females—more than a 25-point gap. What happens when we switch to respondents who reside in the South? Among southerners, the percentage of males who strongly support stricter gun laws, 36.2 percent, is practically the same as that of non-southern males. But among women there is a 10-point drop, from 61.3 percent for non-southern women to 51.9 percent for women in the south. Thus the gender gap is narrower inside the South than outside the South. Clearly, interaction is going on here: The gunlaw2-gender relationship has the same tendency for both values of the control—women are more pro-control than men—but the relationship is stronger for one value of the control (when south = 0) than for the other value of the control (when south = 1). Now let's see what the statistics have to say.

Chi-Square Tests

| south R resides in South | | Value | df | Asymp. Sig. (2-sided) | Exact Sig. (2-sided) | Exact Sig. (1-sided) |
|---|---|---|---|---|---|---|
| 0 Nonsouth | Pearson Chi-Square | 84.181[a] | 1 | .000 | | |
| | Continuity Correction[b] | 83.170 | 1 | .000 | | |
| | Likelihood Ratio | 85.143 | 1 | .000 | | |
| | Fisher's Exact Test | | | | .000 | .000 |
| | Linear-by-Linear Association | 84.117 | 1 | .000 | | |
| | N of Valid Cases | 1315 | | | | |
| 1 South | Pearson Chi-Square | 24.125[c] | 1 | .000 | | |
| | Continuity Correction[b] | 23.493 | 1 | .000 | | |
| | Likelihood Ratio | 24.305 | 1 | .000 | | |
| | Fisher's Exact Test | | | | .000 | .000 |
| | Linear-by-Linear Association | 24.100 | 1 | .000 | | |
| | N of Valid Cases | 981 | | | | |
| Total | Pearson Chi-Square | 101.740[d] | 1 | .000 | | |
| | Continuity Correction[b] | 100.895 | 1 | .000 | | |
| | Likelihood Ratio | 102.681 | 1 | .000 | | |
| | Fisher's Exact Test | | | | .000 | .000 |
| | Linear-by-Linear Association | 101.696 | 1 | .000 | | |
| | N of Valid Cases | 2296 | | | | |

According to the Chi-Square Tests table, the gunlaw2-gender relationship defeats the null hypothesis in the non-South cross-tab (chi-square = 84.181, P-value = .000) and in the South cross-tab (chi-square = 24.125, P-value = .000). Let's see whether the measures of association back up our interaction interpretation. Scroll down to the next two tables, Directional Measures and Symmetric Measures.

Directional Measures

| south R resides in South | | | | Value | Asymp. Std. Error[a] | Approx. T[b] | Approx. Sig. |
|---|---|---|---|---|---|---|---|
| 0 Nonsouth | Nominal by Nominal | Lambda | Symmetric | .221 | .032 | 6.483 | .000 |
| | | | gunlaw2 Fed gov make it more difficult to buy a gun? Dependent | .244 | .035 | 6.081 | .000 |
| | | | gender R gender Dependent | .196 | .038 | 4.696 | .000 |
| | | Goodman and Kruskal tau | gunlaw2 Fed gov make it more difficult to buy a gun? Dependent | .064 | .013 | | .000[c] |
| | | | gender R gender Dependent | .064 | .013 | | .000[c] |
| 1 South | Nominal by Nominal | Lambda | Symmetric | .030 | .046 | .645 | .519 |
| | | | gunlaw2 Fed gov make it more difficult to buy a gun? Dependent | .048 | .052 | .892 | .373 |
| | | | gender R gender Dependent | .012 | .054 | .215 | .829 |
| | | Goodman and Kruskal tau | gunlaw2 Fed gov make it more difficult to buy a gun? Dependent | .025 | .010 | | .000[c] |
| | | | gender R gender Dependent | .025 | .010 | | .000[c] |
| Total | Nominal by Nominal | Lambda | Symmetric | .143 | .027 | 5.157 | .000 |
| | | | gunlaw2 Fed gov make it more difficult to buy a gun? Dependent | .165 | .030 | 5.104 | .000 |
| | | | gender R gender Dependent | .120 | .031 | 3.614 | .000 |
| | | Goodman and Kruskal tau | gunlaw2 Fed gov make it more difficult to buy a gun? Dependent | .044 | .009 | | .000[c] |
| | | | gender R gender Dependent | .044 | .009 | | .000[c] |

Symmetric Measures

| south R resides in South | | | Value | Approx. Sig. |
|---|---|---|---|---|
| 0 Nonsouth | Nominal by Nominal | Phi | -.253 | .000 |
| | | Cramer's V | .253 | .000 |
| | N of Valid Cases | | 1315 | |
| 1 South | Nominal by Nominal | Phi | -.157 | .000 |
| | | Cramer's V | .157 | .000 |
| | N of Valid Cases | | 981 | |
| Total | Nominal by Nominal | Phi | -.211 | .000 |
| | | Cramer's V | .211 | .000 |
| | N of Valid Cases | | 2296 | |

The lambda statistics allow us to put a finer point on the particular form of interaction we saw in the cross-tabulation. Notice that, just as it did with Somers's d, SPSS has reported three lambda statistics: "Symmetric," "gunlaw2 Dependent," and "gender Dependent." Always use the lambda statistic that names the dependent variable in your hypothesis. Because our dependent variable is labeled "gunlaw2," we would use that statistic. The non-South lambda, .244, says that compared with how well we can predict gun control opinions by not knowing gender, we can improve our predictive power by 24.4 percent by taking gender into account. That is a fairly strong relationship—certainly stronger than the gunlaw2-gender relationship in the South, where gender provides a modest 4.8 percent boost in predictive leverage. A comparison of the Cramer's V values, .253 versus .157, also supports our interpretation. Although both relationships "run in the same direction," the gunlaw2-gender relationship is stronger among non-southern respondents than among southern respondents.

## A Problem with Lambda

One final example will emphasize an important limitation of lambda. Here is another gun control hypothesis: In a comparison of individuals, people who reside in "Blue Dog" congressional districts will be less likely to favor tighter restrictions than will people who do not reside in Blue Dog districts. (Blue Dog congressional districts are those that elected a Democratic member to the House of Representatives but also favored the Republican candidate over the Democratic candidate in the presidential contest. The conventional wisdom is that Blue Dog constituencies tend to be conservative on economic and cultural issues.) NES2008 has blue_dog, coded 1 for respondents who reside in Blue Dog districts and coded 0 for those who do not reside in Blue Dog districts. Return to the Crosstabs window for one more analysis, but don't reset the panels. Click gender back into the variable list and click blue_dog into Column(s). Click south back into the variable list, leaving the Layer box empty. Because we just ran a nominal-level analysis—Chi-square, Lambda, Phi and Cramer's V are still selected in Statistics—we should be all set. Click OK. Take a look at the cross-tab and the Chi-Square Tests table.

gunlaw2 Fed gov make it more difficult to buy a gun? * blue_dog Is R in Blue Dog district? Crosstabulation

| | | | blue_dog Is R in Blue Dog district? | | Total |
|---|---|---|---|---|---|
| | | Statistics | 0 No | 1 Yes | |
| gunlaw2 Fed gov make it more difficult to buy a gun? | 1 More difficult | Count | 980 | 114 | 1094 |
| | | % within blue_dog Is R in Blue Dog district? | 49.1% | 37.9% | 47.6% |
| | 2 Same/easier | Count | 1016 | 187 | 1203 |
| | | % within blue_dog Is R in Blue Dog district? | 50.9% | 62.1% | 52.4% |
| Total | | Count | 1996 | 301 | 2297 |
| | | % within blue_dog Is R in Blue Dog district? | 100.0% | 100.0% | 100.0% |

Chi-Square Tests

| | Value | df | Asymp. Sig. (2-sided) | Exact Sig. (2-sided) | Exact Sig. (1-sided) |
|---|---|---|---|---|---|
| Pearson Chi-Square | 13.211[a] | 1 | .000 | | |
| Continuity Correction[b] | 12.765 | 1 | .000 | | |
| Likelihood Ratio | 13.357 | 1 | .000 | | |
| Fisher's Exact Test | | | | .000 | .000 |
| Linear-by-Linear Association | 13.205 | 1 | .000 | | |
| N of Valid Cases | 2297 | | | | |

Based on the cross-tab, the hypothesis clearly has merit. Among non–Blue Dog respondents, 49.1 percent favor stronger gun laws, compared with 37.9 percent of Blue Dog respondents, about an 11-point difference. And according to chi-square (13.211, P-value = .000), we can reject the null hypothesis and infer that, in the population from which the sample was drawn, Blue Dog residents probably are less likely than non–Blue Dog residents to favor stricter controls. Now look at the Directional Measures table and the Symmetric Measures table.

Directional Measures

| | | | Value | Asymp. Std. Error[a] | Approx. T | Approx. Sig. |
|---|---|---|---|---|---|---|
| Nominal by Nominal | Lambda | Symmetric | .000 | .000 | .[b] | .[b] |
| | | gunlaw2 Fed gov make it more difficult to buy a gun? Dependent | .000 | .000 | .[b] | .[b] |
| | | blue_dog Is R in Blue Dog district? Dependent | .000 | .000 | .[b] | .[b] |
| | Goodman and Kruskal tau | gunlaw2 Fed gov make it more difficult to buy a gun? Dependent | .006 | .003 | | .000[c] |
| | | blue_dog Is R in Blue Dog district? Dependent | .006 | .003 | | .000[c] |

**Symmetric Measures**

|  |  | Value | Approx. Sig. |
|---|---|---|---|
| Nominal by Nominal | Phi | .076 | .000 |
|  | Cramer's V | .076 | .000 |
| N of Valid Cases |  | 2297 |  |

Consider the relevant value of lambda, .000, which suggests that blue_dog plays no role in predicting gun opinions. What is going on here? Because of the way lambda is computed, it will sometimes fail to detect a relationship that, by all other evidence, clearly exists. Lambda looks at the modal value of the dependent variable for each category of the independent variable. Lambda can detect a relationship only if the mode is different between categories. In our earlier analysis of the gunlaw2-gender relationship among non-southerners, for example, men and women had different modes on the dependent variable—the modal value for men was code 2 ("same/easier," with 64.1 percent of the males), and the mode for women was code 1 ("more difficult," with 61.3 percent of the females). So lambda picked up the relationship. In the gunlaw2-blue_dog cross-tab, however, both Blue Dog and non–Blue Dog respondents have the same mode—code 2, "same/easier"—the category of 50.9 percent of the non–Blue Dogs and 62.1 percent of the Blue Dogs. Because both categories of the independent variable have the same mode on the dependent variable, lambda returned a value of .000: no relationship detected. In situations like this, which are not uncommon, rely on Cramer's V as a useful gauge of strength. In the current example, Cramer's V registers a value of .076, indicating a weak relationship between the independent and dependent variables.

## EXERCISES

1. (Dataset: States. Variables: abortlaw3, gunlaw_rank3_rev, cook_index3.) Pedantic pontificator is pondering a potential partisan paradox of public policy.

"Think about two sorts of policies that figure prominently in cultural debate: laws restricting abortion and laws restricting guns. For both policies, fewer restrictions mean more choices and greater freedom, while more restrictions mean fewer choices and less freedom. Because choice and freedom are the touchstone values, one would think that partisan elites would be consistent in their positions on these policies. If Republicans favor fewer gun restrictions, then they ought to favor fewer abortion restrictions, too. By the same logic, if Democrats favor fewer abortion restrictions, then they should also support less gun control. As a keen observer of state politics, however, it is my impression that Republican-controlled states have less restrictive gun laws but more restrictive abortion laws. Democrat-controlled states are just the reverse: less restrictive abortion laws and more restrictive gun laws. I am sure that when you analyze the States dataset, you will discover this odd partisan paradox."

The States dataset contains these two policy measures, which will serve as dependent variables: abortlaw3 and gun_rank3_rev. Both variables are identically coded, three-category ordinals. Codes range from 1 (states having fewer restrictions) to 3 (states having more restrictions). Another three-category ordinal, cook_index3, measures states' partisan balance in three codes: 1 (Republican states), 2 (states with an even balance), and 3 (Democratic states). This is the independent variable.

A. If pedantic pontificator is correct, as you compare states across increasing values of cook_index3—from Republican states, to even states, to Democratic states—the percentage of states having more restrictive abortion policies should (circle one)

decrease.     stay the same.     increase.

The percentage of states having more restrictive gun policies should (circle one)

decrease.     stay the same.     increase.

B. If pedantic pontificator is correct, you should find that states having higher codes on cook_index3 will have (circle one)

lower     higher

codes on abortlaw3. You should also find that states having higher codes on cook_index3 will have (circle one)

lower     higher

codes on gun_rank3_rev.

C. Think about how SPSS calculates Somers' d. If pedantic pontificator is correct, the Somers' d statistic for the abortlaw3-cook_index3 relationship will have a (circle one)

negative     positive

sign. The Somers' d statistic for the gunlaw_rank3_rev-cook_index3 relationship will have a (circle one)

negative     positive

sign.

D. Obtain the appropriate cross-tabulation analyses of the abortlaw3-cook_index3 relationship and the gunlaw_rank3_rev-cook_index3 relationship. Make sure to request chi-square and Somers' d. Browse the cross-tabulation results. In the table below, enter the percentage of Democratic states, even states, and Republican states having more restrictive policies. In the abortlaw3 row, for example, record the percentage states having "9-10" abortion restrictions. For the gunlaw_rank3_rev row, enter the percentage of states in the "More restr" category. For each relationship, record chi-square, chi-square's P-value, and Somers' d.

| Dependent variable | More Rep | Even | More Dem | Chi-square | P-value | Somers' d |
|---|---|---|---|---|---|---|
| abortlaw3 % more restrictive | ? | ? | ? | ? | ? | ? |
| gunlaw_rank3_rev % more restrictive | ? | ? | ? | ? | ? | ? |

E. Consider Somers' d for the gunlaw_rank3_rev-cook_index3 relationship. This value of Somers' d means that, compared to how well we can predict gunlaw_rank3_rev without knowing cook_index3 (complete the sentence) _____

_____.

F. Consider the chi-square P-value for the abortlaw3-cook_index3 relationship. This P-value means that, under the assumption that the null hypothesis is correct, (complete the sentence) _____

_____.

Therefore, you should (circle one)

reject     not reject

the null hypothesis.

G. Consider all the evidence from your analysis. The evidence suggests that pedantic pontificator is (circle one)

correct.        incorrect.

Explain your reasoning. _____

_____

_____

_____

_____

2. (Dataset: GSS2008. Variables: abany, femrole2, sex.) Interested student has joined pedantic pontificator in a discussion of the gender gap in U.S. politics.

Interested student: "On what sorts of issues or opinions are men and women most likely to be at odds? What defines the gender gap, anyway?"

Pedantic pontificator: "That's easy. A couple of points seem obvious, to me anyway. First, we know that the conflict over abortion rights is the defining gender issue of our time. Women will be more likely than men to take a strong pro-choice position on this issue. Second—and pay close attention here—on more mundane cultural questions, such as whether women should be homemakers or pursue careers outside the home, men and women will not differ at all."

A. Pedantic pontificator has suggested the following two hypotheses about the gender gap: (check two)

❑ In a comparison of individuals, women will be less likely than men to think that abortion should be allowed.

❑ In a comparison of individuals, women and men will not differ in their abortion opinions.

❑ In a comparison of individuals, women will be more likely than men to think that abortion should be allowed.

❑ In a comparison of individuals, women will be less likely than men to think that women should pursue careers outside the home.

❑ In a comparison of individuals, women and men will not differ in their opinions about female roles outside the home.

❑ In a comparison of individuals, women will be more likely than men to think that women should pursue careers outside the home.

B. Run cross-tabulation analysis to test pedantic pontificator's hypotheses. GSS2008 contains two variables that will serve as dependent variables: abany, which records whether the respondent thinks that a woman should be able to obtain an abortion for any reason ("yes" is coded 1 and "no" is coded 2); and femrole2, which gauges respondents' opinions on the appropriate female role ("home" is coded 1 and "work" is coded 2). The independent variable is sex, coded 1 for males and 2 for females. Request chi-square. Sex is a nominal variable, so be sure to request lambda and Cramer's V. In the abany-sex cross-tabulation, focus on the percentage saying "yes." In the femrole2-sex cross-tabulation, focus on the "work" category. Record your results in the table that follows:

| Dependent variable | Male | Female | Chi-square | P-value | Lambda | Cramer's V |
|---|---|---|---|---|---|---|
| Percent "yes" (abany) | ? | ? | ? | ? | ? | ? |
| Percent "Work" (femrole2) | ? | ? | ? | ? | ? | ? |

C. Based on these results, you may conclude that (check three)

❑ a statistically significant gender gap exists on abortion opinions.

❑ pedantic pontificator's hypothesis about the femrole2-sex relationship is not supported by the analysis.

❑ under the assumption that the null hypothesis is correct, the abany-sex relationship could have occurred by chance more frequently than 5 times out of 100.

❑ pedantic pontificator's hypothesis about the abany-sex relationship is supported by the analysis.

❑ a higher percentage of females than males think that women belong in the workplace.

D. The P-value of the chi-square statistic in the femrole2-sex cross-tabulation tells you that, under the assumption that the null hypothesis is correct (complete the sentence) _____

_____ .

E. The value of lambda for the femrole2-sex relationship tells you that, compared to how well (complete the sentence) _____

_____ .

3. (Dataset: GSS2008. Variables: polview3, racial_liberal3, social_cons3, spend3.) While having lunch together, three researchers are discussing what the terms *liberal, moderate,* and *conservative* mean to most people. Each researcher is touting a favorite independent variable that may explain the way survey respondents describe themselves ideologically.

Researcher 1: "When people are asked a question about their ideological views, they think about their attitudes toward government spending. If people think the government should spend more on important programs, they will respond that they are 'liberal.' If they don't want too much spending, they will say that they are 'conservative.'"

Researcher 2: "Well, that's fine. But let's not forget about social policies, such as abortion and pornography. These issues must influence how people describe themselves ideologically. People with more permissive views on these sorts of issues will call themselves 'liberal.' People who favor government restrictions will label themselves as 'conservative.'"

Researcher 3: "Okay, you both make good points. But you're ignoring the importance of racial issues in American politics. When asked whether they are liberal or conservative, people probably think about their opinions on racial policies, such as affirmative action. People who give racial equality a higher priority will say they are 'liberal,' and people giving racial equality a lower priority will say they are 'conservative.'"

In Chapter 3 you created an ordinal measure of ideology, polview3, which is coded 1 for "liberal," 2 for "moderate," and 3 for "conservative." This is the dependent variable. GSS2008 also contains researcher 1's favorite independent variable, spend3, a three-category ordinal measure of attitudes toward government spending. Higher codes denote more supportive opinions toward spending. Researcher 2's favorite independent variable is social_cons3, a three-category ordinal measure of attitudes on social issues. Higher codes denote less permissive views. Researcher 3's favorite independent variable is racial_liberal3, also a three-category ordinal variable. Higher codes denote more strongly egalitarian opinions on racial issues.

A. Think about how SPSS calculates Somers' d. Assuming that each researcher is correct, SPSS should report (check all that apply)

❑ a negative relationship between polview3 and spend3.

❑  a positive relationship between polview3 and social_cons3.

❑  a negative relationship between polview3 and racial_liberal3.

B.  Run Crosstabs, using polview3 as the dependent variable and spend3, social_cons3, and racial_liberal3 as independent variables. Request chi-square. Obtain Somers's d. Summarize your results in the following table. In the first three columns, enter the percentage of self-identified "conservatives" on polview3 for each value of the independent variable. For example, from the spend3 cross-tab, record the percentage of conservatives among respondents who want to "spend less" (code 0 of spend3), the percentage of conservatives among respondents taking the "middle" position (code 1 on spend3), and the percentage of conservatives among respondents wanting to "spend more" (code 2 on spend3). For each relationship, record chi-square, chi-square's P-value, and Somers' d.

|  | Code on independent variable* | | | Chi-square | P-value | Somers' d |
|---|---|---|---|---|---|---|
|  | 0 | 1 | 2 | | | |
| Percent "conservative" (spend3 cross-tab) | ? | ? | ? | ? | ? | ? |
| Percent "conservative" (social_cons3 cross-tab) | ? | ? | ? | ? | ? | ? |
| Percent "conservative" (racial_liberal3 cross-tab) | ? | ? | ? | ? | ? | ? |

*For spend3, code 0 = "spend less," code 1 = "middle," and code 2 = "spend more." For social_cons3, code 0 = "most permissive," code 1 = "middle," and code 2 = "least permissive." For racial_liberal3, code 0 = "low," code 1 = "middle," and code 2 = "high."

C.  Consider the evidence you have assembled. Your analysis supports which of the following statements? (check three)

❑  As values of spend3 increase, the percentage of respondents describing themselves as conservatives decreases.

❑  As values of social_cons3 increase, the percentage of respondents describing themselves as conservative increases.

❑  The polview3-racial_liberal3 relationship is not statistically significant.

❑  If the null hypothesis is correct, you will obtain the polview3-spend3 relationship less frequently than 5 times out of 100 by chance.

❑  If the null hypothesis is correct, you will obtain the polview3-racial_liberal3 relationship more frequently than 5 times out of 100 by chance.

D.  The value of Somers' d for the polview3-social_cons3 relationship tells you that, compared with how well we can predict polview3 by not knowing (complete the sentence) _____

_____.

E.  The three researchers make a friendly wager. The researcher whose favorite independent variable does the worst job predicting values of the dependent variable has to buy lunch for the other two. Who pays for lunch? (circle one)

Researcher 1      Researcher 2      Researcher 3

4. (Dataset: GSS2008. Variables: muslim_tol, relig_tol3, educ2.) Think for a moment about the general principle of religious tolerance—the willingness of individuals to tolerate diversity in religious faiths and practices. A person who has a high level of religious tolerance would agree that "all religious groups in the U.S. should have equal rights" and that "we must respect all religions." A person with low religious tolerance would tend to disagree with these ideas. GSS2008 contains relig_tol3, which measures respondents' level of agreement with these general principles. Relig_tol3 is coded 0 ("low" tolerance), 1 ("middle"), and 2 ("high" tolerance). Now recall muslim_tol, which you created in Chapter 3. Muslim_tol does not measure general religious tolerance; rather, it measures respondents' level of tolerance toward "anti-American Muslim clergymen." Do people apply general principles of religious tolerance to instances involving radical Muslims? The literature on political tolerance suggests that individuals' willingness or ability to apply general democratic principles to unpopular groups depends on their level of education. Perhaps we would see the same phenomenon in the study of religious tolerance. Consider three hypothetical expectations:

Expectation 1: For low-education individuals, those who have higher levels of general religious tolerance will be significantly more likely to express high levels of tolerance toward anti-American Muslim clergymen than will those who have lower levels of general religious tolerance.

Expectation 2: For high-education individuals, those who have higher levels of general religious tolerance will be significantly more likely to express high levels of tolerance toward anti-American Muslim clergymen than will those who have lower levels of general religious tolerance.

Expectation 3: The relationship between religious tolerance and tolerance toward Muslim clergy will be stronger for high-education individuals than for low-education individuals.

A. Test expectations 1–3 using GSS2008 variables muslim_tol (dependent variable), relig_tol3 (independent variable), and educ2 (control variable). Record the results of your analysis in the table that follows. For respondents at both education levels, write down the percentage with "high" tolerance toward Muslim clergy. Record chi-square and its P-value, and Somers'd.

| | Religious tolerance (relig_tol3) | | | | | |
| | Low | Mid | High | Chi-square | P-value | Somers'd |
|---|---|---|---|---|---|---|
| Low-education respondents:<br><br>Percent "high" tolerance toward Muslim clergy | ? | ? | ? | ? | ? | ? |
| High-education respondents:<br><br>Percent "high" tolerance toward Muslim clergy | ? | ? | ? | ? | ? | ? |

B. Does your analysis support expectation 1?

Yes     No

Please explain your reasoning. _____

_____

_____

_____

C.  Does your analysis support expectation 2?

Yes     No

Please explain your reasoning. _____

_____

_____

_____

D.  Does your analysis support expectation 3?

Yes     No

Please explain your reasoning. _____

_____

_____

_____

5.  (Dataset: World. Variables: protact3, gender_equal3, vi_rel3, pmat12_3.) Ronald Inglehart offers a particularly elegant and compelling idea about the future of economically advanced societies. According to Inglehart, the cultures of many postindustrial societies have been going through a value shift—the waning importance of materialist values and a growing pursuit of postmaterialist values. In postmaterialist societies, economically based conflicts—unions versus big business, rich versus poor—are increasingly supplanted by an emphasis on self-expression and social equality. Postmaterialist societies also are marked by rising secularism and elite-challenging behaviors, such as boycotts and demonstrations. In this exercise you will investigate Inglehart's theory.[4]

The World variable pmat12_3 measures the level of postmaterial values by a three-category ordinal measure: low postmaterialism (coded 1), moderate postmaterialism (coded 2), and high postmaterialism (coded 3). Higher codes denote a greater prevalence of postmaterial values. Use pmat12_3 as the independent variable. Here are three dependent variables, all of which are three-category ordinals: gender_equal3, which captures gender equality (1 = low equality, 2 = medium equality, 3 = high equality); protact3, which measures citizen participation in protests (1 = low, 2 = moderate, 3 = high); and vi_rel3, which gauges religiosity by the percentage of the public saying that religion is "very important" (1 = less than 20 percent, 2 = 20–50 percent, 3 = more than 50 percent). Higher codes on the dependent variables denote greater gender equality (gender_equal3), more protest activity (protact3), and higher levels of religiosity (vi_rel3).

A.  Using pmat12_3 as the independent variable, three postmaterialist hypotheses can be framed:

Gender equality hypothesis (fill in the blank): In a comparison of countries, those with higher levels of postmaterialism will have _____ levels of gender equality than will countries having lower levels of postmaterialism.

Protest activity hypothesis (fill in the blanks): In a comparison of countries, those with _____ levels of postmaterialism will have _____ levels of protest activity than will countries having _____ levels of postmaterialism.

Religiosity hypothesis (complete the sentence): In a comparison of countries, those with _____

_____

_____

_____.

B. Which of the following measures of association is most appropriate for all three relationships? (circle one)

Lambda        Somers' d

C. Consider how the independent variable is coded and how each dependent variable is coded. In the way that SPSS calculates the appropriate measure of association, which one of the three hypotheses implies a negative sign on the measure of association? (check one)

❑  The gender equality hypothesis

❑  The protest activity hypothesis

❑  The religiosity hypothesis

D. Test each hypothesis using cross-tabulation analysis. Obtain Chi-square and the appropriate measure of association. In the table that follows, record the percentages of countries falling into the highest category of each dependent variable. Also, report chi-square statistics, P-values, and measures of association.

| Dependent variable | Level of postmaterialism | | | Chi-square | P-value | Measure of association |
|---|---|---|---|---|---|---|
| | Low | Moderate | High | | | |
| Percentage high gender equality | ? | ? | ? | ? | ? | ? |
| Percentage high protest activity | ? | ? | ? | ? | ? | ? |
| Percentage high religiosity | ? | ? | ? | ? | ? | ? |

E. Which of the following inferences are supported by your analysis? (check all that apply)

❑  The gender equality hypothesis is supported.

❑  Compared with how well we can predict gender equality by not knowing the level of postmaterialism, we can improve our prediction by 20.03 percent by knowing the level of postmaterialism.

❑  The protest activity hypothesis is supported.

❑  If the null hypothesis is correct, the postmaterialism–protest activity relationship would occur, by chance, less frequently than 5 times out of 100.

❑  The religiosity hypothesis is supported.

❑  If the null hypothesis is correct, the postmaterialism-religiosity relationship would occur, by chance, less frequently than 5 times out of 100.

That concludes the exercises for this chapter. Before exiting SPSS, be sure to save your output file.

## NOTES

1. Asymmetry is the essence of hypothetical relationships. Thus we would hypothesize that income causes opinions on abortion policies, but we would not hypothesize that abortion opinions cause income. We would prefer a measure of association that tells us how well income (independent variable) predicts abortion opinions (dependent variable),

not how well abortion opinions predict income. Or, to cite Warner's tongue-in-cheek example: "There are some situations where the ability to make predictions is asymmetrical; for example, consider a study about gender and pregnancy. If you know that an individual is pregnant, you can predict gender perfectly (the person must be female). However, if you know that an individual is female, you cannot assume that she is pregnant." Rebecca M. Warner, *Applied Statistics* (Los Angeles: Sage, 2008), 316.

2. Somers' d may be used for square tables (in which the independent and dependent variables have the same number of categories) and for nonsquare tables (in which the independent and dependent variables have different numbers of categories). Because of its other attractive properties, some methodologists prefer Somers' d to Kendall's tau-b or tau-c. See George W. Bohrnstedt and David Knoke, *Statistics for Social Data Analysis,* 2nd ed. (Itasca, Ill.: Peacock, 1988), 325.

3. Although the "2-sided" part of the P-value's label suggests a two-tailed probability, this number is indeed a one-tailed probability and so permits a one-tailed test of statistical significance.

4. Inglehart has written extensively about cultural change in postindustrial societies. For example, see his *Culture Shift in Advanced Industrial Society* (Princeton: Princeton University Press, 1990).

# 8

# Correlation and Linear Regression

**Procedures Covered**

Analyze → Correlate → Bivariate

Analyze → Regression → Linear

Graphs → Legacy Dialogs → Scatter/Dot

Correlation and regression are powerful and flexible techniques used to analyze interval-level relationships. Pearson's correlation coefficient (Pearson's r) measures the strength and direction of the relationship between two interval-level variables. Pearson's r is not a proportional reduction in error (PRE) measure, but it does gauge strength by an easily understood scale—from –1, a perfectly negative association between the variables, to +1, a perfectly positive relationship. A correlation of 0 indicates no relationship. Researchers often use correlation techniques in the beginning stages of analysis to get an overall picture of the relationships between interesting variables.

Regression analysis produces a statistic, the regression coefficient, that estimates the effect of an independent variable on a dependent variable. Regression also produces a PRE measure of association, R-square, which indicates how completely the independent variable (or variables) explains the dependent variable. In regression analysis the dependent variable is measured at the interval level, but the independent variable can be of any variety—nominal, ordinal, or interval. Regression is more specialized than correlation. Researchers use regression analysis to model causal relationships between one or more independent variables and a dependent variable.

In the first part of this chapter, you will learn to perform correlation analysis using the Correlate procedure, and you will learn to perform and interpret bivariate regression using Regression → Linear. Bivariate regression uses one independent variable to predict a dependent variable. We then will turn to Scatter/Dot, an SPSS graphic routine that yields a scatterplot, a visual depiction of the relationship between two interval-level variables. With generous use of the Chart Editor, you will learn how to add a regression line to the scatterplot and how to edit the graph for elegance and clarity. Finally, you will use Regression → Linear to perform multiple regression analysis. Multiple regression, which uses two or more independent variables to predict a dependent variable, is an essential tool for analyzing complex relationships.

## CORRELATION AND BIVARIATE REGRESSION

Suppose that a student of state politics is interested in the gender composition of state legislatures. Using Descriptives to analyze the States dataset, this student finds that state legislatures range from 10 percent female to 37.5 percent female. Why is there such variation in the female composition of state legislatures? The student researcher begins to formulate an explanation. Perhaps states with lower percentages of college graduates have lower percentages of women legislators than do states with more college-educated residents. And maybe a cultural variable, the percentage of states' residents who frequently attend religious services, plays a role. Perhaps states with higher percentages of frequent attenders have lower percentages of female lawmakers. Correlation analysis would give this researcher an overview of the relationships among these variables. Let's use Correlate and Regression → Linear to investigate.

Open the States dataset. Click Analyze → Correlate → Bivariate. The Bivariate Correlations window is a no-frills interface (Figure 8-1). We are interested in three variables: the percentage of frequent church attenders (attend_pct), the percentage of college graduates (college), and the percentage of female state legislators (womleg_2010). Click each of these variables into the Variables panel, as shown in Figure 8-1.

**Figure 8-1** Bivariate Correlations Window (modified)

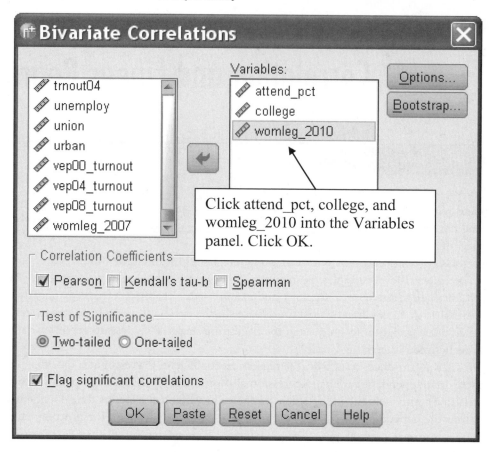

By default, SPSS will return Pearson's correlation coefficients. So the Pearson box, which is already checked, suits our purpose. Click OK. SPSS reports the results in the Viewer.

**Correlations**

| | | attend_pct % freq attend relig serv (Pew) | college Percent of pop w/college or higher | womleg_2010 Percent of state legislators who are women (2010) |
|---|---|---|---|---|
| attend_pct % freq attend relig serv (Pew) | Pearson Correlation | 1 | -.531** | -.648** |
| | Sig. (2-tailed) | | .000 | .000 |
| | N | 50 | 50 | 50 |
| college Percent of pop w/college or higher | Pearson Correlation | -.531** | 1 | .626** |
| | Sig. (2-tailed) | .000 | | .000 |
| | N | 50 | 50 | 50 |
| womleg_2010 Percent of state legislators who are women (2010) | Pearson Correlation | -.648** | .626** | 1 |
| | Sig. (2-tailed) | .000 | .000 | |
| | N | 50 | 50 | 50 |

**. Correlation is significant at the 0.01 level (2-tailed).

The Correlations table, called a correlation matrix, shows the correlation of each variable with each of the other variables—it even shows the correlation between each variable and itself. Each of the correlations in which we are interested appears twice in the table: once above the upper-left-to-lower-right diagonal of 1's, and again below the diagonal. The correlation between womleg_2010 and attend_pct is –.648, which tells us that increasing values of one of the variables is associated with decreasing values of the other variable. So as the percentage of frequent church attenders goes up, the percentage of female legislators goes down. How strong is the relationship? We know that Pearson's r is bracketed by –1 and +1, so we could say that this relationship is a fairly strong negative association. The correlation between womleg_2010 and college, .626, indicates a positive relationship: As states' percentages of college graduates increase, so do their percentages of women legislators. Again, this is a fairly strong association—similar in strength, though not in direction, to the womenleg_2010-attend_pct relationship. Finally, attend_pct and college, with a Pearson's r of –.531, show a moderately strong negative relationship. As the percentage of college graduates increases, the percentage of frequent attenders declines.

Correlation analysis is a good place to start when analyzing interval-level relationships. Even so, a correlation coefficient is agnostic on the question of which variable is the cause and which the effect. Does an increase in the percentage of frequent church attenders somehow cause lower percentages of women in state legislatures? Or do increasing percentages of women in state legislatures somehow cause states to have lower percentages of church attenders? Either way, correlation analysis reports the same measure of association, a Pearson's r of –.648.

Regression is more powerful than correlation, in part because it helps us investigate causal relationships—relationships in which an independent variable is thought to affect a dependent variable. Regression analysis will (1) reveal the precise nature of the relationship between an independent variable and a dependent variable, (2) test the null hypothesis that the observed relationship occurred by chance, and (3) provide a PRE measure of association between the independent variable and the dependent variable. To illustrate these and other points, we will run two separate bivariate regressions. First we will examine the relationship between attend_pct and womleg_2010, and then we will analyze the relationship between college and womleg_2010.

Click Analyze → Regression → Linear. The Linear Regression window appears (Figure 8-2). Click womleg_2010 into the Dependent box. Find attend_pct in the variable list and click it into the Independent(s) box. Click OK.

**Figure 8-2**   Linear Regression Window

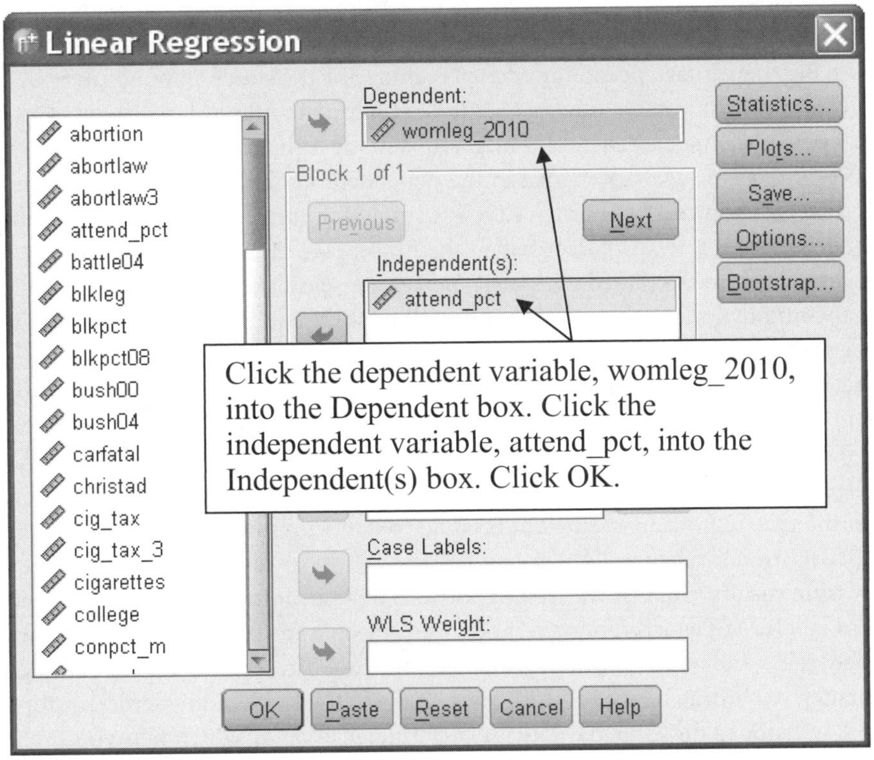

SPSS regression output includes four tables: Variables Entered/Removed, Model Summary, ANOVA (which stands for analysis of variance), and Coefficients. For the regression analyses you will perform in this book, the Model Summary table and the Coefficients table contain the most important information. Let's examine them.

**Model Summary**

| Model | R | R Square | Adjusted R Square | Std. Error of the Estimate |
|---|---|---|---|---|
| 1 | .648[a] | .419 | .407 | 5.2401 |

a. Predictors: (Constant), attend_pct % freq attend relig serv (Pew)

**Coefficients[a]**

| Model | | Unstandardized Coefficients | | Standardized Coefficients | t | Sig. |
|---|---|---|---|---|---|---|
| | | B | Std. Error | Beta | | |
| 1 | (Constant) | 42.435 | 3.198 | | 13.270 | .000 |
| | attend_pct % freq attend relig serv (Pew) | -.470 | .080 | -.648 | -5.889 | .000 |

a. Dependent Variable: womleg_2010 Percent of state legislators who are women (2010)

First, consider the Coefficients table. The leftmost column, under the heading "Model," contains the names of the key elements in the regression equation. "Constant" is the Y-intercept of the regression line, and "% freq attend relig serv" is the label of the independent variable. The numbers along the "Constant" row report statistics about the Y-intercept, and the numbers along the "% freq attend relig serv" row report statistics about the independent variable. Now look at the first column of numbers, which shows the regression coefficient for each parameter. According to these values, the Y-intercept is equal to 42.435, and the regression coefficient is −.470. The regression equation for estimating the effect of attend_pct on womleg_2010, therefore, is as follows (to make the numbers a bit simpler, we will round to two decimal places):

Percent of state legislators who are women = 42.44 − 0.47*attend_pct.

The constant, 42.44, is the estimated value of Y when X equals 0. If you were using this equation to estimate the percentage of women legislators for a state, you would start with 42.44 percent and then subtract .47, or about one-half a percentage point, for each percentage of the state's population who are frequent attenders. So your estimate for a state with, say, 50 percent frequent attenders would be 42.44 − .47*(50) = 42.44 − 23.50 ≈ 19 percent female legislators. The main statistic of interest, then, is the regression coefficient, −.47, which estimates the average change in the dependent variable for each unit change in the independent variable. A regression coefficient of −.47 tells us that, for each one-unit increase in the percentage of frequent attenders, there is a .47-unit decrease in the percentage of female legislators. So a 1-percentage-point increase in attend_pct is associated with a .47-percentage-point decrease in womleg_2010.[1]

What would the null hypothesis have to say about all this? Of course, we are not analyzing a random sample here, since we have information on the entire population of 50 states. But let's assume, for illustrative purposes, that we have just analyzed a random sample and that we have obtained a sample estimate of the effect of attend_pct on womleg_2010. The null hypothesis would say what it always says: In the population from which the sample was drawn, there is no relationship between the independent variable (in this case, the percentage of frequent attenders) and the dependent variable (the percentage of female legislators). In the population the true regression coefficient is equal to 0. Furthermore, the regression coefficient that we obtained, −.47, occurred by chance.

In SPSS regression results, you test the null hypothesis by examining two columns in the Coefficients table—the column labeled "t," which reports t-ratios, and the column labeled "Sig.," which reports P-values. Informally, to safely reject the null hypothesis, you generally look for t-ratios with magnitudes (absolute values) of 2 or greater. According to the results of our analysis, the regression coefficient for attend_pct has a t-ratio of −5.889, well above the informal 2-or-greater rule. A P-value, which tells you the probability of obtaining the results if the null hypothesis is correct, helps you to make more precise inferences about the relationship between the independent variable and the dependent variable. If "Sig." is greater than .05, then the observed results would occur too frequently by chance, and you must not reject the null hypothesis. By contrast, if "Sig." is equal to or less than .05, then the null hypothesis represents an unlikely occurrence and may be rejected. The t-ratio for attend_pct has a corresponding P-value of .000. If the null is correct, then random sampling error would have produced the observed results zero times in a thousand.[2] Reject the null hypothesis. It depends on the research problem at hand, of course, but for most applications you can ignore the t-ratio and P-value for the constant.[3]

How strong is the relationship between attend_pct and womleg_2010? The answer is provided by the R-square statistics, which appear in the Model Summary table. SPSS reports two values, one labeled "R Square," and one labeled "Adjusted R Square." Which one should you use? Most research articles report the adjusted value, so let's rely on adjusted R-square to provide the best overall measure of the strength of the relationship.[4] Adjusted R-square is equal to .407. What does this mean? R-square communicates the proportion of the variation in the dependent variable that is explained by the independent variable. Like any proportion, R-square can assume any value between 0 and 1. Thus, of all the variation in womleg_2010 between states, .407, or 40.7 percent, is explained by attend_pct. The rest of the variation in womleg_2010, 59.3 percent, remains unexplained by the independent variable.

So that you can become comfortable with bivariate regression—and to address a potential source of confusion—let's do another run, this time using college as the independent variable. Click Analyze → Regression → Linear. Leave womleg_2010 in the Dependent box, but click attend_pct back into the Variables list. Click college into the Independent(s) box and click OK. Examine the Coefficients table and the Model Summary table.

**Model Summary**

| Model | R | R Square | Adjusted R Square | Std. Error of the Estimate |
|---|---|---|---|---|
| 1 | .626[a] | .392 | .379 | 5.3645 |

a. Predictors: (Constant), college Percent of pop w/college or higher

**Coefficients[a]**

| Model | | Unstandardized Coefficients | | Standardized Coefficients | t | Sig. |
|---|---|---|---|---|---|---|
| | | B | Std. Error | Beta | | |
| 1 | (Constant) | -.201 | 4.441 | | -.045 | .964 |
| | college Percent of pop w/college or higher | .941 | .169 | .626 | 5.557 | .000 |

a. Dependent Variable: womleg_2010 Percent of state legislators who are women (2010)

The regression equation for the effect of college on womleg_2010 is as follows:

Percent of state legislators who are women = −.20 + 0.94*college.

As is sometimes the case with regression, the constant, −.20, represents an "unreal" situation. For states in which 0 percent of residents have college degrees, the estimated percentage of female legislators is a *minus* .2 percent. Of course, the smallest value of college in the actual data is substantially higher than 0.[5] However, for the regression line to produce the best estimates for real data, SPSS Regression has anchored the line at a Y-intercept. The regression coefficient, .94, says that for each percentage-point increase in college, there is an average increase of .94 of a percentage point in the percentage of female legislators. Again, increase the percentage of college graduates by 1, and the percentage of women legislators goes up by almost 1, on average. In the population, could the true value of the regression coefficient be 0? Probably not, according to the t-ratio (5.557) and the P-value (Sig. = .000). And, according to adjusted R-square, the independent variable does a fair amount of work in explaining the dependent variable. About 38 percent of the variation in womleg_2010 is explained by college. As bivariate regressions go, that's not too bad.

## SCATTERPLOTS

An SPSS graphic routine, Scatter/Dot, adds a visual dimension to correlation and regression and thus can help you paint a richer portrait of a relationship. Consider Figure 8-3, created using Graphs → Legacy Dialogs → Scatter/Dot and edited in the Chart Editor. This graph, generically referred to as a scatterplot, displays the cases in a two-dimensional space according to their values on the two variables. The horizontal axis (X-axis) is defined by the independent variable, college, and the vertical axis (Y-axis) is defined by the dependent variable, womleg_2010. We know from our correlation analysis that Pearson's r for this relationship is .63. We can now see what the correlation "looks like." Based on the figure, states with lower percentages of college graduates tend to have lower percentages of women legislators, with values on the Y-axis that range from 10 percent to about 25 percent or so. The percentages of women legislators for states at the higher end of the X-axis, furthermore, increase 15–20 percent to around 35 percent. So as you move from left to right along the X-axis, values on the Y-axis generally increase, just as the positive correlation coefficient suggested.

The scatterplot has other interesting features. Notice that the dots have been overlaid by the linear regression line obtained from the analysis we just performed:

$$\text{Estimated percentage of women legislators} = -.20 + .94^*\text{college}.$$

Thanks to this visual depiction, we can see that the linear summary of the relationship, while reasonably coherent, is far from perfect. Helpfully, to the lower right-hand side SPSS has supplied the value of R-square, one measure of the perfection (or the completeness, at least) of the relationship. Also, notice the sparse clarity of the graph. The dots are solid, but the background is white, both inside and outside the data space. The axis lines have been removed, leaving only the tick marks, which are labeled in round numbers, without decimal points. The two key data elements—the dots representing each case and the regression line summarizing the relationship—do not compete for our eye with any other lines, colors, or text.

**Figure 8-3**  Scatterplot with Regression Line

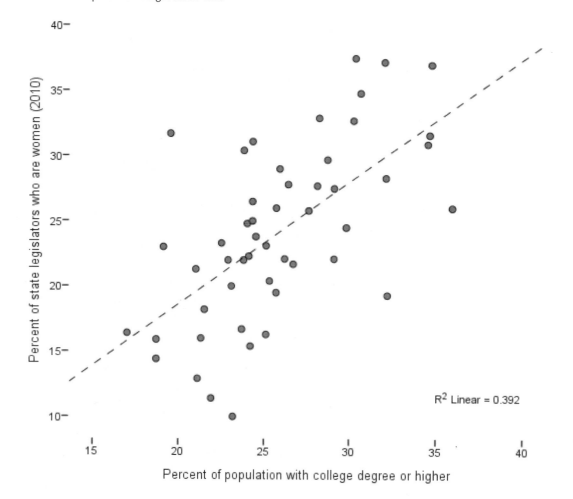

The scatterplot in Figure 8-3 comes close to what Edward R. Tufte calls an "erased" graph, a graph in which nonessential elements have been deleted. In Chapter 5, we touched on Tufte's definition of the data-ink ratio, the proportion of a graph's total ink devoted to depicting the information contained in the data. Tufte, a leading expert on the visual communication of quantitative information, recommends that the greatest share of a graph's elements should be devoted to data ink—graphic features that convey the essence of the relationship.[6] Let's recreate this graphic.

Click Graphs → Legacy Dialogs → Scatter/Dot. In the Scatter/Dot window, select Simple Scatter and click Define, opening the Simple Scatterplot dialog (Figure 8-4). Nothing mysterious in this window. Click the independent variable (college) into the X Axis box, click the dependent variable (womleg_2010) into the Y Axis box, and click OK. SPSS summons its defaults and cranks out a scatterplot (Figure 8-5). This is a good start, but improvement is always possible. Double-click on the image, opening the Chart Editor (Figure 8-6). First we will complement or enhance the data elements—add the regression line, make the dots more prominent—and then we will deemphasize the graph's nondata features by whiting out the scatterplot's fill, border, and axes. Also, we will need to un-bold the axis titles. (In SPSS's default rendition, the bolded axis titles are the first thing one looks at.) Perhaps we also will want to modify the X Axis title to make it more presentable.

To add the regression line, click the Add Fit Line at Total button (see Figure 8-7). SPSS superimposes the line, selects it, and automatically opens the Properties window. The Fit Line tab (the opening tab) does not require our attention. Click the Lines tab. In its solid-black attire, the regression line looks more like a sure thing than a probabilistic estimate. Click the Style drop-down and pick one of the dashed-line options. Click Apply, but be sure to keep the Properties window open.[7] Now click on any one of the hollow circles in the cloud of points or "markers." SPSS selects all the markers (Figure 8-8). In the Color panel of the Marker tab, click Fill. The default setting, a diagonal line through a white background, means "transparent." This won't do. Make a color choice in the palette, and then click Apply. SPSS fills the dots with your selected choice.

**Figure 8-4**   Simple Scatterplot Window (modified)

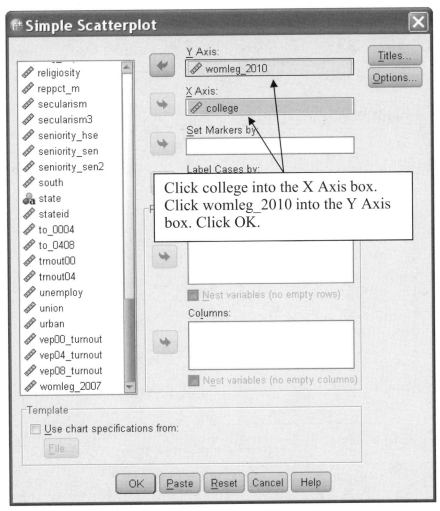

**Figure 8-5**   Unedited Scatterplot in the Viewer

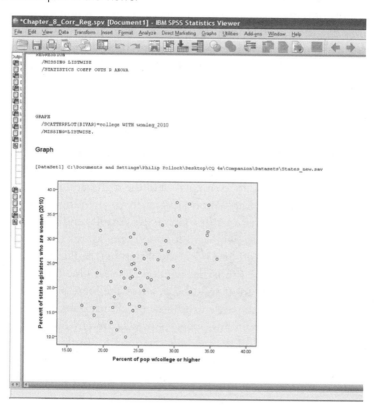

**Figure 8-6**   Scatterplot Ready for Editing

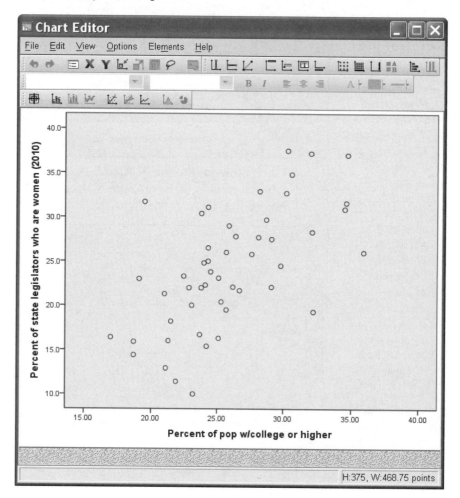

**Figure 8-7**   Adding a Regression Line to the Scatterplot

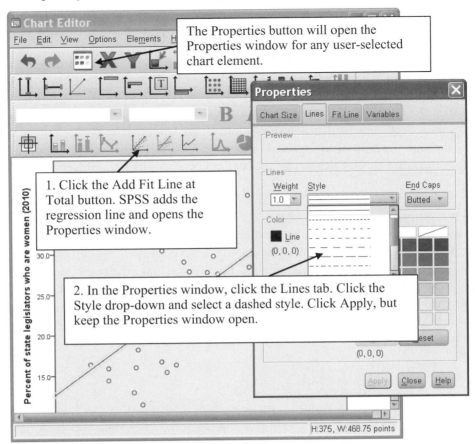

*Note:* To enhance readability, large buttons are shown.

**Figure 8-8**   Adding a Fill Color to Scatterplot Markers

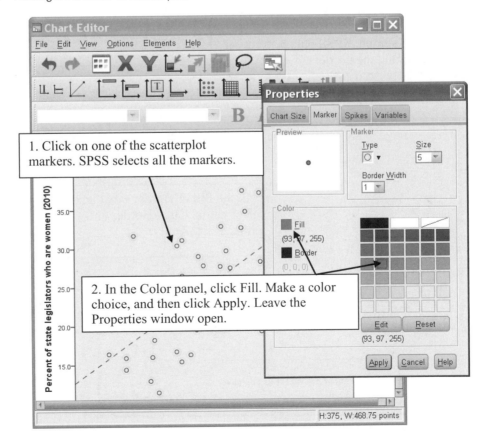

Compared with the effort involved in enhancing the scatterplot's data-related information, the task of erasing the nonessential elements of the graph is rather more labor-intensive. But with a little practice it becomes second nature. To whiten the graphic background (and dim but not whiten the axes), click anywhere on the empty gray space inside the graph (Figure 8-9). The Properties window adapts, telling us which elements are editable. In the Color panel of the Fill & Border tab, click Fill and select white, click Border and select white, and click Apply. Now we will blank out the newly dimmed axes. Carefully click on the X-axis, but make sure not to click on one of the tick labels (Figure 8-10). Depress the Control key and keep it depressed. Click on the Y-axis. Okay, now both axes are selected. In the Color panel of the Lines tab, click Line, select white, and click Apply. While we are here, we also will delete the unnecessary and distracting digits to the right of the decimal points in the axes tick-mark labels. Click the Number Format tab (refer to Figure 8-10). Click in the Decimal Places box (which may be empty) and type 0. Click Apply.

By default, SPSS bolds its graphic axis titles. This draws us away from the data and directs us toward a less important text element. Click on the X-axis title. Hold down the Control key and click on the Y-axis title (Figure 8-11). Both axis titles should now be selected. Click the Preferred Size drop-down and select 10. Click the Style drop-down and select Normal. Click Apply. At long last, you can close the Properties window. One more thing. Click on the X-axis title and modify it to make the graph more presentable (Figure 8-12). Before exiting the Chart Editor, you may want to save your chart preferences as a template, which can be opened and applied to future scatterplot-editing tasks.[8] In any event, if you want a chart that reflects current professional standards of elegant and informative graphic display, the scatterplot you have just created comes very close, indeed. Nicely done.

**Figure 8-9**  Whiting Out the Border and Fill

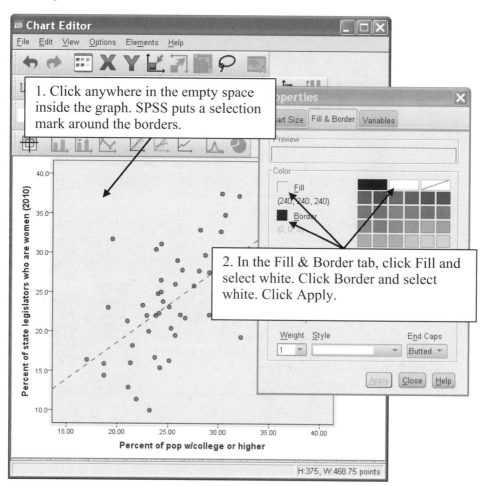

**Figure 8-10**  Editing the Axes of a Scatterplot

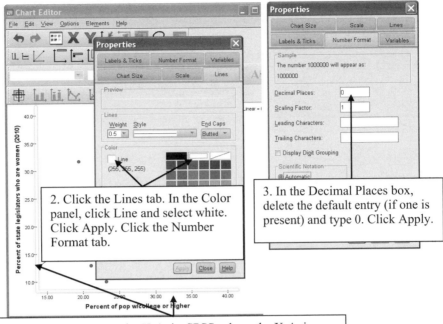

2. Click the Lines tab. In the Color panel, click Line and select white. Click Apply. Click the Number Format tab.

3. In the Decimal Places box, delete the default entry (if one is present) and type 0. Click Apply.

1. Click the mouse on the X-Axis. SPSS selects the X-Axis. While pressing and holding the Control key (Ctrl), click on the Y-Axis. Now both axes are selected.

**Figure 8-11**  Editing the Axis Titles

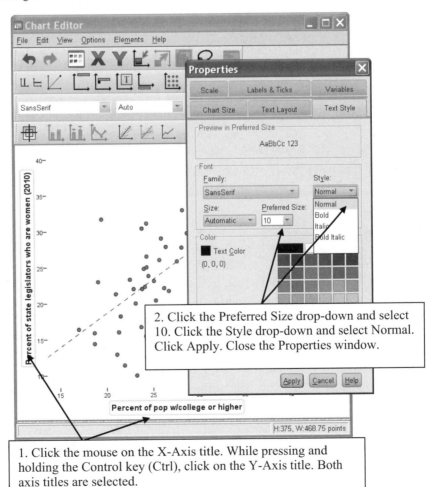

2. Click the Preferred Size drop-down and select 10. Click the Style drop-down and select Normal. Click Apply. Close the Properties window.

1. Click the mouse on the X-Axis title. While pressing and holding the Control key (Ctrl), click on the Y-Axis title. Both axis titles are selected.

**Figure 8-12** Changing an Axis Title

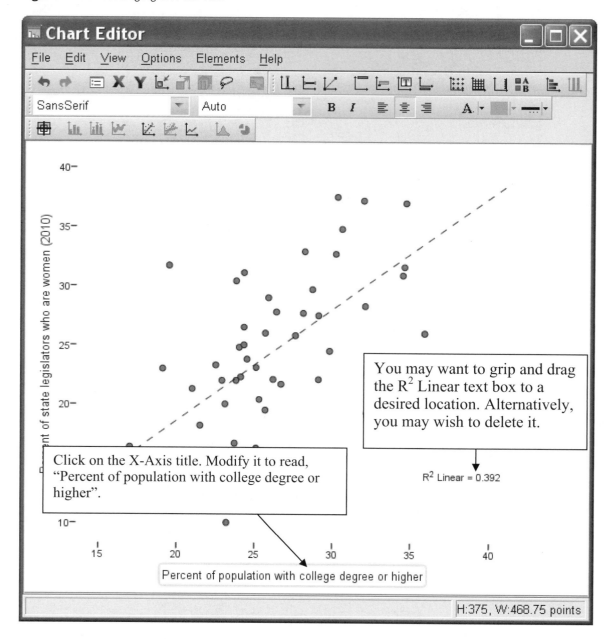

## MULTIPLE REGRESSION

Suppose a policy researcher is investigating factors causally related to motor vehicle deaths in the states. One such factor might be a simple characteristic of states: how densely populated they are. Residents of sparsely populated states, the policy researcher reasons, would typically drive longer distances at higher speeds than would residents of more densely populated states. Plus, a car accident in a thinly populated state would be more likely to be fatal, because "both Good Samaritans and hospitals are more scattered in thinly populated states compared to the denser states."[9] So as density goes up, we should find that fatalities go down. Another variable might be demographic: the proportion of young people in the population. As every insurance agent knows—and as many premium-paying parents will attest—younger drivers are more likely to be involved in automobile accidents than are older drivers. Thus, as the proportion of younger people goes up, fatalities should be found to go up, too.

The States dataset contains three variables: carfatal, the number of motor vehicle deaths per 100,000 residents; density, state population per square mile; and pop_18_24, the percentage of the population between 18 and 24 years of age. Run Analyze → Correlate → Bivariate to obtain a correlation matrix of these three variables.

**Correlations**

|  |  | carfatal Motor vehicle fatalities (per 100,000 pop) | density Population per square mile | pop_18_24 Percent age 18-24 |
|---|---|---|---|---|
| carfatal Motor vehicle fatalities (per 100,000 pop) | Pearson Correlation | 1 | -.574** | .388** |
|  | Sig. (2-tailed) |  | .000 | .005 |
|  | N | 50 | 50 | 50 |
| density Population per square mile | Pearson Correlation | -.574** | 1 | -.478** |
|  | Sig. (2-tailed) | .000 |  | .000 |
|  | N | 50 | 50 | 50 |
| pop_18_24 Percent age 18-24 | Pearson Correlation | .388** | -.478** | 1 |
|  | Sig. (2-tailed) | .005 | .000 |  |
|  | N | 50 | 50 | 50 |

**. Correlation is significant at the 0.01 level (2-tailed).

Note the correlation between each independent variable and the dependent variable. The correlation between density and carfatal is negative, indicating that as density increases, motor vehicle fatalities decrease (r = −.57). The relationship between pop_18_24 and carfatal, as the policy researcher suspected, is positive: As the percentage of young people increases, fatalities also increase (r = .39). But notice, too, that the two independent variables are themselves moderately related (r = −.48). This correlation is negative, suggesting that densely populated states have lower percentages of young people than do sparsely populated states. (This relationship becomes important later on.)

First let's run a simple regression, using carfatal as the dependent variable and density as the independent variable. Click Analyze → Regression → Linear. Put carfatal in the Dependent box and density in the Independent(s) box. Click OK and examine the output.

**Model Summary**

| Model | R | R Square | Adjusted R Square | Std. Error of the Estimate |
|---|---|---|---|---|
| 1 | .574a | .329 | .315 | 4.74790 |

a. Predictors: (Constant), density Population per square mile

**Coefficients<sup>a</sup>**

| Model |  | Unstandardized Coefficients B | Std. Error | Standardized Coefficients Beta | t | Sig. |
|---|---|---|---|---|---|---|
| 1 | (Constant) | 19.939 | .836 |  | 23.847 | .000 |
|  | density Population per square mile | -.013 | .003 | -.574 | -4.855 | .000 |

a. Dependent Variable: carfatal Motor vehicle fatalities (per 100,000 pop)

Consider the Y-intercept (Constant) and the regression coefficient on density. According to these values, the Y-intercept is equal to 19.94, and the regression coefficient is −.013, which rounds to −.01. The regression equation for the effect of density on carfatal, therefore, is

Motor vehicle fatalities per 100,000 pop. = 19.94 − .01*Population per square mile.

What do these coefficients mean? In terms of its magnitude, for example, the regression coefficient seems to be an incredibly small number, and its meaning is not intuitively obvious. Remember to keep the substantive relationship in mind—and focus on the units of measurement. Very thinly populated states will have an estimated fatality rate close to the intercept, or about 20 fatalities per 100,000 population. Alaska, for example, has a population density of just more than 1 person per square mile. So its estimated fatality rate would be close to the intercept of about 20. The regression coefficient tells us that, for each additional person per square mile, the motor vehicle fatality rate drops by .01. New Jersey, for instance, is a very densely populated state, with a density of about 1,200 people per square mile. So New Jersey's estimated fatality rate would be 20 – .01*1200, which is equal to 20 – 12, or about 8 fatalities per 100,000 population. Thus, as density increases, by one person at a time, fatalities decrease by .01 of a fatality per 100,000 population. According to the t-ratio (–4.86) and accompanying P-value (.000), we can safely reject the null hypothesis. SPSS reports an adjusted R-square of .315. Thus, of all the variation among states in automobile fatality rates, about 32 percent is explained by population density.

Now let's run another bivariate regression. We will keep carfatal as the dependent variable, but this time we'll use the percentage of the population aged 18 to 24 (pop_18_24) as the independent variable.

**Model Summary**

| Model | R | R Square | Adjusted R Square | Std. Error of the Estimate |
|---|---|---|---|---|
| 1 | .388[a] | .151 | .133 | 5.34276 |

a. Predictors: (Constant), pop_18_24 Percent age 18-24

**Coefficients[a]**

| Model | | Unstandardized Coefficients | | Standardized Coefficients | t | Sig. |
|---|---|---|---|---|---|---|
| | | B | Std. Error | Beta | | |
| 1 | (Constant) | -10.102 | 9.494 | | -1.064 | .293 |
| | pop_18_24 Percent age 18-24 | 2.729 | .935 | .388 | 2.919 | .005 |

a. Dependent Variable: carfatal Motor vehicle fatalities (per 100,000 pop)

According to the estimated coefficients, the regression line for the effect of pop_18_24 on carfatal is as follows:

$$\text{Motor vehicle fatalities per 100,000 pop.} = -10.10 + 2.73 \times \text{Percent age 18-24.}$$

Again we have a Y-intercept depicting an unreal situation, so let's focus on the regression coefficient, 2.73. This says that for each percentage point increase in pop_18_24, there is a 2.73-unit increase in the motor vehicle fatality rate—2.73 additional fatalities per 100,000 population. As the percentage of younger people in the population increases, so too does the fatality rate. In the population, could the true value of the regression coefficient be 0? Probably not, according to the t-ratio (2.92) and the P-value (.005). And, according to the adjusted R-square of .133, about 13 percent of the variation in carfatal is explained by pop_18_24.

Let's review our analysis so far. In the first bivariate regression, we found that population density has a statistically significant negative effect on motor vehicle fatalities. Low-density states have higher fatality rates than do high-density states. In the second bivariate regression, we found that the percentage of younger people has a significant positive effect on motor vehicle fatalities. States with lower percentages of young people have lower fatality rates than do states with higher percentages of young people. But recall the initial correlation matrix. There we found that the two independent variables are related: As density goes up, the percentage of younger people goes down. So when we compare states with lower percentages of young people with states with higher percentages of young people, we are also comparing high-density states with low-density states. Perhaps states with higher percentages of young people have higher fatality rates not because they have more young people, but because they have lower population densities. Thus the relationship between pop_18_24 and carfatal might be spurious. Then again, it might not be. Unless we reexamine the pop_18_24-carfatal relationship, controlling for density, there is no way to tell.

Multiple regression is designed to estimate the partial effect of an independent variable on a dependent variable, controlling for the effects of other independent variables. Regression → Linear easily allows us to run multiple regression analysis. Let's do such an analysis, again using carfatal as the dependent variable and entering *both* density and pop_18_24 as independent variables. Return to the Regression window. Leaving pop_18_24 in place, click density into the Independent(s) box and click OK.

**Model Summary**

| Model | R | R Square | Adjusted R Square | Std. Error of the Estimate |
|---|---|---|---|---|
| 1 | .588[a] | .346 | .318 | 4.73747 |

a. Predictors: (Constant), density Population per square mile, pop_18_24 Percent age 18-24

**Coefficients[a]**

| Model | | Unstandardized Coefficients | | Standardized Coefficients | t | Sig. |
|---|---|---|---|---|---|---|
| | | B | Std. Error | Beta | | |
| 1 | (Constant) | 9.126 | 9.858 | | .926 | .359 |
| | pop_18_24 Percent age 18-24 | 1.039 | .944 | .148 | 1.101 | .277 |
| | density Population per square mile | -.011 | .003 | -.503 | -3.748 | .000 |

a. Dependent Variable: carfatal Motor vehicle fatalities (per 100,000 pop)

This analysis provides the information we need to isolate the partial effect of each independent variable on the dependent variable. The multiple regression equation is as follows:

$$\text{Motor vehicle fatalities per 100,000 pop.} =$$
$$9.13 + 1.04^* \text{ (Percent age 18-24)} - .01^*(\text{Population per square mile}).$$

Let's focus on the regression coefficients for each of the independent variables. The coefficient for pop_18_24, 1.04, tells us the effect of pop_18_24 on carfatal, controlling for density. Recall that in the bivariate analysis, a 1-percentage-point increase in pop_18_24 was associated with about a 2.73-unit increase in the fatality rate. When we control for density, however, we find a substantial reduction in this effect—to about a 1-unit increase in the fatality rate. What is more, the regression coefficient for pop_18_24, with a P-value of .277, is not statistically significant. Density, on the other hand, retains much of its predictive power. The regression coefficient, −.011, is essentially the same effect we found earlier when we investigated the bivariate relationship between carfatal and density. With a P-value of .000, we can say that, controlling for the percentage of young residents, population density is significantly related to motor vehicle fatalities. It would appear, then, that the carfatal-pop_18_24 relationship is a spurious artifact of differences between states in population density.

In multiple regression, adjusted R-square communicates how well all the independent variables explain the dependent variable. So, by knowing two things about states—the percentage of younger people and the population density—we can account for about 32 percent of the variation in motor vehicle death rates. But notice that this value of adjusted R-square is practically the same as the adjusted R-square we found before, using density by itself to explain carfatal. Clearly, density does the lion's share of explanatory work in accounting for the dependent variable.

## EXERCISES

1. (Dataset: States. Variables: dem_hr09, demstate09, union07.) Consider a plausible scenario for the relationships between three variables: the percentage of a state's U.S. House delegation who are Democrats, the percentage of state legislators who are Democrats, and the percentage of workers in the state who are unionized. We could hypothesize that, compared with states with fewer Democrats in their state legislatures, states having larger percentages of Democratic legislators would also have greater proportions of Democrats in their U.S. House delegations. Furthermore, because unions tend to support Democratic candidates, we would also expect more heavily unionized states to have higher percentages of Democratic legislators at the state legislative and congressional levels. States contains three variables: dem_hr09, the percentage of House members who are Democrats; demstate09, the percentage of state legislators who are Democrats; and union07, the percentage of workers who are union members.

A. Run Correlate to find the Pearson's correlation coefficients among dem_hr09, demstate09, and union07. Fill in the six empty cells of this correlation matrix:

| | | Percent U.S House delegation Democratic | Percent state legislators Democratic | Percent workers who are union members |
|---|---|---|---|---|
| Percent U.S. House delegation Democratic | Pearson's correlation | 1 | ? | ? |
| Percent state legislators Democratic | Pearson's correlation | ? | 1 | ? |
| Percent workers who are union members | Pearson's correlation | ? | ? | 1 |

B. According to the correlation coefficient, as the percentage of Democratic state legislators increases, the percentage of Democratic U.S. representatives (circle one)

increases.     decreases.

C. According to the correlation coefficient, as the percentage of Democratic state legislators decreases, the percentage of Democratic U.S. representatives (circle one)

increases.     decreases.

D. Consider this argument: Unions are more important in U. S. House races than in state legislative races. Does your analysis support this argument? (circle one)

Yes     No

Explain your reasoning. _____

_____

_____

_____

2. (Dataset: States. Variables: conpct_m, cons_hr09.) Two congressional scholars are discussing the extent to which members of the U.S. House of Representatives stay in touch with the voters in their states.

Scholar 1: "When members of Congress vote on important public policies, they are closely attuned to the ideological make-ups of their states. Members from states having lots of liberals will tend to cast votes in the liberal direction. Representatives from states with mostly conservative constituencies, by contrast, will take conservative positions on important policies."

Scholar 2: "You certainly have a naïve view of congressional behavior. Once they get elected, members of Congress adopt a 'Washington, D.C., state of mind,' perhaps voting in the liberal direction on one policy and in the conservative direction on another. One thing is certain: The way members vote has little to do with the ideological composition of their states."

Think about an independent variable that measures the percentage of self-described conservatives among the mass public in a state, with low values denoting low percentages of conservatives and high values denoting high percentages of conservatives. And consider a dependent variable that gauges the degree to which the state's House delegation votes in a conservative direction on public policies. Low scores on this dependent variable tell you that the delegation tends to vote in a liberal direction, and high scores say that the delegation votes in a conservative direction.

A. Below is an empty graphic shell showing the relationship between the independent variable and the dependent variable. Draw a regression line inside the shell that depicts what the relationship should look like if scholar 1 is correct.

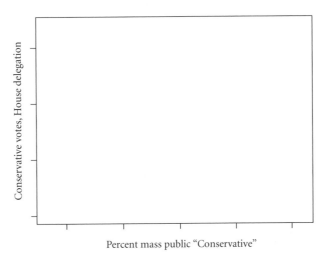

B. Below is another graphic shell showing the relationship between the independent variable and the dependent variable. Draw a regression line inside the shell that depicts what the relationship should look like if scholar 2 is correct.

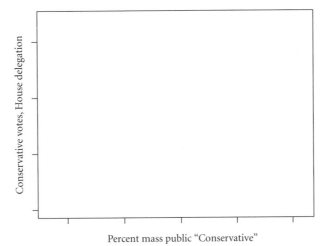

C. States contains the variable conpct_m, the percentage of the mass public calling themselves conservative. This is the independent variable. States also contains cons_hr09, a measure of conservative votes by states' House members. Scores on this variable can range from 0 (low conservatism) to 100 (high conservatism). This is the dependent variable. Run Regression to analyze the relationship between cons_hr and conpct_m.

According to the regression equation, a 1-percentage-point increase in conservatives in the mass public is associated with (check one)

❑  about a 66-point decrease in House conservatism scores.

❑  about a 3-point increase in House conservatism scores.

❑  about a 6-point increase in House conservatism scores.

D. If you were to use this regression to estimate the mean House conservatism score for states having 30 percent conservatives, your estimate would be (circle one)

a score of about 30.        a score of about 45.        a score of about 95.

E. The adjusted R-square for this relationship is equal to (fill in the blank) _____. This tells you that about (fill in the blank) _____ percent of the variation in cons_hr09 is explained by conpct_m.

F. Use Graphs → Legacy Dialogs → Scatter/Dot to create a scatterplot of the relationship between conpct_m (X-axis) and cons_hr09 (Y-axis). Enhance the graph's data-ink ratio by following the procedures described in this chapter for creating an erased graph. Print the graph.

G. Based on your inspection of the graph, the regression line, and adjusted R-square, which congressional scholar is more correct? (Check the correct answer and fill in the appropriate blank)

❑ Scholar 1 is more correct because _____

_____

❑ Scholar 2 is more correct because _____

_____

3. (Dataset: States. Variables: to_0408, obama08). An article of faith among Democratic Party strategists (and a source of apprehension among Republican strategists) is that high voter turnouts help Democratic candidates. Why should this be the case? According to the conventional wisdom, Democratic electorates are less likely to vote than are Republican voters. Thus low turnouts naturally favor Republican candidates. As turnouts push higher, the reasoning goes, a larger number of potential Democratic voters will go to the polls, creating a better opportunity for Democratic candidates. Therefore, as turnouts go up, so should the Democratic percentage of the vote.[10]

A. Use Regression → Linear to test this conventional wisdom. The states dataset contains to_0408, the percentage-point change in presidential election turnout between 2004 and 2008. States in which turnout declined between 2004 and 2008 have negative values on to_0408, while states in which turnout increased have positive values on to_0408. (For example, North Carolina's turnout increased from 57.8 percent to 65.8 percent, giving North Carolina a value of 8.0 on to_0408. Utah's turnout dropped from 58.9 percent to 53.3 percent, giving it a value of –5.6 on to_0408.) To_0408 is the independent variable. Another variable, obama08, the percentage of the vote cast for Democratic candidate Barack Obama, is the dependent variable.

Based on your results, the regression equation for estimating the percentage of votes cast for Obama is (fill in the blanks)

_____ + _____ * to_0408.
(constant)      (regression coefficient)

B. The P-value for the regression coefficient on to_0408 is _____, and adjusted R-square is _____.

C. Consider your findings in A and B. One may conclude that

❑ the conventional wisdom is correct because _____

_____

_____

_____

❑ the conventional wisdom is incorrect because _____

_____

_____

_____

4. (Dataset: States. Variables: abortlaw, permit.) As you are no doubt aware, in its momentous decision in *Roe v. Wade* (1973), the U.S. Supreme Court declared that states may not outlaw abortion. Even so, many state legislatures have enacted restrictions and regulations that, while not banning abortion, make an abortion more difficult to obtain. Other states, however, have few or no restrictions. What factors might explain these differences in abortion laws among the states? We know that the mass public remains divided on this issue. Public opinion in some states is more favorable toward permitting abortion and in other states is less favorable. Does public opinion guide state policy on this issue?

States contains abortlaw, which measures the number of abortion restrictions a state has enacted into law. Values on abortlaw range from 0 (least restrictive) to 10 (most restrictive). This is the dependent variable. States also has the variable permit, the percentage of the mass public saying that abortion should "always" be permitted. This is the independent variable.

A. If you were to use regression analysis to test the idea that public opinion on abortion affects state abortion policy, then you would expect to find (check one)

❑ a negative sign on permit's regression coefficient.

❑ a positive sign on permit's regression coefficient.

B. Analyze the abortlaw-permit relationship using Regression → Linear. According to the results, the regression equation for estimating the number of abortion restrictions is (fill in the blanks)

_____ _____*permit
(constant)        (regression coefficient)

C. The P-value for the regression coefficient is (fill in the blank) _____. The value of adjusted R-square is (fill in the blank) _____.

D. According to States, 48.7 percent of Oregon residents believe that abortion should "always" be permitted. In Oklahoma, by contrast, 25.2 percent of the public holds this view. Based on the regression equation, you would estimate that Oregon would have about (fill in the blanks) _____ abortion restrictions and that Oklahoma would have about _____ abortion restrictions.

E. Use Graphs → Legacy Dialogs → Scatter/Dot to create a scatterplot of the relationship between permit (X-axis) and abortlaw (Y-axis). Enhance the graph's data-ink ratio by following the procedures described in this chapter for creating an erased graph. Print the graph.

5. (Dataset: States. Variables: demstate09, dempct_m, libpct_m.) In Exercise 2 you analyzed the connection between mass political attitudes and congressional voting, and in Exercise 4 you examined the link between public opinion and public policy. In this exercise you will use correlation and multiple regression to examine a set of relationships between mass attitudes and the partisan make-up of state legislatures. State legislatures are remarkably varied in this regard—ranging in partisan composition from about 24 percent Democratic to over 90 percent Democratic. What accounts for this variation? Consider two plausible independent variables: the percentage of a state's citizens who are self-identified Democrats, and the percentage of citizens who are self-described liberals. Each of these variables should have a positive relationship with the percentage of Democrats in the state legislature.

A. States contains these three variables: demstate09, the percentage of state legislators who are Democrats; dempct_m, the percentage of Democrats in the mass electorate; and libpct_m, the percentage of self-described liberals in the mass public. Run Correlate to find the Pearson's correlation coefficients among demstate09, dempct_m, and libpct_m. Fill in the six empty cells of this correlation matrix:

|  |  | demstate09 Percent of state legislators who are Democrats (2009) | dempct_m Percent mass public Democratic | libpct_m Percent mass public Liberal |
|---|---|---|---|---|
| demstate09 Percent of state legislators who are Democrats (2009) | Pearson's correlation | 1 | ? | ? |
| dempct_m Percent mass public Democratic | Pearson's correlation | ? | 1 | ? |
| libpct_m Percent mass public liberal | Pearson's correlation | ? | ? | 1 |

B. According to the correlation coefficient, as the percentage of liberals in the mass public increases, the percentage of Democratic state legislators (circle one)

increases.     decreases.

C. Suppose someone were to make this claim: "Being a Democrat and being a liberal are practically synonymous. The relationship between the percentage of Democratic identifiers and the percentage of liberals, therefore, will be positive and strong." According to the correlation coefficient, this claim is (check the appropriate answer and fill in the appropriate blank)

❑ correct because _____

_____

❑ incorrect because _____

_____

D. Run Regression → Linear to obtain multiple regression estimates for the partial effects of dempct_m and libpct_m on demstate09. Demstate09 is the dependent variable, and dempct_m and libpct_m are the independent variables. Based on your results, the multiple regression for estimating the percentage of Democratic state legislators is (fill in the blanks)

−28.375 + _____*dempct_m + _____*libpct_m.

E. The P-value for the regression coefficient on dempct_m is (fill in the blank) _____, and the P-value for the regression coefficient on libpct_m is (fill in the blank) _____.

F. As you may know, Nebraska's state legislature is unique in two ways: It is unicameral (all other state legislatures are bicameral), and it is nonpartisan. Candidates do not run for the state legislature using party labels, and the legislature is not organized on the basis of party. Thus Nebraska has a missing value on the variable demstate09, and it was not included in the regression analysis you just performed. However, if you were to peruse States, you would find that 29.03 percent of Nebraskans are Democrats and 16.44 percent are self-described liberals.

For the sake of speculation, assume that Nebraska decided that all members of the state legislature should declare a partisan allegiance. Based on your regression model, about what percentage of state legislators would be Democrats? (circle one)

About 30 percent      About 40 percent      About 50 percent

G. Based on your interpretation of the multiple regression output, you can conclude that (check all that apply)

☐ controlling for the percentage of the mass public who are Democratic, a 1-percentage-point increase in the percentage of liberals in the mass public is associated with about a 2.6-percentage-point increase in the percentage of Democratic state legislators.

☐ both independent variables are significantly related to the dependent variable.

☐ taken together, both independent variables explain over one-half of the variation in the dependent variable.

6. (Dataset: GSS2008. Variables: egalit_scale, rincom06, polviews.)

Two scholars are discussing why some people have stronger egalitarian beliefs than other people.

Scholar 1: "The main thing to know about people is how much money they make. As income goes up, egalitarian beliefs decline. I'll bet that if you perform a regression analysis using egalitarianism as the dependent variable and income as the independent variable, the regression coefficient on income will be negative and statistically significant."

Scholar 2: "You are on the right track, but you're not really addressing the question of *why* income and egalitarianism are negatively related. It's a two-step process. First, peoples' incomes shape their ideological worldviews. Lower income people tend to be ideologically liberal, and higher income people tend to be ideologically conservative. Next, ideology affects egalitarianism. Liberals tend to be strong egalitarians, while conservatives have weaker egalitarian beliefs. Go ahead and run your simple bivariate regression using income as the independent variable. Then run a multiple regression, using income *and* ideology as independent variables. The multiple regression will show that ideology is strongly related to egalitarianism. But the coefficient on income will be statistically insignificant.

GSS2008 contains egalit_scale, a metric that runs from 0 (low egalitarianism) to 12 (high egalitarianism). The dataset also has rincom06, which ranges from 1 (low income) to 25 (high income). A familiar measure of self-described ideology, polviews, ranges from 1 (extremely liberal) to 7 (extremely conservative).

A. Run the bivariate regression suggested by scholar 1. Scholar 1's idea is (circle one)

correct.      incorrect.

Explain how you know. _____

_____

_____

B. Run the multiple regression suggested by scholar 2. Scholar 2 claimed that liberals are significantly more egalitarian than conservatives. This claim is (circle one)

correct.      incorrect.

Explain how you know. _____

_____

_____

C.  Scholar 2 claimed that, after controlling for ideology, the egalitarianism-income relationship will be insignificant. This claim is

correct.       incorrect.

Explain how you know. _____

_____

_____

D.  Suppose you wanted to use the multiple regression to estimate the egalitarianism score for the "average" person—that is, for a person having mean levels of income and ideology. Run the appropriate analysis to obtain the mean values of rincom06 and polviews. The mean of rincom06 is equal to _____. The mean of polviews is equal to _____. Your estimate for the average person's egalitarianism score is equal to _____.

That concludes the exercises for this chapter. Before exiting SPSS, be sure to save your output file.

## NOTES

1.  Regression analysis on variables measured by percentages can be confusing. Always stay focused on the exact units of measurement. One percentage point would be 1.00. So if attend_pct increases by 1.00, then womleg_2010 decreases, on average, by .47, or .47 of a percentage point.
2.  SPSS Regression $\rightarrow$ Linear reports two-tailed P-values, not one-tailed P-values. Strictly speaking, then, you may correctly apply the .05 standard by rejecting the null hypothesis for any reported P-value of .10 or less. However, in this book we follow the more conservative practice of rejecting the null hypothesis for P-values of .05 or less.
3.  The t-ratio for the Y-intercept permits you to test the null hypothesis that, in the population, the Y-intercept is 0. In this case we have no interest in testing the hypothesis that states having 0 percent frequent attenders have 0 percent women in their state legislatures.
4.  Most data analysis programs, SPSS included, provide two values of R square—a plain version, which SPSS labels "R Square," and an adjusted version, "Adjusted R Square." Adjusted R Square is often about the same as (but is always less than) plain R Square. What is the difference? Just like a sample mean, which provides an estimate of the unseen population mean, a sample R-square provides an estimate of the true value of R-square in the population. And just like a sample mean, the sample R-square is equal to the population R-square, give or take random sampling error. However, unlike the random error associated with a sample mean, R-square's errors can assume only positive values—squaring any negative error, after all, produces a positive number—introducing upward bias into the estimated value of R-square. This problem, which is more troublesome for small samples and for models with many independent variables, can be corrected by adjusting plain R-square "downward." For a sample of size N and a regression model with k predictors, adjusted R-square is equal to: $1 - (1 - \text{R-square})[(N - 1)/(N - k - 1)]$. See Barbara G. Tabachnick and Linda S. Fidell, *Using Multivariate Statistics*, 3rd ed. (New York: HarperCollins, 1996), 164–165.
5.  If you do a quick Descriptives run, you will find that the lowest value of college is 17 percent.
6.  Edward R. Tufte, *The Visual Display of Quantitative Information*, 2nd ed. (Cheshire, Conn.: Graphics Press, 2001). Tufte's work has inspired other excellent treatments of visual communication. For example, see Stephen Few, *Show Me the Numbers: Designing Tables and Graphs to Enlighten* (Oakland, Calif.: Analytics Press, 2004); Howard Wainer, *Graphic Discovery: A Trout in the Milk and Other Visual Adventures* (Princeton: Princeton University Press, 2005).
7.  You will want to keep the Properties window open for your entire excursion into the Chart Editor. Each time you select a different part of the graph for editing, SPSS automatically adjusts the Properties window to reflect the editable features of the graphic element you have selected. Naturally, you can open the Properties window upon entering the Chart Editor by clicking the Properties button.

8. With the Chart Editor still open, click File → Save Chart Template. In the Save Chart Template window, click in the All Settings box, which selects all chart features. Now uncheck the box next to Text Content. (You don't want SPSS to apply the same axis titles to all of your scatterplots.) Click Continue. Find a good place to save the template (and concoct a descriptive name for the file), which SPSS saves with the .sgt extension. To apply the template to future editing projects: In the Chart Editor, click File → Apply Chart Template, find the .sgt file, and click Open. Experience teaches that SPSS will apply most of the template's features to the new graphic, although some minor editing may still be required.

9. Edward R. Tufte, *Data Analysis for Politics and Policy* (Englewood Cliffs, N.J.: Prentice Hall, 1974), 21. Tufte uses regression analysis to evaluate the effectiveness of motor vehicle inspections, controlling for population density.

10. See Michael D. Martinez and Jeff Gill, "The Effects of Turnout on Partisan Outcomes in U.S. Presidential Elections 1960–2000," *Journal of Politics* 67 (2005): 1248–1274. Martinez and Gill find that the Democratic advantage from higher turnouts has declined over time.

# 9

# Dummy Variables and Interaction Effects

Procedures Covered

Transform → Recode into Different Variables (dummy variables)

Analyze → Regression → Linear (with dummy variables)

Transform → Compute → If (optional case selection condition)

Analyze → Regression → Linear (with interaction variable)

You can adapt regression analysis to different research situations. In one situation you might have nominal or ordinal independent variables. Provided that these variables are dummy variables, you can run a regression analysis, using categorical variables to predict values of an interval-level dependent variable. In this chapter you will learn how to construct dummy variables and how to use them in regression analysis. In a second research situation you might suspect that the effect of one independent variable on the dependent variable is not the same for all values of another independent variable—in other words, that interaction is going on in the data. Provided that you have created an interaction variable, you can use multiple regression to estimate the size and statistical significance of interaction effects. In this chapter you will learn how to create an interaction variable and how to perform and interpret multiple regression with interaction effects.

## REGRESSION WITH DUMMY VARIABLES

A dummy variable can take on only two values, 1 or 0. Each case being analyzed either has the characteristic being measured (a code of 1) or does not have it (a code of 0). For example, a dummy variable for gender might code females as 1 and males as 0. Everybody who is coded 1 has the characteristic of being female, and everybody who is coded 0 does not have that characteristic. To appreciate why this 0 or 1 coding is the essential feature of dummy variables, consider the following regression model. The model, which uses data gathered during the 2008 presidential campaign, is designed to test the hypothesis that women gave Democratic presidential candidate Barack Obama higher feeling thermometer ratings than did men:

Obama feeling thermometer = a + b(female).

In this formulation, gender is measured by a dummy variable, female, which is coded 0 for males and 1 for females. Since males are scored 0 on the dummy, the constant or intercept, a, will tell us the average Obama rating among men. Why so? Substituting 0 for the dummy yields: a + b*0 = a. In the language of dummy variable regression, males are the "omitted" category, the category whose mean value on the dependent variable is captured by the intercept, a. The regression coefficient, b, will tell us how much to adjust the intercept for women—that is, when the dummy switches from 0 to 1. Thus, just as in any regression, b will estimate the average change in the dependent variable for a unit change in the independent variable. Since in this case a unit change in the independent variable is the difference between men (coded 0 on female) and women (coded 1 on female), the regression coefficient will reflect the mean difference in Obama thermometer ratings between males and females.

It is important to be clear on this point: The coefficient, b, does not communicate the mean Obama rating among females. Rather, it estimates the mean difference between males and females. (Of course, an estimated value of the dependent variable among females can be arrived at easily by summing a and b: a + b*1 = a + b.) As with any regression coefficient, we can rely on the coefficient's t-ratio and P-value to test the null hypothesis that there is no statistically meaningful gender difference in thermometer ratings of Obama.

Let's open NES2008 and figure out how to use gender as an independent variable in a regression analysis of Obama thermometer ratings. We'll use obama_therm as the dependent variable. The independent variable, gender, is a nominal-level measure, coded 1 for males and 2 for females. Because of the way it is currently coded, gender could not be used in regression analysis. How can we create a dummy variable, female, coded 0 for males and 1 for females? We could assign these values by using Transform → Recode into Different Variables and applying this recoding scheme:

| Respondent's gender | Old value (gender) | New value (female) |
|---|---|---|
| Male | 1 | 0 |
| Female | 2 | 1 |
| | Missing | Missing |

You know how to use Transform → Recode into Different Variables, so go ahead and create female, which you can label "Female dummy." (Figure 9-1 helps to reacquaint you with the Recode procedure.)

**Figure 9-1**   Recoding to Create a Dummy Variable

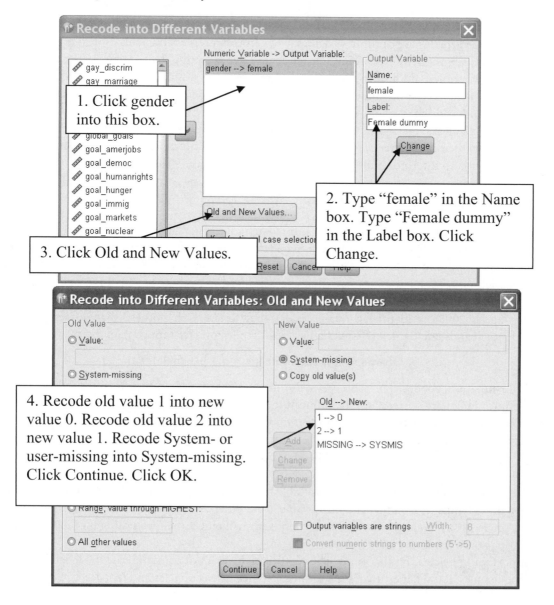

To check your work, run Frequencies on gender and female to ensure that the distributions are the same.

**gender R gender**

|       |          | Frequency | Percent | Valid Percent | Cumulative Percent |
|-------|----------|-----------|---------|---------------|--------------------|
| Valid | 1 Male   | 1048      | 45.1    | 45.1          | 45.1               |
|       | 2 Female | 1275      | 54.9    | 54.9          | 100.0              |
|       | Total    | 2323      | 100.0   | 100.0         |                    |

**female Female dummy**

|       |       | Frequency | Percent | Valid Percent | Cumulative Percent |
|-------|-------|-----------|---------|---------------|--------------------|
| Valid | .00   | 1048      | 45.1    | 45.1          | 45.1               |
|       | 1.00  | 1275      | 54.9    | 54.9          | 100.0              |
|       | Total | 2323      | 100.0   | 100.0         |                    |

The 1,275 respondents coded 2 on gender are coded 1 on female, and the 1,048 respondents coded 1 on gender are coded 0 on female. Return to the Variable View and assign value labels to the dummy variable you have created ("Male" for value 0, "Female" for value 1).

Now let's run linear regression, using the Barack Obama feeling thermometer (obama_therm) as the dependent variable and female as the independent variable. Click Analyze → Regression → Linear. Click obama_therm into the Dependent box, and click female into the Independent(s) box. Click OK.

**Model Summary**

| Model | R      | R Square | Adjusted R Square | Std. Error of the Estimate |
|-------|--------|----------|-------------------|----------------------------|
| 1     | .051[a] | .003     | .002              | 28.388                     |

a. Predictors: (Constant), female Female dummy

**Coefficients[a]**

| Model |                      | Unstandardized Coefficients | | Standardized Coefficients | t | Sig. |
|-------|----------------------|--------|------------|-------|--------|------|
|       |                      | B      | Std. Error | Beta  |        |      |
| 1     | (Constant)           | 56.274 | .880       |       | 63.968 | .000 |
|       | female Female dummy  | 2.891  | 1.189      | .051  | 2.432  | .015 |

a. Dependent Variable: obama_therm Feeling Thermometer: Obama

According to the Coefficients table, the regression equation is as follows:

$$\text{Obama feeling thermometer} = 56.27 + 2.89*\text{Female dummy}.$$

How would we interpret these estimates? As always, the constant estimates the value of the dependent variable when the independent variable is 0. Because males have a value of 0 on female, the mean thermometer rating of Barack Obama for males is 56.27, the intercept. The regression coefficient on female communicates the mean change in the dependent variable for each unit change in the independent variable. So when the dummy switches from 0 to 1, the Obama rating goes up, on average, about 2.9 degrees. We can use this value to estimate the mean rating for females: 56.27 + 2.89 = 59.16. So men rated Obama at about 56 and women rated him at about 59. Was this gender difference produced by random sampling error? Not according to the P-value, .015. Do gender differences account for a big chunk of the variation in Barack Obama thermometer ratings? Not exactly. According to the adjusted R-square, gender alone accounts for two-tenths of 1 percent of the variation in the dependent variable. There must be other variables that contribute to the explanation of Obama's ratings. Let's expand the model.

We would expect partisanship to have a big effect on the Obama thermometer scale. Democrats should score higher on the dependent variable than do independents or Republicans. Plus, we know that women are more likely than men to be Democrats, so the Obama_therm-female relationship might be the spurious result of partisan differences, not gender differences. NES2008 contains partyid3, which codes Democrats as 1, independents as 2, and Republicans as 3. Because partyid3 is a categorical variable, we cannot use it in a regression—not in its present form, anyway. But we can use partyid3 to create a dummy variable for partisanship.

Actually, we need to create not one but two dummy variables from partyid3. Why two? Here is a general rule about dummy variables: If the variable you want to "dummy-ize" has k categories, then you need k–1 dummies to measure the variable. Because partyid3 has three categories, we need two dummy variables. One of these variables, which we will call demdum, is equal to 1 for Democrats and 0 for independents and Republicans. The second dummy variable, repdum, is equal to 1 for Republicans and 0 for Democrats and independents. Independents, then, are uniquely identified by their exclusion from both dummies. Independents have values of 0 on demdum and 0 on repdum. Consider this recoding protocol:

| Party ID: 3 Categories | Old value (partyid3) | New value (demdum) | New value (repdum) |
|---|---|---|---|
| Democrat | 1 | 1 | 0 |
| Independent | 2 | 0 | 0 |
| Republican | 3 | 0 | 1 |
| | Missing | Missing | Missing |

We will create demdum and repdum one at a time. To create demdum, click Transform → Recode into Different Variables. (The gender recode is still in the window, so click Reset.) Follow these steps:

1. Click partyid3 into the Numeric Variable → Output Variable panel.
2. Click in the Name box and type "demdum".
3. Click in the Label box and type "Democrat dummy". Click Change.
4. Click Old and New Values.

In the Recode into Different Variables: Old and New Values window, recode old value 1, the partyid3 code for Democrats, into new value 1, the code for Democrats on demdum. Old values 2 and 3, the partyid3 codes for independents and Republicans, are equal to new value 0, the code for independents and Republicans on demdum. You can use the Range boxes in the Old Value panel to accomplish this change. Make sure that you recode missing values on partyid3 into missing values on demdum. The Old and New Values window should now look like the upper panel of Figure 9-2.

Repeat the recoding procedure, using partyid3 to create repdum. Return to the Transform → Recode into Different Variables window. To avoid confusion, click the Reset button first and then follow these steps:

1. Click partyid3 into the Numeric Variable → Output Variable panel.
2. Type "repdum" in the Name box.
3. Type "Republican dummy" in the Label box and click Change.
4. Click Old and New Values.

This time recode old value 3 on partyid3 into new value 1 on repdum. Old values 1 and 2 become 0 in the new values of repdum. Again, make sure to recode missing values on partyid3 into missing values on repdum (see the lower panel of Figure 9-2).

**Figure 9-2**  Creating Two Dummy Variables from a Three-Category Ordinal

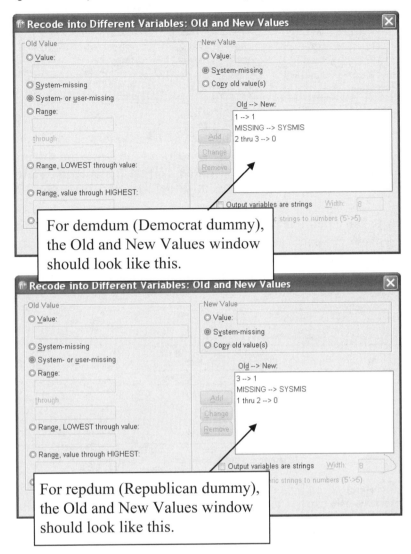

For demdum (Democrat dummy), the Old and New Values window should look like this.

For repdum (Republican dummy), the Old and New Values window should look like this.

Before analyzing these new variables, it would be prudent to check your work. Run a quick Frequencies on partyid3, demdum, and repdum.

partyid3 RECODE of partyid7 (Summary Party ID)

|  |  | Frequency | Percent | Valid Percent | Cumulative Percent |
|---|---|---|---|---|---|
| Valid | 1 Dem | 782 | 33.7 | 34.0 | 34.0 |
|  | 2 Indep | 926 | 39.9 | 40.3 | 74.3 |
|  | 3 Rep | 591 | 25.4 | 25.7 | 100.0 |
|  | Total | 2299 | 99.0 | 100.0 |  |
| Missing | System | 24 | 1.0 |  |  |
| Total |  | 2323 | 100.0 |  |  |

demdum Democrat dummy

|  |  | Frequency | Percent | Valid Percent | Cumulative Percent |
|---|---|---|---|---|---|
| Valid | .00 | 1516 | 65.3 | 66.0 | 66.0 |
|  | 1.00 | 782 | 33.7 | 34.0 | 100.0 |
|  | Total | 2299 | 99.0 | 100.0 |  |
| Missing | System | 24 | 1.0 |  |  |
| Total |  | 2323 | 100.0 |  |  |

*(continues)*

**repdum Republican dummy**

| | | Frequency | Percent | Valid Percent | Cumulative Percent |
|---|---|---|---|---|---|
| Valid | .00 | 1708 | 73.5 | 74.3 | 74.3 |
| | 1.00 | 591 | 25.4 | 25.7 | 100.0 |
| | Total | 2299 | 99.0 | 100.0 | |
| Missing | System | 24 | 1.0 | | |
| Total | | 2323 | 100.0 | | |

According to the distribution of partyid3, NES2008 has 782 Democrats and 591 Republicans. According to the distribution of demdum, 782 respondents are coded 1 on the Democrat dummy. And according to the distribution of repdum, 591 respondents are coded 1 on the Republican dummy. So our recodes check out. Before proceeding, return to the Variable View and label the values for each of these new variables.

Now we are ready to run a multiple regression analysis of obama_therm, using female, demdum, and repdum as independent variables. Click Analyze → Regression → Linear. The variable obama_therm should still be in the Dependent panel and female in the Independent(s) panel. Fine. Click both of the partisanship dummies, demdum and repdum, into the Independent(s) panel. Click OK. Let's see what we have.

**Model Summary**

| Model | R | R Square | Adjusted R Square | Std. Error of the Estimate |
|---|---|---|---|---|
| 1 | .559[a] | .312 | .311 | 23.609 |

a. Predictors: (Constant), repdum Republican dummy, female Female dummy, demdum Democrat dummy

**Coefficients[a]**

| Model | | Unstandardized Coefficients | | Standardized Coefficients | t | Sig. |
|---|---|---|---|---|---|---|
| | | B | Std. Error | Beta | | |
| 1 | (Constant) | 58.185 | .920 | | 63.246 | .000 |
| | female Female dummy | 1.050 | .999 | .018 | 1.051 | .294 |
| | demdum Democrat dummy | 16.343 | 1.158 | .272 | 14.118 | .000 |
| | repdum Republican dummy | -25.071 | 1.250 | -.385 | -20.050 | .000 |

a. Dependent Variable: obama_therm Feeling Thermometer: Obama

The regression equation is as follows (to enhance readability, we will round to two decimals and shorten the variable names to "Female," "Democrat," and "Republican"):

Obama thermometer rating = 58.19 + 1.05*Female + 16.34*Democrat − 25.07*Republican.

First, get oriented by using the constant, 58.19, as a point of reference. Again, because this value estimates the dependent variable when all the independent variables are 0, 58.19 is the mean Obama rating for males who are independents. Why so? Because all the dummies are switched to 0: Female is 0 (that's the "male" part of the intercept) and both the Democrat dummy and the Republican dummy are 0 (that's the "independent" part of the intercept). The regression coefficient on Female tells us how much to adjust the "male" part of the intercept, controlling for partisanship. The regression coefficients on the partisanship dummies tell us how much to adjust the "independent" part of the intercept, controlling for gender. Thus, compared with independents, Democrats average over 16 degrees higher—and Republicans score more than 25 degrees lower—on the Obama thermometer. The partisan coefficients are large and statistically significant, with huge t-ratios and miniscule P-values. What about the effect of gender? The coefficient on

Female, 1.05, tells us that women, on average, score only about 1 degree higher on the Obama scale, controlling for partisanship. This weak effect fails to trump the null hypothesis (t = 1.051 with P-value = .294). In the earlier regression, using the female dummy alone, we found a gender difference of nearly 3 degrees. That regression, of course, didn't account for the fact that women are more likely than men to be Democrats. After taking party differences into account, the gender difference fades to insignificance.

Overall, however, the model performs fairly well. The adjusted R-square value of .311 tells us that all the independent variables, taken together, account for about 31 percent of the variation in the dependent variable. So the "glass is 31 percent full." A skeptic would point out, of course, that the "glass is still 69 percent empty." Before going on to the next section, you may want to exercise your new skills by creating new dummies and further expanding the model. In any event, before proceeding be sure to save the dataset.

## INTERACTION EFFECTS IN MULTIPLE REGRESSION

Multiple regression is a linear and additive technique. It assumes a linear relationship between the independent variables and the dependent variable. It also assumes that the effect of one independent variable on the dependent variable is the same for all values of the other independent variables in the model. In the regression we just estimated, for example, multiple regression assumed that the effect of being female is the same for all values of partisanship—that Democratic females are about 1 degree warmer toward Barack Obama than are Democratic males and that Republican females and independent females also are 1 degree warmer than are their male counterparts. This assumption works fine for additive relationships. However, if interaction is taking place—if, for example, the gap between male and female ratings is significantly larger among Republicans than among Democrats or independents—then multiple regression will not capture this effect. Before researchers model interaction effects by using multiple regression, they have usually performed preliminary analyses that suggest such effects are occurring in the data.

Consider an interesting theory in American public opinion. According to this perspective, which we will call the "polarization perspective," political disagreements are often more intense among people who are interested in and knowledgeable about public affairs than they are among people who are disengaged or who lack political knowledge.[1] For example, it seems reasonable to hypothesize that individuals who think that "protecting the environment is not as important as maintaining jobs and our standard of living" would give the Republican Party higher ratings than would individuals who believe that "it is important to protect the environment, even if it costs some jobs or otherwise reduces our standard of living." So if we were to compare ratings on the Republican Party feeling thermometer for pro-jobs and pro-environment respondents, we should find a higher mean among the pro-jobs group. According to the polarization perspective, however, this relationship will be weaker for people with low political knowledge than for people with higher political knowledge. Among people with lower political knowledge, the mean difference in Republican ratings may be modest at best, with pro-jobs respondents giving the Republican Party somewhat higher average ratings than do pro-environment respondents. As political knowledge increases, however, this mean difference should increase, reflecting greater polarization between the pro-environment and pro-jobs camps. Thus the strength of the relationship between environmental attitudes and evaluations of the Republican Party will depend on the level of political knowledge.

Dataset NES2008 contains rep_therm, which records respondents' feeling thermometer ratings of the Republican Party. Another variable, enviro_jobs, is a 3-point gauge of respondents' environmental attitudes, coded 0 (pro-environment), 1 (middle), and 2 (pro-jobs). A third variable, know_scale3, measures each respondent's political knowledge by three values: 0 (low knowledge), 1 (medium knowledge), or 2 (high knowledge).[2]

First we will perform a preliminary analysis and find out if the polarization perspective has merit. We want to examine the mean of rep_therm (dependent variable) for each value of enviro_jobs (independent variable), controlling for know_scale3. Figure 9-3 provides a review of how to accomplish this. Let's look at the results of this analysis.

**Report**

rep_therm Feeling Thermometer: Republican Party

| know_scal... | enviro_jobs ... | Mean | N |
|---|---|---|---|
| 0 Low | 0 Environ | 49.54 | 71 |
| | 1 Middle | 50.15 | 64 |
| | 2 Jobs | 54.89 | 50 |
| | Total | 51.21 | 185 |
| 1 Med | 0 Environ | 42.81 | 134 |
| | 1 Middle | 47.86 | 117 |
| | 2 Jobs | 53.13 | 144 |
| | Total | 48.07 | 394 |
| 2 High | 0 Environ | 36.26 | 96 |
| | 1 Middle | 46.26 | 80 |
| | 2 Jobs | 59.56 | 119 |
| | Total | 48.35 | 295 |
| Total | 0 Environ | 42.30 | 300 |
| | 1 Middle | 47.93 | 261 |
| | 2 Jobs | 55.85 | 313 |
| | Total | 48.83 | 873 |

**Figure 9-3**  Obtaining Mean Comparisons with a Control Variable

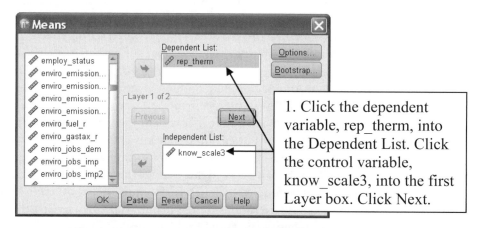

1. Click the dependent variable, rep_therm, into the Dependent List. Click the control variable, know_scale3, into the first Layer box. Click Next.

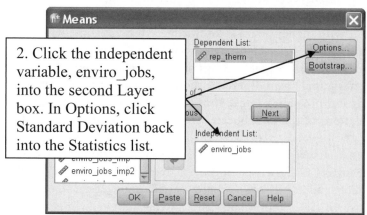

2. Click the independent variable, enviro_jobs, into the second Layer box. In Options, click Standard Deviation back into the Statistics list.

In the way the table is set up, we would assess the effect of environmental attitudes, at each level of political knowledge, by reading down the column within each value of the control. Consider low-knowledge respondents. As we move from pro-environment to pro-jobs, do mean Republican ratings increase? Not by much. The Republicans are rated at 49.54 degrees among pro-environment respondents, 50.15 among those taking the middle position, and 54.89 among pro-jobs respondents. End to end, this is a barely discernible 5-degree increase. The relationship is substantially stronger, however, at medium and high levels of political knowledge. For people with medium knowledge, Republican ratings rise from 42.81 to 53.13, a 10-point increase. And the relationship is stronger still for individuals at the highest knowledge level, for whom the data show about a 23-point difference in ratings of the Republican Party, from 36.26 at the pro-environment end to 59.56 at the pro-jobs end. So it looks like the rep_therm-enviro_jobs relationship does indeed strengthen as political knowledge increases. How would we use regression analysis to estimate the size and statistical significance of these relationships?

We would begin in a familiar way, by estimating the effects of each independent variable, enviro_jobs and know_scale3, on the dependent variable, rep_therm:

$$\text{Republican rating} = a + b1^*\text{enviro\_jobs} + b2^*\text{know\_scale3}.$$

This is a simple additive model. The constant, a, estimates rep_therm for respondents who have a value of 0 on both independent variables—pro-environment respondents who have low political knowledge. The parameter b1 estimates the effect of each unit increase in enviro_jobs, from 0 to 2. The parameter b2 tells us the effect of each unit increase in know_scale3, from 0 to 2.

Think about why the simple additive model does not adequately represent the complex relationships we discovered in the mean comparison analysis. For low-knowledge respondents, for whom know_scale3 is equal to 0, the model reduces to:

$$\text{Republican rating} = a + b1^*\text{enviro\_jobs}.$$

Our previous analysis revealed that, for low-knowledge people, Republican ratings increase from 49.54 (among pro-environment respondents) to 54.89 (among pro-jobs respondents), a weak positive effect. Based on those results, we know that the constant, a, will be equal to about 50 and that the coefficient, b1, will be a small number. Now consider how the simple model would estimate the effect of enviro_jobs for medium-knowledge respondents, for whom know_scale3 is equal to 1:

$$\text{Republican rating} = a + b1^*\text{enviro\_jobs} + b2^*1,$$

which is the same as

$$\text{Republican rating} = (a + b2) + b1^*\text{enviro\_jobs}.$$

The coefficient, b2, adjusts the constant. If b2 is positive, we can say that higher-knowledge people give the Republican Party higher ratings, on average, than do lower-knowledge people. A negative sign on b2 will mean that higher-knowledge people give the Republican Party lower average ratings than do lower-knowledge people. That's fine. But notice that, in the simple additive model, b1 remains unaffected as knowledge goes up. Yet the mean comparison analysis clearly showed that the effect of enviro_jobs on Republican ratings—estimated by b1—strengthens as knowledge increases. We need to add an adjustment to the regression, an adjustment that permits b1 to change as know_scale3 increases.

In multiple regression, this adjustment is accomplished by including an interaction variable as an independent variable. To create an interaction variable, you multiply one independent variable by the other independent variable. Consider how we would create an interaction variable for the problem at hand: enviro_jobs * know_scale3. All respondents who are coded 0 on know_scale3 will, of course, have a value of 0 on the interaction variable. As political knowledge increases, however, so will the magnitude of the interaction variable. Let's include this term in the model just discussed and see what it looks like:

$$\text{Republican rating} = a + b1^*\text{enviro\_jobs} + b2^*\text{know\_scale3} + b3(\text{enviro\_jobs}^*\text{know\_scale3}).$$

The simple additive model did not permit b1 to change as knowledge increased. Consider how the interaction term, enviro_jobs*know_scale3, remedies this situation. Using medium-knowledge respondents (know_scale3 equals 1) to illustrate, the interaction model would be

$$\text{Republican rating} = a + b1*\text{enviro\_jobs} + b2(1) + b3(1*\text{enviro\_jobs}),$$

which is the same as

$$\text{Republican rating} = (a + b2) + (b1+b3)*\text{enviro\_jobs}.$$

As in the simple model, b2 tells us by how much to adjust the constant as knowledge increases. The key difference is the role of b3, which tells us by how much to adjust the *effect* of enviro_jobs as knowledge increases. Because the positive relationship between rep_therm and enviro_jobs gets stronger as know_scale3 increases, we are expecting a positive sign on b3. The t-ratio and P-value on b3 will allow us to test the null hypothesis that the effect of enviro_jobs on rep_therm is the same at all levels of political knowledge.

Let's work through the research problem and get SPSS to estimate the model for us. First we will use Compute to create an interaction variable. Then we will run Regression to estimate the additive effects and the interaction effect.

## USING COMPUTE FOR INTERACTION VARIABLES

Because NES2008 does not have the interaction variable we need for our model, we will use Compute to calculate it. Click Transform → Compute. What to name the variable? In naming interaction variables, it's a good idea to choose a name that implies higher codes on the variables whose product you wish to compute. Higher codes on enviro_jobs denote pro-jobs attitudes. And higher codes on know_scale3 denote greater knowledge. So a name such as "jobs_knowhi" will tell you that respondents who fall into the jobs side of the environmental measure and who have higher levels of knowledge will have higher values on the interaction variable. Type "jobs_knowhi" in the Target Variable box (Figure 9-4). In the Numeric Expression box, type "enviro_jobs*know_scale3." Next, click Type & Label. In the Compute Variable: Type and Label window, type "enviro_jobs*know_scale3" in the Label box. Click Continue, returning to the Compute Variable window.

**Figure 9-4**   Computing an Interaction Variable

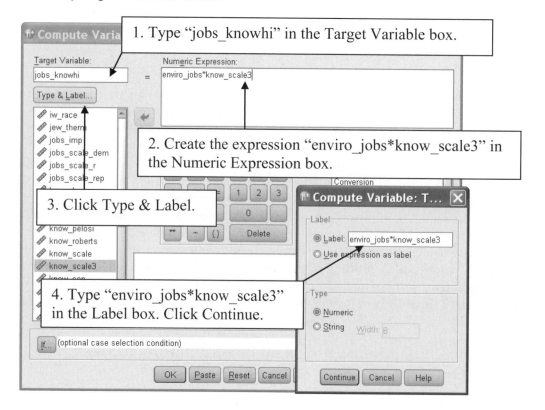

Before clicking OK and computing the variable, there is one more thing to do. Whenever you create a new variable by multiplying one variable by another (as we are doing), and at least one of the variables can take on the value of 0 (as is the case here), you need to make sure that the computation is restricted to cases that have nonmissing values on both variables.[3] In the Compute Variable window, click the button labeled "If (optional case selection condition)," as shown in Figure 9-5. The grayed out Compute Variable: If Cases window appears. Select the radio button next to "Include if case satisfies condition." Doing so wakes up the window. Click in the box and type "not missing(enviro_jobs) & not missing(know_scale3)." Click Continue. Finally set. Click OK. SPSS creates the interaction variable, jobs_knowhi.

Click Analyze → Regression → Linear. Click Reset to clear the panels for our new analysis. Click rep_therm into the Dependent box. Click enviro_jobs, know_scale3, and jobs_knowhi into the Independent(s) box, as shown in Figure 9-6. Click OK.

**Model Summary**

| Model | R | R Square | Adjusted R Square | Std. Error of the Estimate |
|---|---|---|---|---|
| 1 | .263[a] | .069 | .066 | 24.103 |

a. Predictors: (Constant), jobs_knowhi enviro_jobs*know_scale3, know_scale3 RECODE of know_scale, enviro_jobs Environment vs. jobs tradeoff

**Coefficients[a]**

| Model | | Unstandardized Coefficients | | Standardized Coefficients | t | Sig. |
|---|---|---|---|---|---|---|
| | | B | Std. Error | Beta | | |
| 1 | (Constant) | 49.264 | 2.277 | | 21.636 | .000 |
| | enviro_jobs Environment vs. jobs tradeoff | 1.214 | 1.838 | .041 | .660 | .509 |
| | know_scale3 RECODE of know_scale | -6.694 | 1.733 | -.196 | -3.863 | .000 |
| | jobs_knowhi enviro_jobs*know_scale3 | 4.937 | 1.350 | .274 | 3.658 | .000 |

a. Dependent Variable: rep_therm Feeling Thermometer: Republican Party

**Figure 9-5**  Restricting Compute to Nonmissing Cases

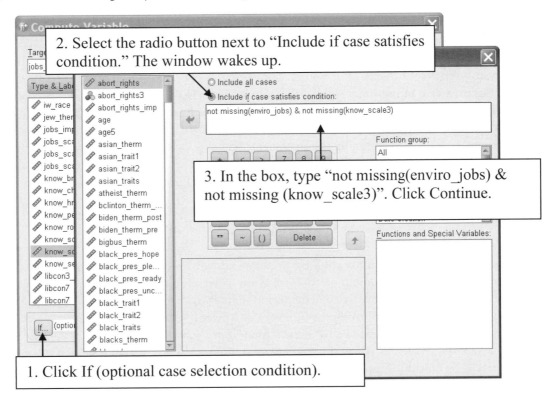

**Figure 9-6**   Requesting Multiple Regression with Interaction Variable

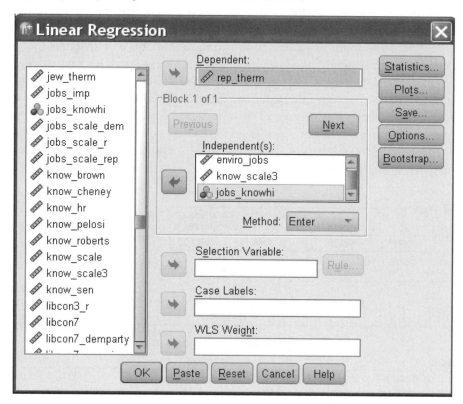

Let's plug the estimates into our model (we'll use the variables' names instead of their lengthy labels):

$$rep\_therm = 49.26 + 1.21*enviro\_jobs - 6.69*know\_scale3 + 4.94*(jobs\_knowhi).$$

Again use the constant, 49.26, to get oriented. This is the estimated mean of rep_therm for respondents who have values of 0 on all the independent variables: pro-environment individuals (code 0 on enviro_jobs) with low political knowledge (a value of 0 on know_scale3). So, this group averages about 49 on the dependent variable. Now consider the small and statistically insignificant coefficient on enviro_jobs, 1.21. This coefficient estimates the effect of enviro_jobs when all the other coefficients are 0—that is, at low levels of knowledge. For respondents taking the middle position on the environmental question, the model returns a predicted Republican rating of 49.26 + 1.21*1 = 50.47. For low knowledge/pro-jobs respondents: 49.26 + 1.21*2 = 51.68. Thus predicted ratings increase slightly, from 49.26 to 51.68. Just as the mean comparison analysis suggested, at low levels of knowledge, environmental opinions have no statistically meaningful effect on ratings of the Republican Party.

Now switch know_scale3 to 1 and light up the coefficients on know_scale3 and the interaction term. The coefficient on know_scale3, –6.69, tells us to subtract 6.69 degrees from the intercept. Performing the arithmetic: 49.26 – 6.69 = 42.57. This number, 42.57, is the estimated mean of rep_therm for pro-environment respondents (enviro_jobs = 0) with medium knowledge (know_scale3 = 1):

$$49.26 + 1.21*0 - 6.69*1 + 4.94*(0*1) = 49.26 - 6.69 = 42.57.$$

A thermometer rating of 42.57 is pretty cold. Do medium-knowledge individuals (know_scale3 = 1) with the strongest pro-jobs opinions (enviro_jobs=2) have a warmer response to the Republicans? Yes, they do, according to the significant positive coefficient on the interaction term, 4.94. As we did with staunch environmentalists, begin by adjusting the intercept downward: 49.26 – 6.69 = 42.57. Now add the (paltry) base effect of enviro_jobs, multiplied by 2, the strongest pro-jobs opinion: 1.21*2 = 2.42. So far, we have: 42.57 + 2.42 = 44.99. Here is where the coefficient on the interaction variable, 4.94, comes into play. This number tells by how much to boost the effect of enviro_jobs as political knowledge goes up. For medium-knowledge, pro-jobs respondents, the boost is equal to 4.94*(2*1) = 9.88. Adding the interaction effect: 42.57 + 2.42 + 9.88 = 54.87. For medium-knowledge respondents, then, the model's estimates range from 42.57 for pro-environment respondents to 54.87 for pro-jobs respondents. These estimates are a fairly accurate fit to the actual data, which range from 42.81 to 53.13. If you were to work out the estimates for high-knowledge individuals (know_scale3 = 2), you would arrive at 35.88 for pro-environment respondents and 58.06 for pro-jobs respondents, which match the data closely (36.26 and 59.56). Thus our regression model has captured the interaction effect quite nicely.

## EXERCISES

1. (Dataset: World. Variables: fhrate08_rev, gdp_cap3.) In one of the exercises in Chapter 5, you used World to investigate the relationship between economic development and democracy. In this exercise you will use multiple regression to reanalyze this relationship, using an interval-level measure of democracy (fhrate08_rev) and a set of dummy variables that you will create from gdp_cap3, a 3-category ordinal measure of per-capita gross domestic product (GDP).

   A. World's gdp_cap3 is coded 1 for countries with "low" GDP per capita, 2 for countries in the "middle" category, and 3 for countries with "high" GDP per capita. Use gdp_cap3 to create two dummy variables, one named gdp_mid and labeled "mid-gdp dummy," and the other named gdp_high and labeled "high-gdp dummy." Follow this recoding scheme:

   | GDP per capita 3 Categories | Old value (gdp_cap3) | New value (gdp_mid) | New value (gdp_high) |
   |---|---|---|---|
   | Low | 1 | 0 | 0 |
   | Middle | 2 | 1 | 0 |
   | High | 3 | 0 | 1 |
   | | Missing | Missing | Missing |

   Check your recoding work by running Frequencies on gdp_cap3, gdp_mid, and gdp_high. In the table that follows, write the number of cases (raw frequencies) in the cells that have question marks:

   | GDP per capita: 3 cats | Frequency | gdp_mid | Frequency | gdp_high | Frequency |
   |---|---|---|---|---|---|
   | Low | 59 | 0 | ? | 0 | ? |
   | Middle | ? | 1 | ? | 1 | ? |
   | High | ? | | | | |
   | Valid total | 177 | | | | |

   B. Imagine running a regression using gdp_mid and gdp_high to estimate a dependent variable: Dependent variable = Constant + b1*gdp_mid + b2*gdp_high. Complete the matching exercise below by drawing a line connecting the desired estimate on the left to the appropriate coefficient (or combination of coefficients) on the right.

   | Your estimate of the . . . | Would be provided by (the) . . . |
   |---|---|
   | mean difference between countries with the lowest gdp and the highest gdp . . . | constant |
   | mean of the dependent variable for the highest gdp countries . . . | b1 |
   | mean of the dependent variable for the lowest gdp countries . . . | constant + b2 |
   | mean difference between the lowest gdp and the middle gdp countries . . . | b2 |

C. World contains fhrate08_rev, a measure of democratic freedoms. The variable fhrate08_rev measures countries on a scale from 0 (least free) to 12 (most free). Run Regression → Linear, using fhrate08_rev as the dependent variable and gdp_mid and gdp_high as independent variables. The regression equation for estimating fhrate08_rev is as follows (fill in the blanks, putting the constant in the first blank):

fhrate08_rev = _____ + _____*gdp_mid + _____*gdp_high.

D. Use the regression coefficients to arrive at estimated mean values of fhrate08_rev for countries at each level of economic development. Write the estimates in the table that follows:

| GDP per capita: 3 cats | Estimated mean on democratic freedoms scale |
|---|---|
| Low | ? |
| Middle | ? |
| High | ? |

E. Examine the t-ratio and P-value on gdp_mid. Do the middle GDP per capita countries score significantly higher on fhrate08_rev than do the lowest GDP per capita countries? Circle one

Yes     No

Briefly explain. _____

_____

_____

F. Examine the t-ratio and P-value on gdp_high. Do the highest GDP per capita countries score significantly higher on fhrate08_rev than do the lowest GDP per capita countries? Circle one

Yes     No

Briefly explain. _____

_____

_____

G. According to Adjusted R-square, GDP per capita accounts for _____ percent of the variation in fhrate08_rev.

2. (Dataset: World. Variables: gini08, hi_gdp, democ_regime, rich_democ.) As a country becomes richer, do more of its citizens benefit economically? Or do economic resources become distributed inequitably across society? The answer may depend on the type of regime in power. Democratic regimes, which need to appeal broadly for votes, may adopt policies that redistribute wealth. Dictatorships, by contrast, are less concerned with popular accountability, and so might hoard economic resources among the ruling elite, creating a less equitable distribution of wealth. This explanation suggests a set of interaction relationships. It suggests that, when we compare poorer democracies with richer democracies, richer democracies will have a more equitable distribution of wealth. However, it also suggests that, when we compare poorer dictatorships with richer dictatorships, richer dictatorships will have a less equitable distribution of wealth. In this exercise you will investigate this set of relationships.

World contains the variable gini08, which measures the extent to which wealth is distributed inequitably in society. The variable can take on any value between 0 (equal distribution of wealth) and 100 (unequal distribution of wealth). So, lower values of gini08 denote less economic inequality and higher values of gini08 denote greater economic inequality. The dependent variable is gini08. World also has a dummy variable, hi_gdp, that classifies each country as "low GDP" (coded 0) or "high GDP" (coded 1). Hi_gdp will serve as the measure of the independent variable, level of wealth. Another dummy, democ_regime, which categorizes each country as a democracy (coded 1 and labeled "yes" on democ_regime) or dictatorship (coded 0 and labeled "no" on democ_regime), is the control variable.

A. Exercise a skill you learned in Chapter 5. To see whether interaction is occurring, obtain a multiple line chart of the gini08-hi_gdp-democ_regime relationships. Click Graphs → Legacy Dialogs → Line → Multiple. Select Other statistic and click gini08 into the Variable box in the Lines Represent panel. (SPSS will offer to graph the mean of gini08, which suits your purpose.) The independent variable, hi_gdp, will go in the Category Axis box, and the control variable, democ_regime, will go in the Define Lines by box. Edit the graph for clarity. For example, you will want the line styles to distinguish clearly between dictatorships and democracies. Print the multiple line chart you created.

B. Examine the chart you just created. It would appear that interaction (circle one)

is          is not

occurring in the data.

Explain your reasoning. _____

_____

_____

_____

C. World contains rich_democ, an interaction variable computed by the expression hi_gdp*democ_regime. Rich_democ takes on the value of 1 for high-GDP democracies and the value of 0 for all other countries.

Run Regression → Linear, using gini08 as the dependent variable and hi_gdp, democ_regime, and rich_democ as independent variables. The regression equation for estimating gini08 is as follows (fill in the blanks, putting the constant in the first blank):

gini08 = _____ + _____*hi_gdp + _____*democ_regime + _____*rich_democ.

D. Use the regression to arrive at estimated mean values of gini08 for low-GDP and high-GDP democracies and dictatorships. Write your estimates in the table that follows:

| Country GDP and regime | Estimated mean of gini08 |
|---|---|
| Low-GDP democracies | ? |
| Low-GDP dictatorships | ? |
| High-GDP democracies | ? |
| High-GDP dictatorships | ? |

E. Suppose someone claimed that, from the standpoint of statistical significance, low-GDP dictatorships have a significantly more equitable distribution of wealth than do low-GDP democracies. This claim is (circle one)

correct.          incorrect.

Explain your reasoning. _____

_____

_____

_____

F. Suppose someone claimed that, as GDP increases, wealth becomes significantly more equitably distributed in democracies and significantly less equitably distributed in dictatorships. This claim is (circle one)

correct.          incorrect.

Explain your reasoning. _____

_____

_____

_____

Before proceeding to the next exercise, be sure to save the World dataset.

3. (Dataset: GSS2008. [If you are running Student Version, use GSS2008B_student for Exercises 3 and 5. Exercise 4 does not perform well for GSS2008B_student.] Variables: abortion_scale, dogmatism, educ2.) The abortion issue is a perennially conflictual debate in U.S. politics. What factors divide people on this issue? Because opposition to abortion is often deeply rooted in religious convictions, you could hypothesize that individuals having dogmatic religious convictions will be more likely to oppose abortion than will those with less dogmatic beliefs. (If this hypothesis seems too commonplace to test, be patient. It gets more interesting below.) GSS2008 contains the variable abortion_scale, which records the percentage of conditions under which respondents believe an abortion should be allowed. Scores can assume values between 0 (abortion should be allowed under none, or 0 percent, of the conditions) and 100 (abortion should be allowed under all, or 100 percent, of the conditions). So higher scale scores denote a stronger pro-abortion stance. GSS2008 also has dogmatism, which measures religious dogmatism along a scale from 0 (very low dogmatism) to 12 (very high dogmatism). If the dogmatism-abortion_scale hypothesis is correct, then higher scores on dogmatism should produce lower scores on abortion_scale.

A. Run Regression to test the hypothesis that people with more dogmatic religious beliefs are more opposed to abortion than are people with less dogmatic beliefs. Fill in the blank and circle the correct term:

For each one-unit increase in dogmatism, there is a _____ (circle one) increase      decrease

in the abortion scale.

B. The analysis (circle one)

supports the hypothesis.          does not support the hypothesis.

Explain your reasoning. _____

_____

_____

_____

A critic, upon examining your results, might reasonably respond, "Did you control for education? It could be that individuals with dogmatic beliefs have lower levels of education than do the less dogmatic. If education is also related to abortion opinions, then you might be confusing the effect of religious dogmatism with the effect of education. After you control for education, dogmatism's effect on abortion opinions will weaken substantially."

GSS2008 contains educ2, a dummy variable scored 0 for respondents with lower education and 1 for respondents with higher education. Run the regression again, using dogmatism and educ2 as independent variables.

C.  Fill in the blanks and circle the correct terms:

Controlling for education, for each one-unit increase in dogmatism, there is a _____ (circle one)

increase          decrease

in the abortion scale.

Controlling for dogmatism, people with higher levels of education score _____ points (circle one)

higher          lower

on the abortion scale than do people with lower levels of education.

D.  The analysis (circle one)

supports          does not support

the critic's argument.

Briefly explain your reasoning. _____

_____

_____

4.  (Dataset: GSS2008. [GSS2008B_student does not perform well for this exercise.] Variables: abortion_scale, dogmatism, educ2.) One of the examples in this chapter discussed the polarization perspective—the idea that political conflict is more pronounced among people who are more knowledgeable about politics than it is among less knowledgeable people. Perhaps the same pattern applies to the relationship between religious dogmatism and abortion opinions. That is, it could be that dogmatism has a strong effect on abortion attitudes among politically knowledgeable people but that this effect is weaker for people who have lower knowledge about politics. We can use the education dummy (educ2) as a surrogate for political knowledge because we can reasonably assume that people with more education will be more politically knowledgeable than will less educated people. In this exercise you will compute an interaction variable. You will then run and interpret a multiple regression that includes the interaction variable you created.

A. Use Transform → Compute to create a new variable, educH_dog, by multiplying dogmatism by educ2. Make sure to restrict the computation of educH_dog to respondents having nonmissing values on dogmatism and educ2. (If you are unsure how to do this, review this chapter's "Using Compute for Interaction Variables" section.) Think about educH_dog, the interaction variable you computed.

A respondent with a lower level of education (coded 0 on educ2) has what value on educH_dog? (circle one)

    A value of 0    A value of 1    A value equal to his or her score on the dogmatism variable

A respondent with a higher level of education (coded 1 on educ2) has what value on educH_dog? (circle one)

    A value of 0    A value of 1    A value equal to his or her score on the dogmatism variable

B. Run a multiple regression, using abortion_scale as the dependent variable and dogmatism, educ2, and educH_dog as the independent variables. According to the Coefficients table, the multiple regression equation for estimating scores on the abortion scale is as follows (fill in the blanks, putting the constant in the first blank):

_____ + _____*dogmatism + _____*educ2 + _____*educH_dog.

C. Use the regression coefficients to estimate scores on the abortion scale for respondents having different values on the dogmatism scale and the education dummy. Calculate the estimates for low- and high-education people with very low dogmatism (a score of 0 on the dogmatism scale), moderate dogmatism (a score of 5), and very high dogmatism (a score of 12). For example, to arrive at the estimate for less educated respondents with very low dogmatism, dogmatism will take on a value of 0 and educ2 will take on a value of 0. Record the abortion_scale estimates next to the question marks in the table that follows:

| | Level of dogmatism | | |
|---|---|---|---|
| Level of education (educ2) | Very low dogmatism (dogmatism = 0) | Moderate dogmatism (dogmatism = 5) | Very high dogmatism (dogmatism = 12) |
| Lower education (educ2 = 0) | ? | ? | ? |
| Higher education (educ2 = 1) | ? | ? | ? |

D. Examine the table of estimates. Think about the polarization perspective. Does the analysis support the idea that religious dogmatism has a larger effect on abortion opinions among people with more education than among people with less education?

                              Yes     No

Explain your reasoning. _____

_____

_____

_____

_____

_____

5. (Dataset: GSS2008. Variables: polviews, race2, homosex2.) If one were trying to predict ideological self-identification on the basis of opinions on social issues, such as homosexuality, one would expect most African Americans to be conservatives. Indeed, blacks are considerably more likely to oppose homosexuality than are whites. According to the GSS2008 data, for example, over 70 percent of blacks say that homosexuality is "always wrong," compared with 50 percent of whites. Yet only about 20 percent of blacks call themselves "conservative," compared with 40 percent of whites. Why? A commonly accepted view is that social issues lack *salience* for African Americans. Issues such as homosexuality may matter for whites—whites who object to homosexuality are more likely to be self-described conservatives than are whites who do not object—but they have no effect for blacks. According to this argument, blacks who think homosexuality is wrong are no more likely to call themselves conservatives than are blacks who do not think homosexuality is wrong. Or so the familiar argument goes. Is this idea correct? More research needs to be done on this question.[4]

You can model salience with interaction variables. Consider the 7-point ideological scale (polviews) as a dependent variable, ranging from "extremely liberal" at 1 to "extremely conservative" at 7. Now bring in two independent variables: a dummy variable for race (race2, with blacks scored 1 and whites scored 0) and a dummy variable gauging opposition to homosexuality (homosex2, scored 1 if the respondent said homosexuality is "always wrong" and 0 for "not always wrong"). Finally, think about (but don't compute yet) an interaction variable, black_wrong, created by multiplying race2 and homosex2. Examine the regression model that follows:

$$polviews = a + b1{*}race2 + b2{*}homosex2 + b3{*}black\_wrong.$$

A. The interaction variable, black_wrong, will take on a value of 1 for (check one)

❑ blacks who think that homosexuality is "always wrong."

❑ blacks who think that homosexuality is "not always wrong."

❑ all respondents.

B. To gauge the effect of homosex2 among whites, you would need to compare values of polviews for "not always wrong" whites and "always wrong" whites.

Which of the following will estimate polviews for "not always wrong" whites? (circle one)

a      a + b1      a + b2

Which of the following will estimate polviews for "always wrong" whites? (circle one)

a      a + b1      a + b2

C. Remember that higher scores on polviews denote stronger conservative self-identifications.

If the salience argument is correct—the idea that heightened opposition to homosexuality leads to stronger conservative ideological leanings among whites but not blacks—then the sign on the coefficient, b2, will be (circle one)

negative.      positive.      zero.

If the salience argument is correct, then the sign on the coefficient, b3, will be (circle one)

negative.      positive.      zero.

D. Use Compute to create black_wrong. Remember to use the If cases option to restrict computation to respondents having nonmissing values on homosex2 and race2.

Run Regression → Linear to obtain estimates for the model. The regression equation for estimating polviews is as follows (fill in the blanks, putting the constant in the first blank):

$$polviews = \underline{\hspace{2cm}} + \underline{\hspace{2cm}} *race2 + \underline{\hspace{2cm}}$$
$$*homosex2 + \underline{\hspace{2cm}} *black\_wrong.$$

E. Which of the variables in the model have statistically significant effects on polviews? (check all that apply)

❑ race2

❑ homosex2

❑ black_wrong

F. Use the model to estimate polviews for "not always wrong" whites and "always wrong" whites. For "not always wrong" whites you obtain _____, and for "always wrong" whites you obtain _____.

G. Use the model to estimate polviews for "not always wrong" blacks and "always wrong" blacks. For "not always wrong" blacks you obtain _____, and for "always wrong" blacks you obtain _____.

H. Consider all the evidence you have adduced. Based on the evidence, the salience idea appears to be (circle one)

correct.     incorrect.

Explain your answer. _____

_____

_____

_____

_____

_____

That concludes the exercises for this chapter. Before exiting SPSS, be sure to save your data and output files.

## NOTES

1. See John R. Zaller's influential work, *The Nature and Origins of Mass Opinion* (New York: Cambridge University Press, 1992).
2. Enviro_jobs is based on the 2008 NES variable V083154, a 7-point scale that gauges opinion from 1 ("Protect environment, even if it costs jobs and standard of living") to 7 ("Jobs and standard of living more important than environment"). Enviro_jobs recodes points 1–3 ("Environment"), point 4 ("Middle"), and points 5–7 ("Jobs"). Know_scale is based on respondents' ability to recognize the office held by Nancy Pelosi (V085120a), Dick Cheney (V085121a), Gordon Brown (V085122a), and John Roberts (V085123a). Each failure was coded 0 and each success coded 1. The 0–4 scale, created by summing responses, was recoded: 0 ("Low"), 1-2 ("Med"), and 3-4 ("High").

3. In calculating a multiplicative product, SPSS will assign a valid code of 0 to any case that has a missing value on one of the variables and a value of 0 on the other variable. For example, a respondent who has a missing value on enviro_jobs and who has a value of 0 on know_scale3 will be assigned a valid, analyzable value on the interaction variable—a value of 0. This respondent should be treated as missing but instead ends up in the analysis. SPSS also returns a valid code of 0 for any expression that divides 0 by a missing value. Avoid the goofy quirk by restricting the Compute procedure to cases having nonmissing values on both variables. See *SPSS 16.0 Command Syntax Reference* (Chicago: SPSS, Inc., 2007), 116.

4. See Quentin Kidd, Herman Diggs, Mehreen Farooq, and Megan Murray, "Black Voters, Black Candidates, and Social Issues: Does Party Identification Matter?" *Social Science Quarterly* 88 (March 2007): 165–176.

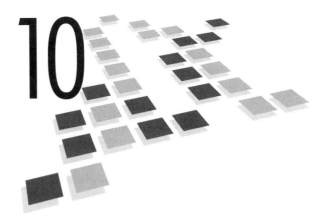

# 10

# Logistic Regression*

| Procedures Covered | |
|---|---|
| | Analyze → Regression → Binary Logistic |
| | Transform → Compute (Predicted probabilities) |
| | Graphs → Legacy Dialogs → Line (Multiple/Summaries of separate variables) |

You now have an array of SPSS skills that enable you to perform the appropriate analysis for just about any situation you will encounter. To analyze the relationship between two categorical variables—variables measured at the nominal or ordinal level—you would enlist Crosstabs. If the dependent variable is an interval-level scale and the independent variable is categorical, then mean comparison analysis would be one way to go. Alternatively, you might create a dummy variable (or variables), specify a linear regression model, and run Regression → Linear to estimate the effects of the categorical variable(s) on the dependent variable. Finally, if both the independent and dependent variables are interval level, then SPSS Correlate or Regression → Linear would be appropriate techniques. There is, however, a common research situation that you are not yet equipped to tackle.

In its most specialized application, logistic regression is designed to analyze the relationship between an interval-level independent variable and a binary dependent variable. A binary variable, as its name suggests, can assume only two values. Binary variables are just like the dummy variables you created and analyzed earlier in this book. Either a case has the attribute or behavior being measured or it does not. Voted/did not vote, married/not married, favor/oppose gay marriage, and South/non-South are examples of binary variables.

Consider a binary dependent variable of keen interest to students of political behavior: whether people voted in an election. This variable, of course, has only two values: Either individuals voted (coded 1 on the binary variable) or they did not vote (coded 0). Now think about an interval-level independent variable often linked to turnout, years of education. As measured by the General Social Survey, this variable ranges from 0 (no formal schooling) to 20 (20 years of education). We would expect a positive relationship between the independent and dependent variables. As years of education increase, the probability of voting should increase as well. So people with fewer years of schooling should have a relatively low probability of voting, and this probability should increase with each additional year of education. Now, we certainly can conceptualize this relationship as positive. However, for statistical and substantive reasons, we cannot assume that it is linear—that is, we cannot assume that a 1-year change in education occasions a consistent increase in the probability of voting. Garden-variety regression, often called ordinary least squares or OLS regression, assumes a linear relationship between the independent and dependent variables.[1] Thus we cannot use Regression → Linear to analyze the relationship between education and the probability of

---

*For this chapter you will need access to a full-version SPSS installation that includes the SPSS Regression Models module. The full version of SPSS Base, by itself, does not permit the user to perform logistic regression. The SPSS Student Version does not contain the Regression Models module.

voting. But as luck and statistics would have it, we can assume a linear relationship between education and the logged odds of voting. Let's put the relationship into logistic regression form and discuss its special properties:

$$\text{Logged odds (voting)} = a + b(\text{years of education}).$$

This logistic regression model is quite OLS-like in appearance. Just as in OLS regression, the constant or intercept, a, estimates the dependent variable (in this case, the logged odds of voting) when the independent variable is equal to 0—that is, for people with no formal education. And the logistic regression coefficient, b, will estimate the change in the logged odds of voting for each 1-year increase in education. What is more, the analysis will produce a standard error for b, permitting us to test the null hypothesis that education has no effect on turnout. Finally, SPSS output for logistic regression will provide R-square-type measures, giving us an idea of the strength of the relationship between education and the likelihood of voting. In all these ways, logistic regression is comfortably akin to linear regression.

However, logistic regression output is more difficult to interpret than are OLS results. In ordinary regression, the coefficients of interest, the constant (a) and the slope (b), are expressed in actual units of the dependent variable. If we were to use OLS to investigate the relationship between years of education (X) and income in dollars (Y), the regression coefficient on education would communicate the dollar-change in income for each 1-year increase in education. With OLS, what you see is what you get. With logistic regression, by contrast, the coefficients of interest are expressed in terms of the logged odds of the dependent variable. The constant (a) will tell us the logged odds of voting when education is 0, and the regression coefficient (b) will estimate the change in the logged odds for each unit change in education. Logged odds, truth be told, have no intuitive appeal. Thus we often must translate logistic regression results into language that makes better intuitive sense.

## USING REGRESSION → BINARY LOGISTIC

Let's run the voting-education analysis and clarify these points. GSS2008 contains voted04, coded 0 for respondents who did not vote in the 2004 election and coded 1 for those who voted.[2] GSS2008 also has educ, which records the number of years of schooling for each respondent. Click Analyze → Regression → Binary Logistic, opening the Logistic Regression window (Figure 10-1). Find voted04 in the variable list and click it into the Dependent box. Click educ into the Covariates box. (In logistic regression, independent

**Figure 10-1** The Logistic Regression Window

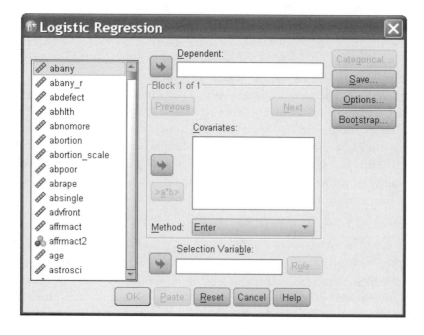

**Figure 10-2** Requesting Logistic Regression with Iteration History

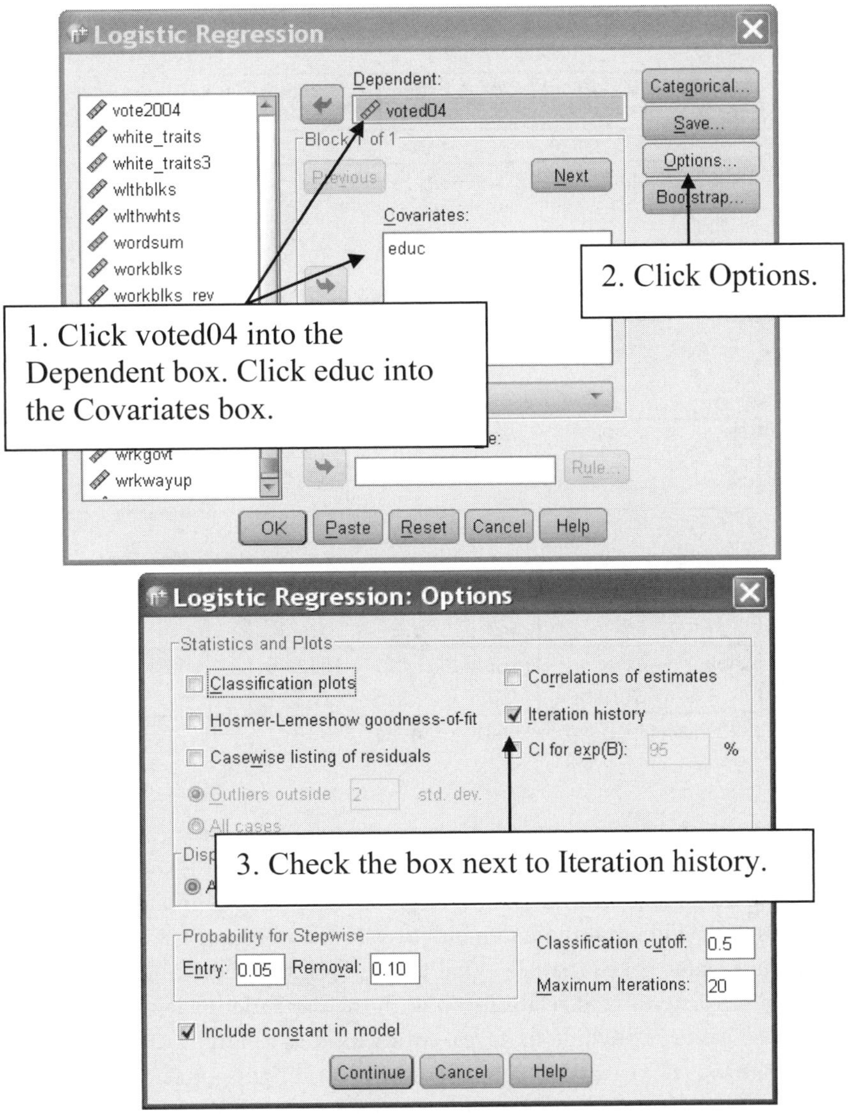

variables are often called covariates.) For this run, we will do one additional thing. In the Logistic Regression window, click Options. The Logistic Regression: Options window opens (Figure 10-2). Click the box next to "Iteration history." This option will produce output that helps to illustrate how logistic regression works. Click Continue, returning to the main Logistic Regression window. Click OK.

In typical fashion, SPSS has given us a wealth of information. Eleven tables now populate the Viewer. Happily, for the essential purposes of this book, you need to be conversant with only three or four of these tables. Scroll to the bottom of the output, to the table labeled "Variables in the Equation." Here you will find the main results of the voted04-educ analysis (Figure 10-3).

Just as in Regression → Linear, the numbers in the column labeled "B" are the estimates for the constant and the regression coefficient. Plug these estimates into our model:

$$\text{Logged odds (voting)} = -2.434 + .246(\text{educ}).$$

What do these coefficients tell us? Again, the constant says that, for people with no education, the estimated logged odds of voting is equal to –2.434. And the logistic regression coefficient on educ says that the logged odds of voting increases by .246 for each 1-year increase in education. So, as expected, as the independent variable increases, the likelihood of voting increases, too. Does education have a statistically significant effect on the likelihood of voting? In OLS regression, SPSS determines statistical significance by calculating

**Figure 10-3** Logistic Regression Output with One Independent Variable: Variables in the Equation and Model Summary

**Model Summary**

| Step | -2 Log likelihood | Cox & Snell R Square | Nagelkerke R Square |
|------|-------------------|----------------------|---------------------|
| 1 | 1982.332[a] | .091 | .129 |

a. Estimation terminated at iteration number 4 because parameter estimates changed by less than .001.

**Classification Table[a]**

| | | | Predicted | | |
|---|---|---|---|---|---|
| | | | voted04 Did R vote? | | Percentage Correct |
| Observed | | | 0 Did not vote | 1 Voted | |
| Step 1 | voted04 Did R vote? | 0 Did not vote | 77 | 458 | 14.3 |
| | | 1 Voted | 33 | 1177 | 97.3 |
| | Overall Percentage | | | | 71.9 |

a. The cut value is .500

**Variables in the Equation**

| | | B | S.E. | Wald | df | Sig. | Exp(B) |
|---|---|---|------|------|-----|------|--------|
| Step 1[a] | educ | .246 | .021 | 138.545 | 1 | .000 | 1.278 |
| | Constant | -2.434 | .276 | 77.721 | 1 | .000 | .088 |

a. Variable(s) entered on step 1: educ.

a t-statistic and an accompanying P-value. In logistic regression, SPSS calculates a Wald statistic (which is based on chi-square) and reports a P-value for Wald. Interpretation of this P-value, displayed in the column labeled "Sig.," is directly analogous to ordinary regression. If the P-value is greater than .05, then do not reject the null hypothesis. Conclude that the independent variable does not have a significant effect on the dependent variable. If the P-value is less than or equal to .05, then reject the null hypothesis and infer that the independent variable has a significant relationship with the dependent variable. In our output, the P-value for educ is .000, so we can conclude that, yes, education has a significant effect on voting turnout.

Now let's return to the logistic regression coefficient, .246, and figure out how to make it more meaningful. Consider the rightmost column of the Variables in the Equation table, the column labeled "Exp(B)." Here SPSS has reported the value 1.278 for the independent variable, educ. Where did this number originate? SPSS obtained this number by raising the natural log base *e* (approximately equal to 2.72) to the power of the logistic regression coefficient, .246. This procedure translates the logged odds regression coefficient into an *odds ratio*. An odds ratio tells us by how much the odds of the dependent variable change for each unit change in the independent variable. An odds ratio of less than 1 says that the odds decrease as the independent variable increases (a negative relationship). An odds ratio equal to 1 says that the odds do not change as the independent variable increases (no relationship). And an odds ratio of greater than 1 says that the odds of the dependent variable increase as the independent variable increases (a positive relationship). An odds ratio of 1.278 means that respondents at a given level of education are 1.28 times more likely to have voted than are respondents at the next lower level of education. So people with, say, 10 years of education are 1.28 times more likely to have voted than are people with 9 years of education, people with 14 years are 1.28 times more likely to have voted than people with 13 years, and so on.

The value of Exp(B) is sometimes used to obtain an even more understandable estimate, the *percentage change in the odds* for each unit change in the independent variable. Mercifully, simple arithmetic accomplishes this task. Subtract 1 from Exp(B) and multiply by 100. In our current example: (1.28 − 1) * 100 = 28. We can now say that each 1-year increment in education increases the odds of voting by 28 percent. As you can see, when the relationship is positive—that is, when the logistic regression coefficient is greater than 0 and the odds ratio is greater than 1—figuring out the percentage change in the odds requires almost no thought. Just subtract 1 from Exp(B) and move the decimal point two places to the right. But be alert for negative relationships, when the odds ratio is less than 1. (In the exercises at the end of this chapter, you will interpret negative relationships.) Suppose, for example, that Exp(B) were equal to .28, communicating a negative relationship between the independent variable and the probability of the dependent variable. The percentage change in the odds would be equal to $(.28 - 1) * 100 = -72.0$, indicating that a one-unit change in the independent variable decreases the odds of the dependent variable by 72 percent.

How strong is the relationship between years of education and the likelihood of voting? Consider the table labeled "Model Summary," also shown in Figure 10-3. OLS researchers are quite fond of R-square, the overall measure of strength that gauges the amount of variation in the dependent variable that is explained by the independent variable(s). For statistical reasons, however, the notion of "explained variation" has no direct analog in logistic regression. Even so, methodologists have proposed various "pseudo R-square" measures that gauge the strength of association between the dependent and independent variables, from 0 (no relationship) to 1 (perfect relationship). SPSS reports two of these: the Cox and Snell R-square and the Nagelkerke R-square. Cox-Snell is the more conservative measure—that is, its maximum achievable value is less than 1. The Nagelkerke measure adjusts for this, and so it generally reports a higher pseudo R-square than does Cox-Snell.[3] These two measures are never wildly different, and they do give the researcher a ballpark feel for the strength of the relationship. With values in the range of .091 to .129, you could conclude that education, though related to voting, by itself provides a less-than-complete explanation of it.

One other measure is reported in the Model Summary table, "–2 Log likelihood," equal to 1982.332. In some ways this is the most important measure of strength produced by logistic regression. By itself, however, the magnitude of –2 log likelihood doesn't mean very much. But scroll up a bit, so that you can view the tables labeled "Iteration History" and "Omnibus Tests of Model Coefficients" together on your screen (Figure 10-4).[4]

**Figure 10-4**  Logistic Regression Output with One Independent Variable: Iteration History and Omnibus Test of Model Coefficients

**Iteration History**[a,b,c,d]

| Iteration | | -2 Log likelihood | Constant | educ |
|---|---|---|---|---|
| Step 1 | 1 | 1993.637 | -1.778 | .188 |
| | 2 | 1982.420 | -2.368 | .240 |
| | 3 | 1982.332 | -2.433 | .246 |
| | 4 | 1982.332 | -2.434 | .246 |

a. Method: Enter
b. Constant is included in the model.
c. Initial -2 Log Likelihood: 2149.589
d. Estimation terminated at iteration number 4 because parameter estimates changed by less than .001.

This final-step "–2 Log likelihood" tells us how well we can predict voting *by using* education as a predictive tool.

This "Initial –2 Log likelihood" tells us how well we can predict voting *without using* education as a predictive tool.

**Omnibus Tests of Model Coefficients**

| | | Chi-square | df | Sig. |
|---|---|---|---|---|
| Step 1 | Step | 167.257 | 1 | .000 |
| | Block | 167.257 | 1 | .000 |
| | Model | 167.257 | 1 | .000 |

This Chi-square statistic is the difference between the initial log likelihood and the final-step log likelihood.

**Model Summary**

| Step | -2 Log likelihood | Cox & Snell R Square | Nagelkerke R Square |
|---|---|---|---|
| 1 | 1982.332[a] | .091 | .129 |

a. Estimation terminated at iteration number 4 because parameter estimates changed by less than .001.

In figuring out the most accurate estimates for the model's coefficients, logistic regression uses a technique called maximum likelihood estimation (MLE). When it begins the analysis, MLE finds out how well it can predict the observed values of the dependent variable without using the independent variable as a predictive tool. So MLE first determined how accurately it could predict whether individuals voted by not knowing how much education they have. The number labeled "Initial –2 Log Likelihood" (equal to 2149.589 and found beneath the Iteration History table) summarizes this "know-nothing" prediction. MLE then brings the independent variable into its calculations, running the analysis again—and again and again—to find the best possible predictive fit between years of education and the likelihood of voting.

According to the Iteration History table, SPSS ran through four iterations, finally deciding that it had maximized its ability to predict voting by using education as a predictive instrument. This final-step log likelihood, 1982.332, is recorded in the Iteration History table and it appears, as well, in the Model Summary table. The amount of explanatory leverage gained by including education as a predictor is determined by subtracting the final-step –2 log likelihood (1982.332) from the initial –2 log likelihood (2149.589). If you performed this calculation by hand, you would end up with 167.257, which appears in the Omnibus Tests of Model Coefficients table next to "Model." This number, which could be more accurately labeled "Change in –2 log likelihood," is a chi-square test statistic. In the "Sig." column of the Omnibus Tests of Model Coefficients table, SPSS has reported a P-value of .000 for this chi-square statistic. Conclusion: Compared with how well we can predict voting without knowing education, including education as a predictor significantly enhances the performance of the model.

By now you are aware of the interpretive challenges presented by logistic regression analysis. In running good old Regression → Linear, you had a mere handful of statistics to report and discuss: the constant, the regression coefficient(s) and accompanying P-value(s), and adjusted R-square. That's about it. With Regression → Binary Logistic, there are more statistics to record and interpret. Below is a tabular summary of the results of the voted04-educ analysis. You could use this tabular format to report the results of any logistic regressions you perform:

| Model estimates and model summary: Logged odds (voting) = a + b(educ) | | | | |
|---|---|---|---|---|
| Model estimates | Coefficient | Significance | Exp(B)* | Percentage change in odds |
| Constant | −2.434 | | | |
| Education | .246 | .000 | 1.278 | 27.8 |
| Model summary | Value | Significance | | |
| Chi-square** | 167.257 | .000 | | |
| Cox-Snell R-square | .091 | | | |
| Nagelkerke R-square | .129 | | | |

\* Alternatively, this column could be labeled "Odds ratio."
\*\* Alternatively, this row could be labeled "Change in –2 log likelihood."

## LOGISTIC REGRESSION WITH MULTIPLE INDEPENDENT VARIABLES

The act of voting might seem simple, but we know that it isn't. Certainly, education is not the only characteristic that shapes the individual's decision whether to vote or to stay home. Indeed, we have just seen that years of schooling, although clearly an important predictor of turnout, returned so-so pseudo-R square statistics, indicating that other factors might also contribute to the explanation. Age, race, marital status, strength of partisanship, political efficacy—all these variables are known predictors of turnout. What is more, education might itself be related to other independent variables of interest, such as age or race. Thus you might reasonably want to know the partial effect of education on turnout, controlling for the effects of these other independent variables. When performing OLS regression, you can enter multiple independent variables into the model and estimate the partial effects of each one on the dependent variable. Logistic regression, like OLS regression, can accommodate multiple predictors of a binary dependent variable. Consider this logistic regression model:

$$\text{Logged odds (voting)} = a + b_1(\text{educ}) + b_2(\text{age}).$$

Again we are in an OLS-like environment. As before, educ measures number of years of formal education. The variable age measures each respondent's age in years, from 18 to 89. From a substantive standpoint, we would again expect $b_1$, the coefficient on educ, to be positive: As education increases, so too should the logged odds of voting. We also know that older people are more likely to vote than are younger people. Thus we should find a positive sign on $b_2$, the coefficient on age. Just as in OLS, $b_1$ will estimate the effect of education on voting, controlling for age, and $b_2$ will estimate the effect of age on the dependent variable, controlling for the effect of education. Finally, the various measures of strength—Cox-Snell, Nagelkerke, –2 log likelihood—will give us an idea of how well both independent variables explain turnout.

Let's see what happens when we add age to our model. Click Analyze → Regression → Binary Logistic. Everything is still in place from our previous run: voted04 is in the Dependent box and educ is in the Covariates box. Good. Now locate age in the variable list and click it into the Covariates box. Click OK to run the analysis. Now scroll to the bottom of the output and view the results displayed in the Variables in the Equation and Model Summary tables.

**Model Summary**

| Step | -2 Log likelihood | Cox & Snell R Square | Nagelkerke R Square |
|---|---|---|---|
| 1 | 1774.028[a] | .191 | .270 |

a. Estimation terminated at iteration number 5 because parameter estimates changed by less than .001.

**Variables in the Equation**

| | | B | S.E. | Wald | df | Sig. | Exp(B) |
|---|---|---|---|---|---|---|---|
| Step 1[a] | educ | .311 | .024 | 171.525 | 1 | .000 | 1.365 |
| | age | .053 | .004 | 168.780 | 1 | .000 | 1.055 |
| | Constant | -5.738 | .405 | 200.508 | 1 | .000 | .003 |

a. Variable(s) entered on step 1: educ, age.

Plug these estimates into our model:

$$\text{Logged odds (voting)} = -5.738 + .311(\text{educ}) + .053(\text{age}).$$

Interpretation of these coefficients follows a straightforward multiple regression protocol. The coefficient on educ, .311, tells us that, controlling for age, each additional year of education increases the logged odds of voting by .311. And notice that, controlling for education, age is positively related to the likelihood of voting. Each 1-year increase in age produces an increase of .053 in the logged odds of voting. According to Wald and accompanying P-values, each independent variable is significantly related to the dependent variable.

Now consider SPSS's helpful translations of the coefficients, from logged odds to odds ratios, which are displayed in the "Exp(B)" column. Interestingly, after controlling for age, the effect of education is somewhat stronger than its uncontrolled effect, which we analyzed earlier. Taking respondents' age differences into account, we find that each additional year of schooling increases the odds ratio by 1.365 and boosts the odds of voting by 36.5 percent: $(1.365 - 1) * 100 = 36.5$.[5] For age, too, the value of Exp(B), 1.055, is greater than 1, again communicating the positive relationship between age and the likelihood of voting. If you were to compare two individuals having the same number of years of education but who differed by 1 year in age, the older person would be 1.055 times more likely to vote than the younger person. Translating 1.055 into a percentage change in the odds: $(1.055 - 1) * 100 = 5.5$. Conclusion: Each additional year in age increases the odds of voting by 5.5 percent.[6]

According to Cox-Snell (.191) and Nagelkerke (.270), adding age to the model increased its explanatory power, at least when compared with the simple analysis using education as the sole predictor. The value of –2 log likelihood, 1774.028, is best viewed through the lens of the chi-square test, which you will find by scrolling up to the tables labeled "Omnibus Tests of Model Coefficients" and "Iteration History."

**Iteration History**[a,b,c,d]

| Iteration | | -2 Log likelihood | Coefficients | | |
|---|---|---|---|---|---|
| | | | Constant | educ | age |
| Step 1 | 1 | 1814.732 | -3.734 | .205 | .036 |
| | 2 | 1775.333 | -5.350 | .290 | .050 |
| | 3 | 1774.031 | -5.722 | .310 | .053 |
| | 4 | 1774.028 | -5.738 | .311 | .053 |
| | 5 | 1774.028 | -5.738 | .311 | .053 |

a. Method: Enter
b. Constant is included in the model.
c. Initial -2 Log Likelihood: 2143.117
d. Estimation terminated at iteration number 5 because parameter estimates changed by less than .001.

**Omnibus Tests of Model Coefficients**

| | | Chi-square | df | Sig. |
|---|---|---|---|---|
| Step 1 | Step | 369.089 | 2 | .000 |
| | Block | 369.089 | 2 | .000 |
| | Model | 369.089 | 2 | .000 |

MLE's initial know-nothing model—estimating the likelihood of voting without using education or age as predictors—returned a –2 log likelihood of 2143.117. After bringing the independent variables into play and running through five iterations, MLE settled on a –2 log likelihood of 1774.028, an improvement of 369.089. This value, which is a chi-square test statistic, is statistically significant ("Sig." = .000). This tells us that, compared with the know-nothing model, both independent variables significantly improve our ability to predict the likelihood of voting.

## WORKING WITH PREDICTED PROBABILITIES: MODELS WITH ONE INDEPENDENT VARIABLE

You now know how to perform basic logistic regression analysis, and you know how to interpret the logistic regression coefficient in terms of an odds ratio and in terms of a percentage change in the odds. No doubt, odds ratios are easier to comprehend than are logged odds. And percentage change in the odds seems more understandable still. Not surprisingly, most researchers prefer to think in terms of probabilities. One might reasonably ask, "What is the effect of a 1-year increase in education on the probability of voting?" Inconveniently, with logistic regression the answer is always, "It depends."

In the first analysis we ran, which examined the voting-education relationship, logistic regression assumed that a linear relationship exists between years of education and the logged odds of voting. This linearity assumption permitted us to arrive at an estimated effect that best fits the data. However, the technique also assumed a nonlinear relationship between years of education and the probability of voting. That is, it assumed that for people who lie near the extremes of the independent variable—respondents with either low or high levels of education—a 1-year increase in education will have a weaker effect on the probability of voting than will a 1-year increase for respondents in the middle range of the independent variable. Because people with low education are unlikely to vote, a 1-year change should not have a huge effect on this likelihood. Ditto for people with many years of schooling. They are already quite likely to vote, and a one-unit increase should not greatly enhance this probability. By contrast, in the middle range of the independent variable, education should have its most potent marginal impact, pushing individuals over the decision threshold from "do not vote" to "vote." So the effect of a 1-year change in education is either weaker or stronger, depending on where respondents "are" on the education variable.

In logistic regression models having more than one independent variable, such as the voted04-educ-age analysis, working with probabilities becomes even more problematic. The technique assumes that the independent variables have additive effects on the logged odds of the dependent variable. Thus for any combination of values of the independent variables, we arrive at an estimated value of the logged odds of the dependent variable by adding up the partial effects of the predictor variables. However, logistic regression also assumes that the independent variables have interactive effects on the probability of the dependent variable. For example, in the case of younger respondents (who have a lower probability of voting), the technique might estimate a large effect of education on the probability of voting. For older respondents (who have a higher probability of voting), logistic regression might find a weaker effect of education on the probability of voting. So the effect of each independent variable on the probability of the dependent variable will depend on the values of the other predictors in the model.

Let's explore these issues one at a time, beginning with the simple model that used education alone to predict voting. Even though we cannot identify a single coefficient that summarizes the effect of education on the probability of voting, we can use SPSS to calculate a predicted probability of voting for respondents at each level of education. How does this work? Recall the logistic regression equation SPSS estimated in our first analysis:

$$\text{Logged odds (voting)} = -2.434 + .246(\text{educ}).$$

SPSS would use this logistic regression model to obtain an estimated logged odds of voting for each respondent. It would plug in each respondent's education level, do the math, and calculate an estimated value of the dependent variable, the logged odds of voting. SPSS would then use the following formula to convert the estimated logged odds of voting into a predicted probability of voting:

$$\text{Probability of voting} = \text{Exp}(\text{Logged odds of voting})/(1 + \text{Exp}(\text{Logged odds of voting})).$$

According to this formula, we retrieve the probability of voting by first raising the natural log base $e$ to the power of the logged odds of voting. We then divide this number by the quantity one plus $e$ raised to the power of the logged odds of voting.[7] Clear as mud.

To get an idea of how SPSS calculates predicted probabilities, let's work through an example. Consider respondents who have a high school education: 12 years of schooling. Using the logistic regression equation obtained in the first guided example, we find the logged odds of voting for this group to be $-2.434 + .246(12) = -2.434 + 2.952 = .518$. What is the predicted probability of voting for people with 12 years of education? It would be $\text{Exp}(.518)/(1 + \text{Exp}(.518)) = 1.679/2.679 \approx .63$. So for respondents with a high school education, the estimated probability of voting is .63. At the user's request, SPSS will follow this procedure to calculate predicted probabilities for individuals at all values of education, and it will save these predicted probabilities as a new variable in the dataset.

Let's run the voted04-educ analysis again and request that SPSS calculate and save the predicted probability of voting for each respondent. Click Analyze → Regression → Binary Logistic. All the variables are still in place from our previous run. This time, however, we want to use only one independent variable, educ. Select age with the mouse and click it back into the variable list, leaving educ in the Covariates box and voted04 in the Dependent box. Now click the Save button in the Logistic Regression window. This opens the Logistic Regression: Save window (Figure 10-5). In the Predicted Values panel, click the Probabilities box. Click Continue, which returns you to the Logistic Regression window. One more thing. We won't be discussing iteration history on this run, so click Options and uncheck the Iteration history box. Click Continue. Ready to go. Click OK.

**Figure 10-5** Requesting Predicted Probabilities

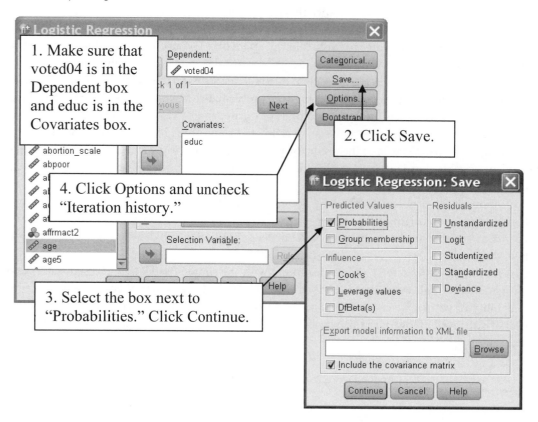

SPSS generates output that is identical (except for the iteration history) to our earlier run. So where are the predicted probabilities that we requested? Because we just ran the analysis, SPSS has taken us to the Viewer. Return to the Variable View of the Data Editor. Scroll to the bottom of the Variable View. As you know, this is where SPSS puts the new variables that you create using Recode or Compute. There you will find a new variable bearing the innocuous name "PRE_1" and the equally bland label "Predicted probability" (Figure 10-6). SPSS has performed just as requested. It ran the analysis, generated the logistic regression output, and discreetly saved a new variable, the predicted probability of voting for each case in the dataset. We will want to have a look at PRE_1. But first we need to give it a more

**Figure 10-6** Predicted Probability Saved as a New Variable in the Data Editor

| | Name | Type | Width | Decimals | Label | Values | N |
|---|---|---|---|---|---|---|---|
| 326 | 77_Student | Numeric | 1 | 0 | filter_$ = 1 (FILTER) | {0, Not Sele... | Non |
| 327 | | | | | Homosexuality always ... | {0, Not alwa... | Non |
| 328 | | | | | Strong relig affil | {0, Not stro... | Non |
| 329 | | | | | Female dummy | {0, Male}... | Non |
| 330 | | | | | Religious dogmatism | None | Non |
| 331 | | | | | | None | Non |
| 332 | black_wrong | Numeric | 8 | 2 | | None | Non |
| 333 | voted04 | Numeric | 8 | 0 | Did R vote? | {0, Did not ... | Non |
| 334 | PRE_1 | Numeric | 11 | 5 | Pred prob:educ-voted04 | None | Non |
| 335 | | | | | | | |

Click in the Label cell and type a more informative label, such as "Pred prob:educ-voted04."

descriptive label. Click in the Label cell and type a more informative variable label, such as "Pred prob: educ-voted04."

In what ways can this new variable, PRE_1, help us to describe changes in the estimated probability of voting as education increases? Remember, SPSS now has a predicted probability of voting for respondents at each value of the education variable, from 0 years to 20 years. So there are two complementary ways to describe the relationship between education and PRE_1, the predicted probability of voting. First, we can perform Analyze → Compare Means → Means, asking SPSS to calculate the mean values of PRE_1 (dependent variable) for each value of educ (independent variable). This would show us by how much the estimated probability of voting increases between groups of respondents having different numbers of years of schooling. Second, we can obtain a line chart of the same information. To obtain a line chart, click Graphs → Legacy Dialogs → Line → Simple and click educ into the "Category Axis" box. Then select "Other statistic," and click PRE_1 into the Line Represents panel. This allows us to visualize the nonlinear relationship between education and the predicted probability of voting.

To you, both of these modes of analysis are old hat, so go ahead and perform the analyses. In the mean comparison results (Figure 10-7), the values of educ appear in ascending order down the left-hand column, and mean predicted probabilities (somewhat distractingly, to 7-decimal-point precision) are reported in the column labeled "Mean." The line chart (Figure 10-8) adds clarity and elegance to the relationship. To get a feel for what is going on, scroll back and forth between the tabular analysis and the graphic output. What happens to the predicted probability of voting as education increases? Notice that, in the lower range of the independent variable, between 0 years and about 6 years, the predicted probabilities are quite low (between

**Figure 10-7** Mean Comparison Table for Predicted Probabilities

PRE_1 Pred prob:educ-voted04

| educ Highest y... | Mean | N |
|---|---|---|
| 0 | .0806444 | 6 |
| 2 | .1253748 | 2 |
| 3 | .1548680 | 5 |
| 4 | .1897936 | 12 |
| 5 | .2304478 | 5 |
| 6 | .2768348 | 34 |
| 7 | .3285724 | 14 |
| 8 | .3848337 | 30 |
| 9 | .4443531 | 48 |
| 10 | .5055133 | 64 |
| 11 | .5665090 | 132 |
| 12 | .6255539 | 558 |
| 13 | .6810839 | 178 |
| 14 | .7319081 | 280 |
| 15 | .7772813 | 90 |
| 16 | .8168964 | 304 |
| 17 | .8508176 | 47 |
| 18 | .8793825 | 121 |
| 19 | .9031009 | 36 |
| 20 | .9225659 | 50 |
| Total | .6831059 | 2017 |

Notice that, between 9 and 10 years of education, the predicted probability switches from less than .5 to greater than .5.

**Figure 10-8**  Line Chart for Predicted Probabilities

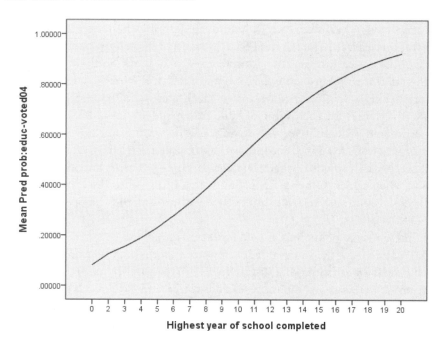

.08 and about .28) and these probabilities increase on the order of .03 to .04 for each increment in education. Now shift your focus to the upper reaches of education and note much the same thing. Beginning at about 14 years of schooling, the estimated probability of voting is at or above about .73—a high likelihood of turning out—and so increments in this range have weaker effects on the probability of voting. In the middle range, from 7 to 13 years, the probabilities increase at a "faster" marginal rate, and within this range the graphic curve shows its steepest slope.

Although most political researchers like to get a handle on predicted probabilities, as we have just done, there is no agreed-upon format for succinctly summarizing logistic regression results in terms of probabilities. One commonly used approach is to report the so-called full effect of the independent variable on the probability of the dependent variable. The full effect is calculated by subtracting the probability associated with the lowest value of the independent variable from the probability associated with the highest value of the independent variable. According to our Compare Means analysis, the predicted probability of voting for people with no formal schooling is about .08, and the predicted probability for those with 20 years of education is .92. The full effect would be .92 − .08 = .84. So, measured across its full range of observed values, education boosts the probability of voting by a healthy .84.

Another way of summarizing a relationship in terms of probabilities is to report the interval of the independent variable that has the biggest impact on the probability of the dependent variable. Suppose that you had to pick the 1-year increment in education that has the largest impact on the probability of voting. What would that increment be? Study the output and think about the phenomenon you are analyzing. Remember that voting is an up or down decision. A person either decides to vote or decides not to vote. But between which two values of education does a "vote" decision become more likely than a "do not vote" decision? You may have noticed that, between 9 years and 10 years, the predicted probabilities increase from .444 to .506, a difference of .062 and the largest marginal increase in the data. And it is between these two values of education that, according to the analysis, the binary decision shifts in favor of voting—from a probability of less than .50 to a probability of greater than .50. So the interval between 9 years and 10 years is the "sweet spot"—the interval with the largest impact on the probability of voting, and the interval in which the predicted probability switches from less than .50 to more than .50.[8]

## WORKING WITH PREDICTED PROBABILITIES: MODELS WITH MULTIPLE INDEPENDENT VARIABLES

Saving predicted probabilities using the Logistic Regression: Save option works fine for simple models with one independent variable. By examining these predicted probabilities, you are able to summarize the full effect of the independent variable on the dependent variable. Furthermore, you can describe the interval of the independent variable having the largest impact on the probability of the dependent variable. Of course, SPSS will also gladly save predicted probabilities for logistic regression models having more than one independent variable. With some specialized exceptions, however, these predicted probabilities are not very useful for summarizing the effect of each independent variable on the probability of the dependent variable, controlling for the other independent variables in the model. As noted earlier, although logistic regression assumes that the independent variables have an additive effect on the logged odds of the dependent variable, the technique also assumes that the independent variables have an interactive effect on the probability of the dependent variable. Thus the effect of, say, education on the probability of voting will be different for younger people than for older people. And the effect of age will vary, depending on the level of education being analyzed. How can we summarize these interaction effects? In dealing with logistic regression models with multiple independent variables, many researchers use the *sample averages method* for presenting and interpreting probabilities. Another approach, the *probability profile method*, permits the researcher to compare probabilities across groups. Let's take a look at both of these methods.

### The Sample Averages Method

In the sample averages approach, the analyst examines the effect of each independent variable while holding the other independent variables constant at their sample means. For example, we would ask and answer these questions: "For people of 'average' age, what effect does education have on the probability of voting?" and "For respondents with 'average' levels of education, what effect does age have on the probability of voting?" In this way, we can get an idea of the effect of each variable on individuals who are "average" on all the other variables being studied. Unfortunately, Regression → Binary Logistic will not calculate the predicted probabilities associated with each value of an independent variable while holding the other variables constant at their sample means.[9] That's the bad news. The good news is that the desired probabilities can be obtained using Transform → Compute, and they are readily analyzed using Compare Means.

Here is the logistic regression model that SPSS estimated for the voted04-educ-age relationships:

$$\text{Logged odds (voting)} = -5.738 + .311(\text{educ}) + .053(\text{age}).$$

We can enlist this equation for two tasks. First, we can plug in the sample mean of age and calculate the full effect of educ on the probability of voting. Second, we can plug in the sample mean of educ and calculate the full effect of age on the probability of voting. Here we will work through the first task only—figuring out the full effect of education on the probability of voting for people of average age. Before proceeding, of course, we need to obtain the sample mean of age. A quick Descriptives run reveals that age has a mean value of 46.22 years. The following equation would permit us to estimate the logged odds of voting at any value of educ, holding age constant at its mean:

$$\text{Logged odds (voting)} = -5.738 + .311(\text{educ}) + .053(46.22).$$

We have already seen that probabilities may be retrieved from logged odds via this conversion:

$$\text{Probability of voting} = \text{Exp(Logged odds of voting)}/(1 + \text{Exp(Logged odds of voting)}).$$

**Figure 10-9** Computing a Predicted Probability for Different Values of an Independent Variable at the Mean Value of Another Independent Variable

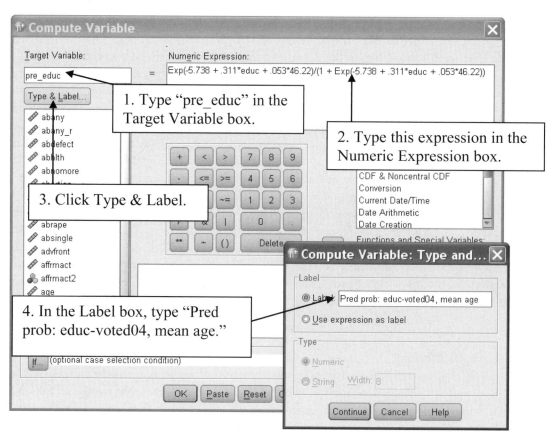

The following equation, therefore, would convert a logged odds of voting into a predicted probability of voting for any plugged-in value of educ, holding age constant at its mean of 46.22:

$$\text{Probability of voting} = \text{Exp}(-5.738 + .311 \text{*educ} + .053 \text{*}46.22) /$$
$$(1 + \text{Exp}(-5.738 + .311 \text{*educ} + .053 \text{*}46.22)).$$

Of course, we could figure all this out using a hand calculator—first finding the predicted probability of voting for individuals having no formal schooling (educ = 0) and then calculating the predicted probability of people with 20 years of education (educ = 20). By subtracting the first probability (when educ = 0) from the second probability (when educ = 20), we would arrive at the full effect of educ at the mean value of age. But let's ask SPSS to do the work for us. Click Transform → Compute. What do we want SPSS to do? We want it to calculate the predicted probability of voting for respondents at each level of education, holding age constant at its sample mean. Because we are holding age constant but allowing educ to vary, we will name this variable "pre_educ." Type "pre_educ" in the Target Variable box. In the Numeric Expression box, type this expression: "Exp(−5.738 + .311*educ + .053*46.22)/(1 + Exp(−5.738 + .311*educ + .053*46.22))," as shown in Figure 10-9.[10] Click Type & Label. In the Label box, give pre_educ the descriptive label "Pred prob: educ-voted04, mean age." Click OK. SPSS computes a new variable, pre_educ, and enters this variable into the dataset.

Finally we have the estimates that permit us to examine the effect of education on the probability of voting for respondents of average age. Click Analyze → Compare Means → Means. Click our newly computed variable, pre_educ, into the Dependent List, and click educ into the Independent List. Click OK.

pre_educ Pred prob: educ-voted04,
mean age

| educ Highest y... | Mean | N |
|---|---|---|
| 0 | .0360 | 6 |
| 2 | .0650 | 2 |
| 3 | .0866 | 5 |
| 4 | .1146 | 12 |
| 5 | .1502 | 5 |
| 6 | .1943 | 34 |
| 7 | .2476 | 14 |
| 8 | .3100 | 30 |
| 9 | .3800 | 48 |
| 10 | .4555 | 64 |
| 11 | .5331 | 132 |
| 12 | .6091 | 558 |
| 13 | .6802 | 178 |
| 14 | .7438 | 280 |
| 15 | .7985 | 90 |
| 16 | .8439 | 304 |
| 17 | .8807 | 47 |
| 18 | .9097 | 121 |
| 19 | .9322 | 36 |
| 20 | .9494 | 50 |
| Total | .6795 | 2017 |

What is the full effect of education? Notice that people with 0 years of education have a probability of voting equal to about .04 (4 chances in 100 that the individual voted), compared with a probability of about .95 for individuals with 20 years of education (95 chances in 100 that the individual voted). Thus holding age constant at its sample mean, we find that the full effect of education is equal to .95 − .04 = .91. Note that the largest marginal effect of education, a boost of nearly .08 in the probability of voting, occurs between 10 and 11 years of schooling.

### The Probability Profile Method

As you can see, the sample averages method is a convenient way of summarizing the effect of an independent variable on the probability of a dependent variable. A more sophisticated approach, the probability profile method, affords a closer and more revealing look at multivariate relationships. In the probability profile method, we compare the effects of one independent variable at two (or more) values of another independent variable. Such comparisons can provide a clearer picture of the interaction relationships between the independent variables and the probability of the dependent variable. In estimating the voted04-educ-age model, for example, logistic regression assumed that the effect of education on the probability of voting is not the same for people of all ages. But just how much does the effect of education vary across age groups? If we were to estimate and compare the effects of education among two groups of respondents—those who are, say, 26 years of age, and those who are 65 years of age— what would the comparison reveal? Once again, here are the logistic regression estimates for the voted04-educ-age model:

$$\text{Logged odds (voting)} = -5.738 + .311(\text{educ}) + .053(\text{age}).$$

In the sample averages method, we used these estimates to compute the probability of voting at each value of education while holding age constant at its sample mean. In the probability profile method, we can ask SPSS to compute two estimated probabilities for the effect of education on voting—one while holding

**Figure 10-10** Computing a Predicted Probability for Different Values of an Independent Variable at a Fixed Value of Another Independent Variable

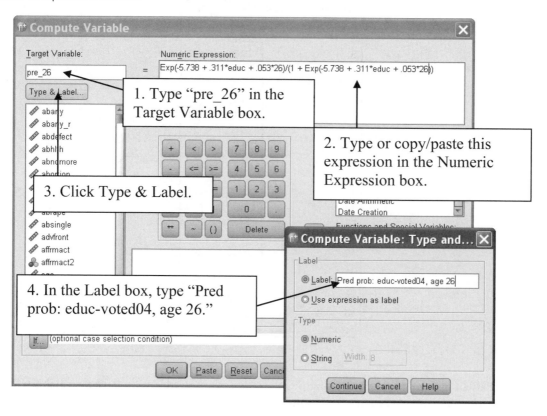

age constant at 26 years, and one while holding age constant at 65 years. Consider the two numeric expressions that follow:

Probability of voting, age 26:
$$\text{pre\_26} = \text{Exp}(-5.738 + .311^*\text{educ} + .053^*26)/(1 + \text{Exp}(-5.738 + .311^*\text{educ} + .053^*26)).$$

Probability of voting, age 65:
$$\text{pre\_65} = \text{Exp}(-5.738 + .311^*\text{educ} + .053^*65)/(1 + \text{Exp}(-5.738 + .311^*\text{educ} + .053^*65)).$$

The first command will estimate the probability of voting at each value of education, while holding age constant at 26 years. And it asks SPSS to save these predicted probabilities in a new variable, pre_26. The second statement estimates the probability of the dependent variable at each value of education, holding age constant at 65 years, and it too will save a new variable, pre_65. To obtain these estimates, we'll need to make two circuits through Transform → Compute, each requiring a fair amount of typing in the Numeric Expression box—although you may have already figured out a useful shortcut that greatly reduces the keyboard drudgery.[11] Figure 10-10 shows the Compute Variable window for obtaining pre_26. To get pre_65, modify the relevant details of the window: Change the Target Variable name to "pre_65," replace "26" with "65" in the Numeric Expression box, and alter Label to read "Pred prob: educ-voted04, age 65." Go ahead and run the Computes.

Run Means → Compare Means, putting educ in the Independent List and both pre_26 and pre_65 in the Dependent List. To enhance readability, in Options click Standard Deviation and Number of Cases back into the Statistics list. SPSS will return a bare-bones mean comparison table:

| educ Highest year of school completed | pre_26 Pred prob: educ-voted04, age 26 | pre_65 Pred prob: educ-voted04, age 65 |
|---|---|---|
| 0 | .0126 | .0917 |
| 2 | .0232 | .1583 |
| 3 | .0315 | .2042 |
| 4 | .0425 | .2594 |
| 5 | .0571 | .3234 |
| 6 | .0763 | .3948 |
| 7 | .1013 | .4710 |
| 8 | .1333 | .5486 |
| 9 | .1735 | .6239 |
| 10 | .2227 | .6936 |
| 11 | .2811 | .7555 |
| 12 | .3480 | .8083 |
| 13 | .4214 | .8520 |
| 14 | .4985 | .8871 |
| 15 | .5757 | .9147 |
| 16 | .6493 | .9360 |
| 17 | .7165 | .9523 |
| 18 | .7752 | .9646 |
| 19 | .8248 | .9738 |
| 20 | .8653 | .9807 |
| Total | .4598 | .8334 |

Consider the dramatically different effects of education for these two age groups. To be sure, the full effect of education is about the same for 26 year olds (.87 − .01 = .86) and 65 year olds (.98 − .09 = .89). But the patterns of marginal effects are not the same at all. For younger people with low levels of education (between 0 and 10 years of schooling), the probability of voting is extraordinarily low, in the range of .01–.22. Indeed, the educational increment with the largest marginal effect—the increment in which the probability of voting switches from less than .50 to more than .50—occurs at a fairly high level of educational attainment, between 14 and 15 years of schooling. Compare the probability profile of younger respondents—sluggish marginal effects in the lower range of education, a high "switchover" threshold—with the probability profile of older respondents. Does education work the same way as we read down the column labeled "pre_65"? Here the probabilities start at a higher level (about .09) and build quite rapidly, in increments of .04 to .05, crossing the .50 threshold at a fairly low level of education, between 7 and 8 years of schooling. So 65 year olds with a junior high school education are almost as likely to vote as are 26 year olds with three years of college.[12]

When you use the probability profile method to explore complex relationships, you will want to complement your analyses with appropriate graphic support. Consider Figure 10-11, a multiple line chart that has spent some "erasing" time in the Chart Editor. This chart instantly communicates the remarkably different ways in which education affects turnout for 26 year olds (dashed line) and 65 year olds (solid line). With one minor exception, the skills you developed earlier in this book will allow you to obtain an unedited version of this graphic. By using the editing skills you already have—and acquiring additional skills through practice and experimentation—you can create the edited version. Click Graphs → Legacy Dialogs → Line (Figure 10-12). In the Line Chart window, select Multiple. Here is something new: Instead of the default setting, "Summaries for groups of cases," select the radio button next to "Summaries of separate variables," as shown in Figure 10-12. Click Define. The Summaries of Separate Variables window is a reasonably famil-iar-looking sight (Figure 10-13). The variable whose effects we want to display, educ, goes in the Category Axis box. The two predicted-probability variables, pre_26 and pre_65, go in the Lines Represent panel. SPSS offers to graph mean values of pre_26 and pre_65, which fits our purpose. Nothing more to it. Clicking OK produces the requested graphic result (Figure 10-14). This is a line chart with a lot of potential. Before moving on to the exercises, see what improvements you can make.[13]

**Figure 10-11**   Edited Multiple Line Chart of Two Logistic Regression Curves

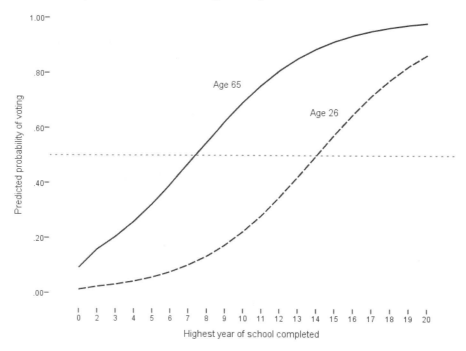

**Figure 10-12**   Changing the Defaults in the Line Chart Windows

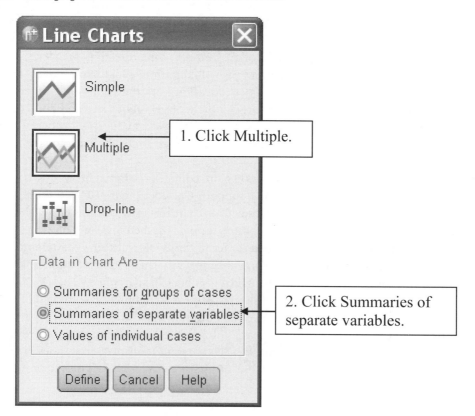

**Figure 10-13**   Define Multiple Line: Summaries of Separate Variables Window (modified)

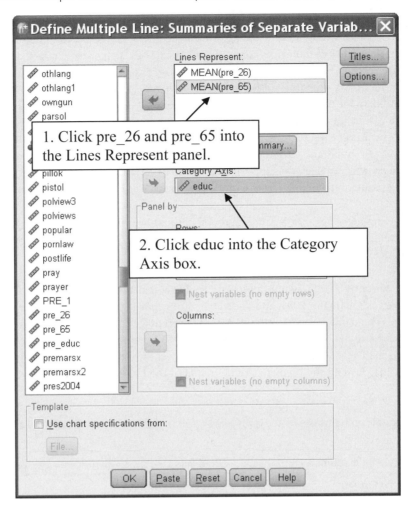

**Figure 10-14**   Multiple Line Chart of Two Logistic Regression Curves

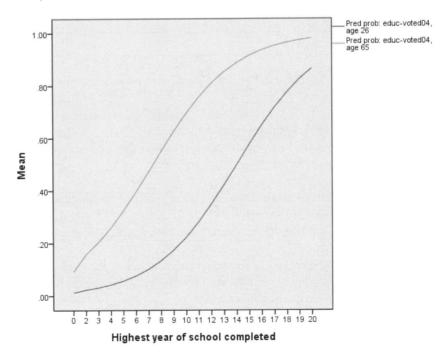

# EXERCISES

1. (Dataset: States. Variables: obama_win08, attend_pct, gunlaw_rank.) As you know, presidential elections in the United States take place within an unusual institutional context. Naturally, candidates seek as many votes as they can get, but the real electoral prizes come in winner-take-all, state-sized chunks: The plurality-vote winner in each state receives all the electoral college votes from that state. Cast in logistic regression terms, each state represents a binary outcome—it goes either Democratic or Republican. What variables shape this outcome? As a candidate for the Democratic nomination, Barack Obama pointed toward several plausible variables. During a campaign appearance, Obama suggested that many voters, frustrated by the disappearance of economic opportunities, "cling to guns or religion or antipathy to people who aren't like them or anti-immigrant sentiment or anti-trade sentiment as a way to explain their frustrations." Obama said that these factors helped explain why his electoral support was weaker in certain geographical areas of the country.[14] These remarks were controversial, to be sure. Controversy aside, was candidate Obama onto something?

Consider the following hypothesis: In a comparison of states, those with higher percentages of residents who frequently attend religious services were less likely to have been won by Obama than were states having lower percentages of residents who frequently attend religious services.

The States dataset contains obama_win08, a binary variable coded 1 if the state's electoral vote went to Democratic candidate Barack Obama in 2008, and coded 0 if the state went to Republican John McCain. This is the dependent variable. The dataset also has attend_pct, the percentage of state residents who frequently attend religious services. Attend_pct is the independent variable.[15] Run Regression → Binary Logistic, clicking obama_win08 into the Dependent box and clicking attend_pct into the Covariates box. In Options, request iteration history. Click Save. In the Predicted Values panel of the Logistic Regression: Save window, select Probabilities.

A. The following table contains seven question marks. Fill in the correct value next to each question mark.

| Model estimates | Coefficient | Significance |
|---|---|---|
| Constant | ? | |
| attend_pct | ? | ? |
| Model summary | Value | Significance |
| Chi-square | ? | ? |
| Cox-Snell R-square | ? | |
| Nagelkerke R-square | ? | |

Before proceeding to part B, review the procedure for converting an odds ratio into a percentage change in the odds for negative relationships.

B. For the variable attend_pct, Exp(B) is equal to (fill in the blank) _____. After converting this number to a percentage change in the odds of an Obama win, you can say that a one-unit increase in attend_pct decreased the odds of an Obama win by _____ percent.

C. Run Analyze → Compare Means → Means, obtaining mean values of the predicted probability of obama_win08 (dependent variable, which SPSS saved as PRE_1) for each value of attend_pct (independent variable). Print the mean comparison table. Use the results to answer parts D, E, and F.

D. For the state having the lowest percentage of frequent church attenders, the predicted probability of an Obama win was equal to _____. For the state having the highest percentage of frequent attenders, the predicted probability of an Obama win was equal to _____. What is the full effect of attend_pct on the probability of obama_win08? The full effect is equal to _____. (*Hint:* Because the probability of an Obama win decreases as attend_pct increases, the full effect will be a negative number.)

E.  A Democratic strategist must decide in which states to concentrate her limited campaign resources. To achieve maximum effect, this strategist should concentrate her campaign on (check one)

❏  a state in which 31 percent of its residents frequently attend religious services.

❏  a state in which 41 percent of its residents frequently attend religious services.

❏  a state in which 50 percent of its residents frequently attend religious services.

F.  Briefly explain your reasoning in part E. _____

_____

_____

_____

Suppose you want to improve the predictive performance of the obama_win08-attend_pct model by adding a second independent variable, the Brady campaign's ranking of states' gun laws (gunlaw_rank). Higher scores on gunlaw_rank denote fewer restrictions on firearms ownership. You reason that states having fewer firearms restrictions were less likely to have been won by Obama than states having more restrictions. You also think that the controlled effects of both variables, attend_pct and gunlaw_rank, will be statistically significant. However, a critic suggests plausibly that the two independent variables, attend_pct and gunlaw_rank, are themselves related—states with many frequent churchgoers will have fewer gun restrictions than will states with fewer frequent churchgoers. This critic argues that only one of the variables, either attend_pct or gunlaw_rank, will be significantly related to obama_win08.

G.  Run the obama_win08 analysis again, using attend_pct and gunlaw_rank as independent variables. Click Save and uncheck Probabilities. In Options, uncheck Iteration history. Refer to the output. For the variable gunlaw_rank, Exp(B) is equal to (fill in the blank) _____. Convert this number to a percentage change in the odds of an Obama win. Controlling for attend_pct, you can say that a one-unit increase in gunlaw_rank decreased the odds of an Obama win by _____ percent. Controlling for gunlaw_rank, a one-unit increase in attend_pct decreased the odds of an Obama win by _____ percent.

H.  Recall the critic's suggestion that only one independent variable, either attend_pct or gunlaw_rank, will be significantly related to obama_win08. Your analysis shows that this critic is (circle one)

correct.       incorrect.

2.  (Dataset: World. Variables: democ_regime, frac_eth, gdp_10_thou.) In Chapter 5 you tested this hypothesis: In a comparison of countries, those having lower levels of ethnic heterogeneity will be more likely to be democracies than will those having higher levels of ethnic heterogeneity. This hypothesis says that, as heterogeneity goes up, the probability of democracy goes down. You then reran the analysis, controlling for a measure of countries' economic development, per-capita gross domestic product (GDP). For this independent variable, the relationship is thought to be positive: As economic development increases, so does the likelihood that a country will be democratic. In the current exercise, you will reexamine this set of relationships, using interval-level independent variables and a more powerful method of analysis, logistic regression.

World contains these three variables: democ_regime, frac_eth, and gdp_10_thou. Democ_regime is coded 1 if the country is a democracy and coded 0 if it is not a democracy. This is the dependent variable. One of the independent variables, frac_eth, can vary between 0 (denoting low heterogeneity) and 1 (high heterogeneity). The other independent variable, gdp_10_thou, measures per-capita GDP in units of $10,000.

A. Run Regression → Binary Logistic, clicking democ_regime into the Dependent box and clicking frac_eth and gdp_10_thou into the Covariates box. In Options, request iteration history. (For this exercise, you will not be saving predicted probabilities.) Click OK to run the analysis. The following table contains eight question marks. Fill in the correct value next to each question mark.

| Model estimates | Coefficient | Significance | Exp(B) |
|---|---|---|---|
| Constant | .842 | | |
| frac_eth | −1.592 | ? | ? |
| gdp_10_thou | .713 | ? | ? |
| Model summary | Value | Significance | |
| Chi-square | ? | ? | |
| Cox-Snell R-square | ? | | |
| Nagelkerke R-square | ? | | |

B. Use each value of Exp(B) to calculate a percentage change in the odds. Controlling for gdp_10_thou, a one-unit change in frac_eth, from low heterogeneity to high heterogeneity, (check one)

❏ increases the odds of democracy by about 20 percent.

❏ decreases the odds of democracy by about 20 percent.

❏ decreases the odds of democracy by about 80 percent.

Controlling for frac_eth, each $10,000 increase in per-capita GDP (check one)

❏ increases the odds of democracy by about 104 percent.

❏ increases the odds of democracy by about 204 percent.

❏ increases the odds of democracy by about 40 percent.

To respond to parts C, D, E, and F, you will need to use Compute to calculate a new variable, which you will name "pre_frac" and label "Pred prob: frac_eth-democ, mean gdp." Pre_frac will estimate the probability of democracy for each value of frac_eth, holding gdp_10_thou constant at its mean. *Useful fact:* The mean of gdp_10_thou is equal to .602. *Helpful hint:* The numeric expression for computing the predicted probability of democracy for each value of frac_eth is "Exp(.842 − 1.592*frac_eth + .713*.602)/(1 + Exp(.842 − 1.592*frac_eth + .713*.602))." After computing pre_frac, run Compare Means → Means, entering pre_frac as the dependent variable and frac_eth as the independent variable. (*Note:* This long tabular output will have 185 rows. Recent releases of SPSS show only the first 100 rows in the Viewer. To see the rest of the Means output, double-click on the table, which reopens it in a separate window. Once in the new window, you can use the navigation buttons to click through the table, or use the View → Navigate Rows drop-down. You do not need to print the means table. However, you will need to refer to it for parts C, D, and E.)

C. As an empirical matter, the most homogeneous country in the World dataset has a value of 0 on frac_eth, and the most heterogeneous country has a value of .93 on frac_eth. The predicted probability of democracy for a highly homogeneous country (frac_eth = 0) with an average level of gdp_10_thou is equal to (fill in the blank) _____. The predicted probability of democracy for a highly heterogeneous country (frac_eth = .93) with an average level of gdp_10_thou is equal to (fill in the blank) _____.

D. As frac_eth increases, from low heterogeneity to high heterogeneity, the predicted probability of democracy (circle one)

decreases.     does not change.     increases.

E. At mean levels of gdp_10_thou, the full effect of frac_eth (from 0 to .93) on the probability of democracy is equal to (fill in the blank) _____.

F. Imagine a country that has average per-capita GDP and also has an average level of ethnic fractionalization: .44 on the frac_eth scale. This country (circle one)

is probably not a democracy.     is probably a democracy.

3. (Dataset: NES2008. Variables: voted, educ_r, age.) In this chapter's guided examples, you used GSS2008 to analyze the relationship between education, age, and turnout in the 2004 election. (The 2008 General Social Survey concluded its last interview in September, so interviewers could not ask about voting behavior in the upcoming election.) Recall that both independent variables, education and age, were significantly related to the probability of voting. Recall also the markedly different effects of education for younger voters and older voters. In this exercise you will use NES2008 to reexamine this set of relationships and see whether the same pattern of effects occurred in 2008. NES2008 contains voted (coded 1 for voters and 0 for nonvoters), educ_r (number of years of schooling), and age (respondent age in years).

A. Run the analysis. Write the correct values next the question marks in the following table:

| Model estimates | Coefficient | Significance | Exp(B) |
|---|---|---|---|
| Constant | ? | | |
| educ_r | ? | ? | ? |
| age | ? | ? | ? |
| Model summary | Value | Significance | |
| Chi-square | ? | ? | |
| Cox-Snell R-square | ? | | |
| Nagelkerke R-square | ? | | |

B. Use the estimates in part A to calculate the effect of education on the probability of voting for voters who are 26 years old and for those who are 65 years old. You will need to process two Compute expressions and create two new variables. Just as you did in the guided examples, name the new variables "pre_26" and "pre_65." Run a means analysis, using educ_r as the independent variable and pre_26 and pre_65 as the dependent variables. In Options, click Standard Deviation and Number of Cases back into the Statistics list. Print the means table.

C. Your GSS2008 analyses showed that, for older voters, the probabilities switched in favor of voting (from less than .5 to more than .5) between 7 years and 8 years of education. For younger voters, the switchover occurred at a substantially higher level of education—between 14 years and 15 years. In what ways are your NES2008 findings similar to the GSS2008 findings? In what ways are the two sets of findings different? Write a short paragraph describing the similarities and differences. _____

_____

_____

_____

_____

_____

D. Use Graphs → Legacy Dialogs → Line to create a multiple line chart showing the relationship between education and predicted probabilities of voting for 26 year olds and for 65 year olds. In the Line Charts window, make sure to request summaries of separate variables. Edit the graph for clarity. Strive for presentation quality. Print the line graph you created.

That concludes the exercises for this chapter. Before exiting SPSS, be sure to save your output file.

## NOTES

1. In arriving at the estimated effect of the independent variable on the dependent variable, linear regression finds the line that minimizes the square of the distance between the observed values of the dependent variable and the predicted values of the dependent variable—predicted, that is, on the basis of the independent variable. The regression line is often referred to as the "least squares" line or "ordinary least squares" line.

2. For all guided examples and exercises in this chapter, the binary dependent variables are naturally coded 0 or 1. To get logistic regression to work, SPSS must have 0/1 binaries. However, here is a bit of SPSS trivia. In running Regression → Binary Logistic, SPSS will check to make sure that the dependent variable has only two values. The values could be 0 and 1, 3 and 5, 2 and 6, or any two (but only two) unique values. If the two values are not 0 and 1, then SPSS will temporarily recode the variable for the immediate purposes of the analysis, encoding one value of the dependent as 0 and the other as 1. SPSS output informs you which natural code it changed to 0 and which it changed to 1. The encoding does not alter your permanent dataset codes.

3. Cox-Snell's maximum achievable value depends on the analysis at hand, but it can never exactly equal 1. For a binary dependent in which the probabilities of 0 and 1 are equal (probability of 0 = .5 and probability of 1 = .5), Cox-Snell reaches a maximum of only .75 for a model in which all cases are predicted perfectly. Nagelkerke's adjustment divides the calculated value of Cox-Snell by the maximum achievable value of Cox-Snell, returning a coefficient that varies between 0 and 1. See D. R. Cox and E. J. Snell, *The Analysis of Binary Data* (London: Chapman and Hall, 1989); N. J. D. Nagelkerke, "A Note on a General Definition of the Coefficient of Determination," *Biometrika* 78 (September 1991): 691–692.

4. When you request iteration history, SPSS will by default produce two histories—one appearing near the beginning of the output beneath the label "Block 0: Beginning Block" and one appearing later beneath the label "Block 1: Method = Enter." In most situations, all of the information you will need can be found under the Block 1 entry. Figure 10-4 portrays the information contained in the Block 1 entry.

5. Why is the controlled effect of education (a 36.5-percent increase in the odds of voting) somewhat greater than its uncontrolled effect (a 27.8-percent increase in the odds of voting)? Running Analyze → Correlate → Bivariate for educ and age provides an important clue: educ and age are negatively correlated (r = −.044). Thus in the earlier analysis, in which we compared respondents having less education with respondents having more education (but in which we did not control for age), we were also comparing older respondents (who, on average, have fewer years of schooling) with younger respondents (who, on average, have more years of schooling). Because younger people are less likely to vote than are older people, the uncontrolled effect of age weakens the zero-order relationship between educ and voted04. In a situation like this, age is said to be a *suppressor variable,* because it suppresses or attenuates the true effect of education on turnout.

6. When using interval-level independent variables with many values, you will often obtain logistic regression coefficients and odds ratios that appear to be quite close to null hypothesis territory (coefficients close to 0 and odds ratios close to 1) but that nonetheless trump the null hypothesis. Remember that logistic regression, like OLS, estimates the marginal effect of a one-unit increment on the logged odds of the dependent variable. In the current example, logistic regression estimated the effect of a 1-year change in age (from, say, an age of 20 years to 21 years) on the logged odds of voting. The researcher may describe the relationship in terms of larger increments. Thus, if a 1-year increase in age (from 20 years to 21 years) increases the odds of voting by an estimated 5.5 percent, then a 10-year increase in age (from 20 years to 30 years) would produce a 55-percent increase in the odds of voting.

7. The expression "Exp(Logged odds of voting)" translates logged odds into odds: Exp(Logged odds of voting) = Odds of voting. You get from an odds to a probability by dividing the odds by the quantity one plus the odds: Probability of voting = Odds of voting/(1 + Odds of voting). Thus the formula for the probability of voting, "Exp(Logged odds of voting)/(1 + Exp(Logged odds of voting))," is equivalent to the formula "Odds of voting/(1 + Odds of voting)."

8. The largest marginal effect of the independent variable on the probability of the dependent variable is sometimes called the *instantaneous effect.* In our example, the instantaneous effect is equal to .062, and this effect occurs between 9 years and 10 years of education. The effect of a one-unit change in the independent variable on the probability of the dependent variable is always greatest for the interval containing a probability equal to .5. The

instantaneous effect, calculated by hand, is equal to b*.5*(1 − .5), in which b is the value of the logistic regression coefficient. For a discussion of the instantaneous effect, see Fred C. Pampel, *Logistic Regression: A Primer,* Sage University Papers Series on Quantitative Applications in the Social Sciences, series no. 07-132 (Thousand Oaks, Calif.: Sage Publications, 2000), 24–26.

9. In calculating predicted probabilities for multivariate logistic regression models, SPSS returns estimated probabilities for subjects having each combination of values on the independent variables. It does not calculate the probabilities associated with each value of a given independent variable while holding the other predictors constant.

10. SPSS has a large repertoire of canned statistical functions. The function Exp(numerical expression) returns the natural log base *e* raised to the power of the numerical expression. This is precisely what we want here, because the estimated probability of voting is equal to Exp(Logged odds of voting)/(1 + Exp(Logged odds of voting)).

11. The Compute Variable window's micro-font Numeric Expression box is a tedious and error-prone place to work, especially if you wish to create long, nested expressions. On the plus side, the box accepts cut-and-paste editing, and it's not choosy about word processing software. In a word processor, you can type an expression just as you want it to appear in the Numeric Expression box, and then copy and paste it into the box. Because the current example's two expressions differ in only one detail—the pre_26 expression requires a "26" and the pre_65 expression requires a "65"—we could type and copy-paste the pre_26 statement, and then compute pre_26. We would then return to Transform → Compute, change the Target Variable to pre_65, replace "26" with "65" in the Numeric Expression box, and compute pre_65.

12. See Raymond E. Wolfinger and Steven J. Rosenstone's classic study of turnout, *Who Votes?* (New Haven: Yale University Press, 1980). Using probit analysis, a technique that is very similar to logistic regression, Wolfinger and Rosenstone explored the effects of a range of demographic characteristics on the likelihood of voting.

13. You may want to experiment with a few choices in the Chart Editor's Options menu: Y Axis Reference Line, Text Box, and Hide Legend. SPSS sometimes produces charts that are too "square." A chart with a width of between 1.3 and 1.5 times its height may be more pleasing to the eye. In any event, you can alter the aspect ratio (the ratio of width to height) in the Chart Size tab of the Properties window. After unchecking the box next to "Maintain aspect ratio," click in the Width box and type a number that is between 1.3 and 1.5 times the value appearing in the Height box.

14. Obama reportedly made this remark at a fundraising event in San Francisco on April 6, 2008. See http://www.cnbcfix.com/obama-cling-guns-religion.html.

15. Attend_pct, which is drawn from the Pew Forum on Religion and Public Life, is one of the four factors Pew uses in its measure of secularism. See http://pewforum.org/docs/?DocID=504. The States dataset contains the Pew secularism scale.

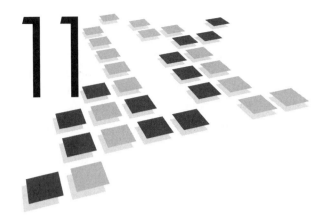

# Doing Your Own Political Analysis

In working through the guided examples in this book, and in performing the exercises, you have developed some solid analytic skills. The datasets you have analyzed here could, of course, become the raw material for your own research. You would not be disappointed, however, if you were to look elsewhere for excellent data. High-quality social science data on a wide variety of phenomena and units of analysis—individuals, census tracts, states, countries—are increasingly accessible via the Internet and might serve as the center-piece for your own research. Your school, for example, may be a member of the Inter-university Consortium for Political and Social Research (ICPSR), the premier organizational clearinghouse for datasets of all kinds.[1] In this chapter we will take a look at various sources of available data and provide practical guidance for inputting it into SPSS. In the process, we also cover situations in which you are confronted with raw, uncoded data—perhaps from an original questionnaire you developed and administered—that you need to code and analyze.

To get you thinking about doing your own research, we begin by laying out the stages of the research process and by offering some manageable ideas for original analysis. We then consider different data sources and procedures for inputting the data into SPSS's Data Editor. Finally, we describe a serviceable format for an organized and presentable research paper.

## FIVE DOABLE IDEAS

Let's begin by describing an ideal research procedure and then discuss some practical considerations and constraints. In an ideal world you would

1. observe an interesting behavior or relationship and frame a research question about it;
2. develop a causal explanation for what you have observed and construct a hypothesis;
3. read and learn from the work of other researchers who have tackled similar questions;
4. collect and analyze the data that will address the hypothesis; and
5. write a research paper or article in which you present and interpret your findings.

In this scenario the phenomenon that you observe in stage 1 drives the whole process. First, think up a question, and then research it and obtain the data that will address it. As a practical matter, the process is almost never this clear cut. Often someone else's idea or assertion may pique your interest. For example, you might read articles or attend lectures on a variety of topics—democratization in developing countries, global environmental issues, ideological change in the Democratic or Republican Party, the effect of election laws on turnout and party competition, and so on—that suggest hypotheses you would like to examine. So you may begin the process at stage 3 and then return to stage 1 and refine your own ideas. Furthermore, the availability of relevant data, considered in stage 4, almost always plays a role in the sorts of questions we address. Suppose, for example, that you want to assess the organizational efforts to mobilize African Americans in your state in the last presidential election. You want precinct-level registration data, and you need to compare these numbers with the figures from previous elections. You would soon become an expert in the bureaucratic hassles and expense involved in dealing with county governments, and you might have to revise your research agenda. Indeed, for professional researchers and academics, data collection in itself can be a full-time job. For students who wish to produce a competent and manageable project, the so-called law of available data can be a source of frustration and discouragement.

A doable project often requires a compromise between stage 1 and stage 4. What interesting question can you ask, given the available data? Fortunately, this compromise need not be as restrictive as it sounds. Consider five possibilities: political knowledge, economic performance and election outcomes, state courts and criminal procedure, electoral turnout in comparative perspective, and Congress.

### Political Knowledge

As you may have learned in other political science courses, scholars continue to debate the levels of knowledge and political awareness among ordinary citizens. Do citizens know the length of a U.S. senator's term of office? Do they know what constitutional protections are guaranteed by the First Amendment? Do people tend to know more about some things—Internet privacy or abortion policy, for example—and less about other things, such as foreign policy or international politics? Political knowledge is a promising variable because the researcher is likely to find some people who know a lot about politics, some who know a fair amount, and others who know very little. One could ask, "What causes this variation?" Imagine constructing a brief questionnaire that asks 8 or 10 multiple-choice questions about basic facts and is tailored to the aspects of political knowledge you find most thought provoking.[2] After including questions that gauge some potentially important independent variables (partisanship, gender, liberalism/conservatism, college major, class standing), you could conduct an exploratory survey among perhaps 50 or 100 of your classmates.

### Economic Performance and Election Outcomes

Here is one of the most widely discussed ideas in political science: The state of the economy before an election has a big effect on the election result. If the economy is strong, the candidate of the incumbent party does well, probably winning. If the economy is performing poorly, the incumbent party's nominee pays the price, probably losing. This idea has a couple of intriguing aspects. For one thing, it works well—but not perfectly. (The 2000 presidential election is a case in point.) Moreover, the economy-election relationship has several researchable layers. Focusing on presidential elections, you can imagine a simple two-category measure of the dependent variable—the incumbent party wins or the incumbent party loses. Now consider several stints in the reference section of the library, collecting information on some potential independent variables for each presidential election year: inflation rates, unemployment, economic growth, and so on. Alternatively, you could look at congressional or state-level elections, or elections in several different countries. Or you could modify and refine the basic idea, as many scholars have done, by adding additional noneconomic variables you believe to be important. Scandal? Foreign policy crises? With some hands-on data collection and guidance from your instructor, you can produce a well-crafted project.

### State Courts and Criminal Procedure

To what extent does a justice's partisanship (or political ideology) affect his or her ruling in a case? This is a perennial question in the annals of judicial research. Again, the 2000 election comes to mind. The U.S. Supreme Court based its pivotal decision on judicial principles, but the Court split along partisan lines. And, given the level of partisan acrimony that accompanies the nominations of would-be federal judges, members of the U.S. Senate behave as if political ideology plays a role in judicial decision making. Original research on judicial proceedings, particularly at the federal level, is among the most difficult to conduct, even for seasoned scholars. But consider state judicial systems. Using an online resource available through most university servers, you could collect information about a large number of, say, criminal cases heard on appeal by the highest court in your state.[3] You could record whether the criminal defendant won or lost, and then determine the party affiliations of the justices. Additionally, you might compare judicial decision making in two states—one in which judges are appointed and one in which they are elected. You could make this comparison at the individual justice level at one point in time. Or you could look at the same set of courts over time, using aggregate units of analysis.

### Electoral Turnout in Comparative Perspective

The record of voter turnout in American presidential elections, while showing encouraging reversals in 2004 and 2008, is relatively low. The situation in other democratic countries is strikingly different. Turnouts in some Western European countries average well above 70 percent. Why? More generally, what causes turnout to vary between countries? Some scholars have focused on legal factors. Unlike the United States, some countries may not require their citizens to register beforehand, or they may penalize citizens for not voting. Other scholars look at institutional differences in electoral systems. Many countries, for example, have systems of proportional representation in which narrowly focused parties with relatively few supporters nonetheless can gain representation in the legislature. Are citizens more likely to be mobilized to vote under such institutional arrangements? Using data sources available on the Internet,[4] you could gather information on a number of democratic countries. You could then look to see if different legal requirements and institutional arrangements are associated with differences in turnout. This area of research might also open the door for some informed speculation on your part. What sort of electoral reforms, if instituted in the United States, might enhance electoral turnout? What other (perhaps unintended) consequences might such reforms have?

### Congress

Political scholars have long taken considerable interest in questions about the U.S. Congress. Some researchers focus on internal dynamics: the role of leadership, the power of party ties versus the pull of constituency. Others pay attention to demographics: Have the numbers of women and minorities who serve in Congress increased in the recent past? Still others look at ideology: Are Republicans, on average, becoming more conservative and Democrats more liberal in their congressional voting? The great thing about Congress is the rich data that are available. The U.S. House and the U.S. Senate are among the most-studied institutions in the world. Several annual or biannual publications chronicle and report a large number of attributes of members of the House and Senate.[5] And the Internet is rife with information about current and past Congresses. Liberal groups, such as Americans for Democratic Action, conservative groups, such as the American Conservative Union, and nonpartisan publications, such as the *National Journal*, regularly rate the voting records of elected officials and post these ratings on their Web sites.[6]

## INPUTTING DATA

Each of these five possibilities represents a practical compromise between posing an interesting question, obtaining available data, and using SPSS to perform the analysis. However, as you will no doubt discover, data sources vary in their "input friendliness"—some data are easy to input into SPSS, and other data require more typing. This section reviews different data sources and input procedures.

### SPSS Formatted Datasets

The least labor-intensive sources provide SPSS datasets that are ready to download and analyze. One such source, the ICPSR's data clearinghouse at the University of Michigan (www.icpsr.umich.edu), was mentioned at the beginning of this chapter. But many other sites exist, often maintained by scholars, academic departments, and private foundations. For example, if you are interested in international relations or comparative politics, visit Pippa Norris's Web site at Harvard's John F. Kennedy School of Government (www.pippanorris.com). For links to a number of SPSS datasets having a particular emphasis on Latino politics, see Prof. Matt A. Barreto's site at the University of Washington (http://faculty.washington.edu/mbarreto/data/index.html). Are you interested in the political beliefs and civic behavior of young people? The Center for Information and Research on Civic Learning and Engagement (CIRCLE) provides excellent data in SPSS format (www.civicyouth.org/research/products/data.htm). More generally, UC Berkeley's SDA Web site—a clearinghouse for the General Social Surveys, the American National Election Studies, and Census Microdata—allows you to download customized datasets and codebooks in a variety of formats, including SPSS (http://sda.berkeley.edu/archive.htm).

### *Microsoft Excel Datasets*

Internet data often are not SPSS-ready but, rather, are available in spreadsheet form, predominately Microsoft Excel format. In these situations, you can copy/paste the data from Excel into the SPSS Data Editor. There are a few caveats to keep in mind, however. To illustrate, consider a typical U.S. Census site, www. census.gov/compendia/statab/, which contains an Excel dataset that records consumer complaints of fraud and identity theft, by state (Figure 11-1). This set provides an instructive example of a common "gotcha" in transferring data from Excel to SPSS.

**Figure 11-1**    Opening an Excel Dataset and Evaluating Its SPSS-Friendliness

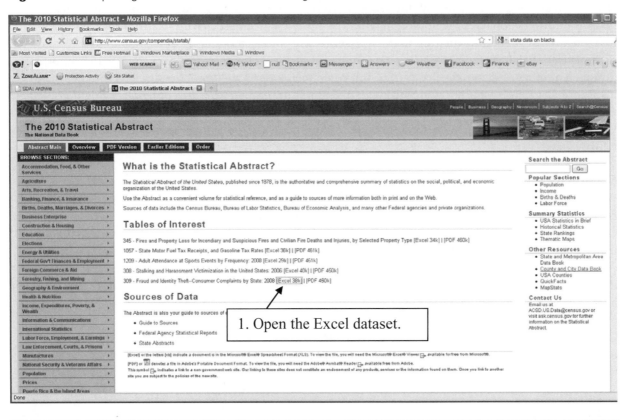

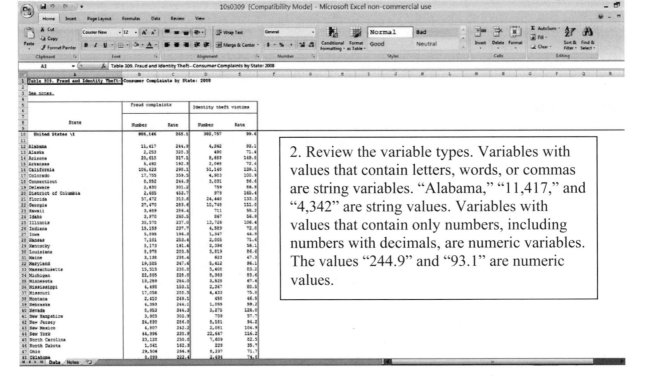

**Figure 11-2**   Removing Commas from Data Values Using Excel

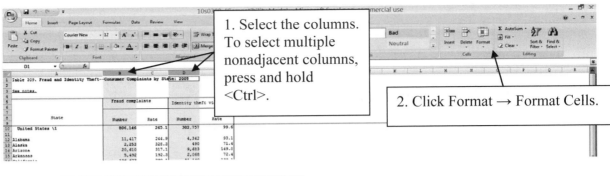

1. Select the columns. To select multiple nonadjacent columns, press and hold <Ctrl>.

2. Click Format → Format Cells.

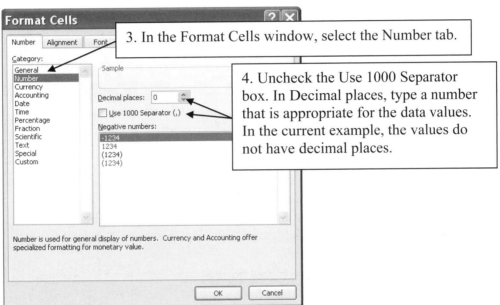

3. In the Format Cells window, select the Number tab.

4. Uncheck the Use 1000 Separator box. In Decimal places, type a number that is appropriate for the data values. In the current example, the values do not have decimal places.

SPSS recognizes two basic forms of data, numeric and string. Numeric data contain only numbers, including numbers with decimals. String data contain letters, words, symbols, or commas. Although some string data are essential—case identifiers, such as state or country names, are obvious examples—SPSS much prefers to analyze numerics, not strings. In the current example, the data in the "Rate" columns of the fraud and identity theft dataset are numeric. But note that the data in the two "Number" columns contain commas and, so, will be recognized by SPSS as strings. SPSS would be happy to let you paste these values into the Data Editor, but it would not be at all happy to analyze them for you. To remove the commas, and thereby convert the data from string to numeric, follow these steps, which are illustrated in Figure 11-2.

1. Select the columns you wish to edit by clicking the column header. To select multiple nonadjacent columns, select the first column, press and hold the Control key, and then select the second column.
2. On Excel's main menu bar, click Format → Format Cells.
3. In the Category pane of the Format Cells window, select the Number tab.
4. In the Number tab, you will always want to uncheck the Use 1000 Separator box. Depending on the exact character of the data, you may want to modify the value in the Decimal places box. If the data contain decimals, then specify the number of decimal places. In the current example, the numbers in the edited columns do not contain decimals, so we would type "0" in the Decimal places box.

To copy/paste the edited Excel data into the SPSS Data Editor, follow the four steps listed at the bottom of page 237. They are illustrated in Figure 11-3.

**Figure 11-3**  Copy/Pasting from Excel into the SPSS Data Editor

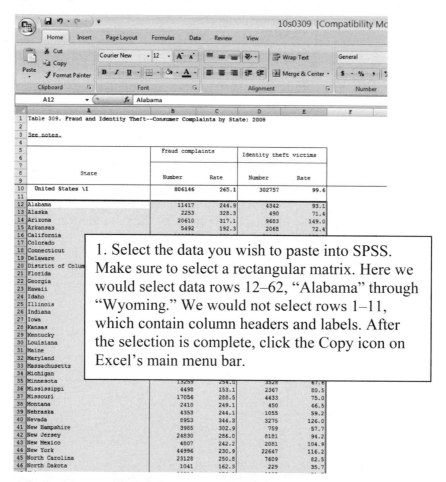

**Figure 11-3**    Copy/Pasting from Excel into the SPSS Data Editor (continued)

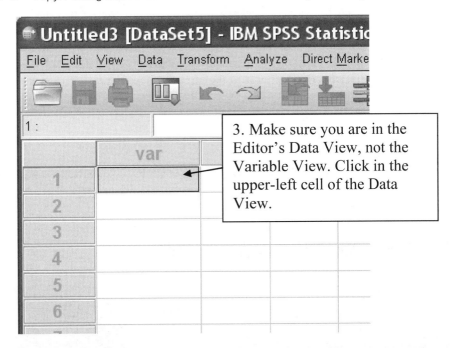

3. Make sure you are in the Editor's Data View, not the Variable View. Click in the upper-left cell of the Data View.

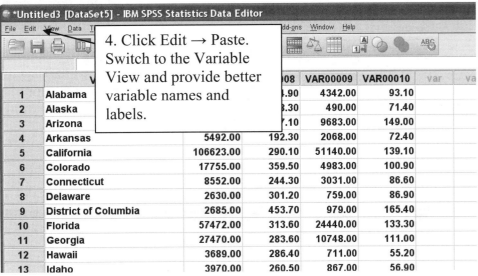

4. Click Edit → Paste. Switch to the Variable View and provide better variable names and labels.

1.  Select the Excel data rows that you want to paste into SPSS. Make sure that the selection is rectangular—that is, ensure that each row contains the same number of columns. Avoid selecting column headers and labels. Also, do not use Excel's row-number markers to make the selection. (This copies the desired columns, plus a number of empty columns.) Rather, select the data by clicking inside the matrix. In the current example, we would begin the selection by clicking on "Alabama," selecting the four columns to the right, and selecting down through the last state, "Wyoming." After completing the selection, click the Copy icon on the Excel menu bar.

2.  On the SPSS menu bar, click File → New → Data.

3.  If SPSS opens in the Variable View, select the Data View tab. Click in the upper-left cell of the Data View.

4.  Click Edit → Paste. SPSS dumps the Excel data into the Data Editor and supplies generic variable names. Switch to the Variable View and provide descriptive names and labels for the variables.

As we have just seen, if the data are available in Excel format, it is a relatively simple matter to copy/paste into SPSS. If the data are in HTML format, it is also a relatively simple matter to copy/paste into Excel. Note, however, that the HTML-to-Excel procedure works with Internet Explorer, but it will not work with Mozilla Firefox or Google Chrome. By way of illustration, consider the *National Journal*'s data on members of the U.S. House (www.nationaljournal.com/njmagazine/nw_20100227_7237.php). As shown in Figure 11-4, we would select the data, copy it to the clipboard, and paste into Excel. Once in

**Figure 11-4** Data in HTML Format

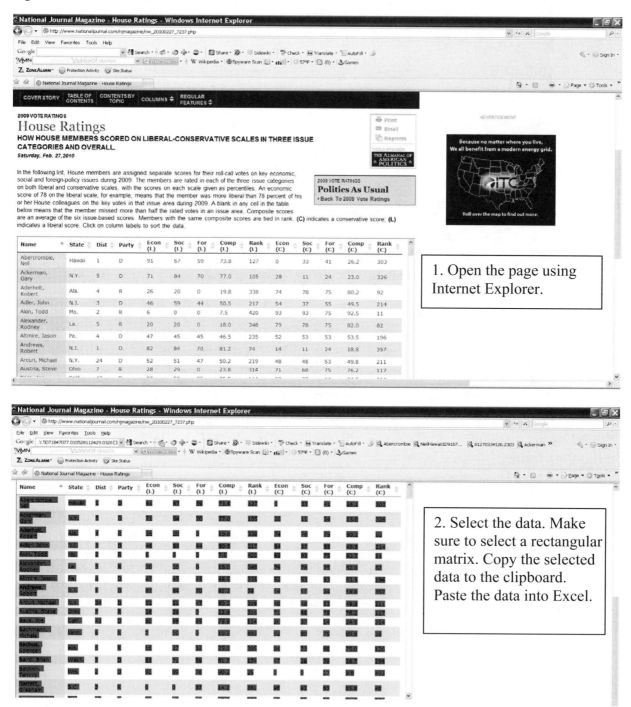

**Figure 11-5**   Editing HTML Data in Excel Before Copy/Pasting into SPSS

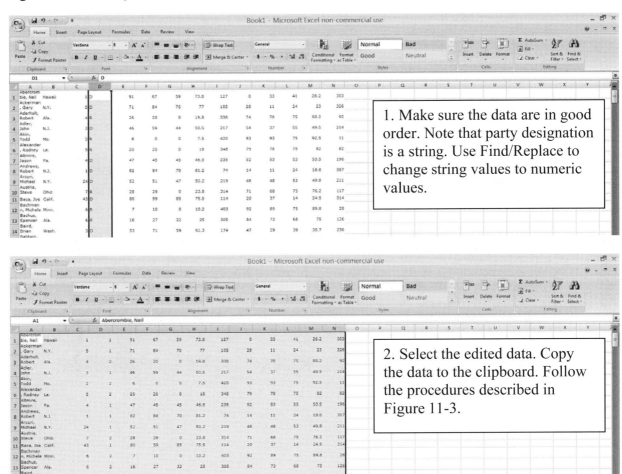

Excel (see Figure 11-5), inspect the data values for unwanted strings. In the current example, each House member's party affiliation is recorded as a "D" or an "R." Numeric codes are preferred—such as code 1 for Democrat and code 2 for Republican. To make the change, select the column, and then use Find & Replace → Replace to replace the Ds with 1s and the Rs with 2s. Once you are satisfied that things look okay, follow the earlier-described procedure for copying Excel spreadsheets into the SPSS Data Editor (see Figure 11-3).

### PDF Format or Hand-coded Data

How do you proceed when your data are in paper-and-pencil form (a stack of completed questionnaires, for example), or are downloadable in print-only format, such as PDF? You could code the data into Excel and copy/paste into SPSS. Better still, you could enter the data directly into the SPSS Data Editor. Figure 11-6 displays a partial Web page from the Americans for Democratic Action's site, which reports ratings of the 2009 Senate. The ADA is a liberal group that uses a series of key votes to rate members of congress, awarding a plus sign (+) for a liberal vote and a minus sign (–) for a conservative vote. In 2009 the ADA monitored 20 key votes, denoted by the unhelpful numbers 1 through 20 on the Web page. Elsewhere on the ADA site we find that the first key vote involved wage discrimination legislation, which the ADA supported. Any senator who voted for the legislation is coded as "+," and those opposed are coded "–." To the right of the page we find ADA's overall liberalism score for 2009, under the heading "LQ," which stands for "Liberal Quotient." ADA arrives at the Liberal Quotient by determining the percentage of votes on which the senator supported ADA's position. Scores can range from 0% (most

**Figure 11-6** Data in PDF Format

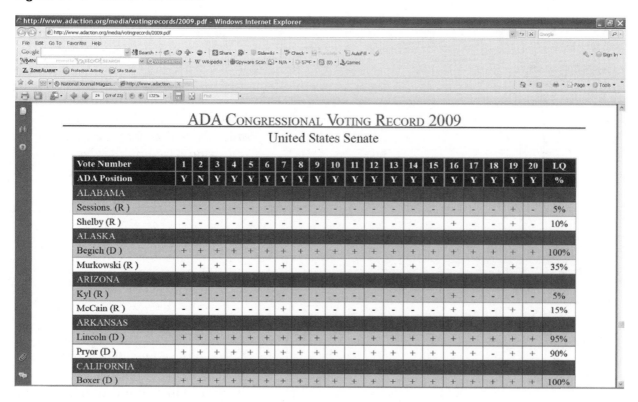

conservative) to 100% (most liberal). How would you take the information from this site and put it into a form that SPSS can analyze?

Here, in raw form, is the first data line, for Senator Sessions of Alabama:

Sessions. (R) – – – – – – – – – – – – – – – – – – + – 5%

Every data value in this line of code is a string. If you were coding these data into SPSS, you would first need to convert all the values, except the senators' names, to numerics. An intuitive scheme might be to code party affiliation as 1 for Democrats and 2 for Republicans, to code minus signs as zeros and plus signs as 1s, and to remove the string symbol "%" from the LQ scores. Here is Senator Sessions's information, converted to numerics:

Sessions 2 0 0 0 0 0 0 0 0 0 0 0 0 0 0 0 0 0 0 1 0 5

Once the coding protocol is established, it is a simple (if tedious) matter to type the data directly into the SPSS Data Editor (see Figure 11-7). Type the code directly into each cell, using the arrow keys to navigate between cells. Use the Variable View to supply more descriptive variable names and labels. And don't forget to click the Save File icon frequently.

## WRITING IT UP

Several of the datasets described thus far would provide great raw material for analysis. After inputting your data, you can let the creative juices flow—describing the variables, performing cross-tabulation and mean comparison analyses, running linear regression and logistic regression models. Rewarding findings are guaranteed. Yet at some point the analysis ends, and the writing must begin. It is at this point, as well, that two contradictory considerations often collide. On one hand, you have an embarrassment of riches. You have worked on your research for several weeks, and you know the topic well—better, perhaps, than does anyone who will read the paper. There may be a large amount of material that you want to include in your paper. On the other hand, you want to get it written, and you do not want to write a book.

**Figure 11-7**    Entering Data Directly into the Data Editor

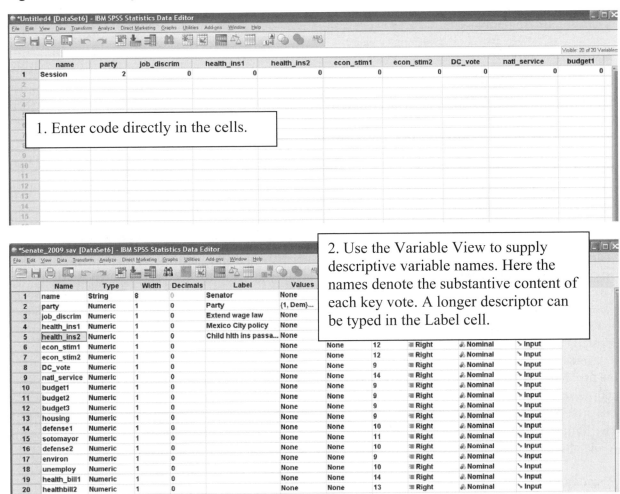

Viewed from an instructor's perspective, the two questions most frequently asked by students are, "How should my paper be organized?" and "How long should it be?" (The questions are not necessarily asked in this order.)

Of course, different projects and instructors might call for different requirements. But here is a rough outline for a well-organized paper of 16–24 double-spaced pages (in 12-point font).

I. The research question (3–4 pages)
   A. Introduction to the problem (1 page)
   B. Theory and process (1–2 pages)
   C. Propositions (1 page)
II. Previous research (2–4 pages)
   A. Descriptive review (1–2 pages)
   B. Critical review (1–2 pages)
III. Data and hypotheses (3–4 pages)
   A. Data and variables (1–2 pages)
   B. Measurement (1 page)
   C. Hypotheses (1 page)
IV. Analysis (5–8 pages, including tables)
   A. Descriptive statistics (1–2 pages)
   B. Bivariate comparisons (2–3 pages)
   C. Controlled comparisons (2–3 pages)

V.   Conclusions and implications (3–4 pages)
   A.   Summary of findings (1 page)
   B.   Implications for theory (1–2 pages)
   C.   New issues or questions (1 page)

### The Research Question

Because of its rhetorical challenges, the opening section of a paper is often the most difficult to write. In this section the writer must both engage the reader's interest and describe the purpose of the research. Here is a heuristic device that may be useful: In the first page of the write-up, place the specific research problem in the context of larger, clearly important issues or questions. For example, suppose your research is centered on the landmark health care legislation passed by Congress in 2010. A narrowly focused topic? Yes. A dry topic? Not at all. The opening page of this paper could frame larger questions about the sometimes conflicting roles of congressional party leadership and constituency interests in shaping the behavior of representatives and senators. Thus your analysis will advance our knowledge by illuminating one facet of a larger, more complex question.

Following the introduction, begin to zero in on the problem at hand. The "theory and process" section describes the logic of the relationships you are studying. Many political phenomena, as you have learned, have competing or alternative explanations. You should describe these alternatives, and the tension between them, in this section. Although a complete description of previous research does not appear in this section, you should give appropriate attribution to the most prominent work. These references tie your work to the scholarly community, and they raise the points you will cover in a more detailed review.

You should round out the introductory section of your paper with a brief statement of purpose or intent. Think about it from the reader's perspective. Thus far you have made the reader aware of the larger context of the analysis, and you have described the process that may explain the relationships of interest. If this process has merit, then it should submit to an empirical test of some kind. What test do you propose? The "Propositions" section serves this role. Here you set the parameters of the research—informing the reader about the units of analysis, the concepts to be measured, and the type of analysis to be performed.

### Previous Research

In this section you provide an intellectual history of the research problem, a description and critique of the published research on which the analysis is based. You first would describe these previous analyses in some detail. What data and variables were used? What were the main findings? Did different researchers arrive at different conclusions? Political scientists who share a research interest often agree on many things. Yet knowledge is nourished through criticism, and in reviewing previous work you will notice key points of disagreement—about how concepts should be measured, what are the best data to use, or which variables need to be controlled. In the latter part of this section of the paper, you would review these points and perhaps contribute to the debate. A practical point: The frequently asked question "How many articles and books should be reviewed?" has no set answer. It depends on the project. However, here is an estimate: A well-grounded yet manageable review should discuss at least four references.

### Data, Hypotheses, and Analysis

Together, the sections "Data and Hypotheses" and "Analysis" form the heart of the project, and they have been the primary concerns of this book. By now you are well versed in how to describe your data and variables and how to frame hypotheses. You also know how to set up a cross-tabulation or mean comparison table, and you can make controlled comparisons and interpret your findings.

In writing these sections, however, bear in mind a few reader-centered considerations. First, assume that the reader might want to replicate your study—collect the data you gathered, define and measure the concepts as you have defined and measured them, manipulate the variables just as you have computed and recoded them, and produce the tables you have reported. By explaining precisely what you did, your write-up should provide a clear guide for such a replication. Second, devote some space to a statistical description of the variables. Often you can add depth and interest to your analysis by briefly presenting the frequency distributions of the variables, particularly the dependent variable. Finally, exercise care in constructing readable tables. You can select, copy, and paste the tables generated by SPSS directly into a word processor, but they always require further editing for readability.

### *Conclusions and Implications*

No section of a research paper can write itself. But the final section comes closest to realizing this optimistic hope. Here you discuss the analysis on three levels. First, you provide a condensed recapitulation. What are the main findings? Are the hypotheses borne out? Were there any unexpected findings? Second, you describe where the results fit in the larger fabric of scholarly research on the topic. In what ways are the findings consistent with the work of previous researchers? Does your analysis lend support to one scholarly perspective as opposed to another? Third, research papers often include obligatory "suggestions for further research." Indeed, you might have encountered some methodological problems that still must be worked out, or you might have unearthed a noteworthy substantive relationship that could bear future scrutiny. You should describe these new issues or questions. Here, too, you are allowed some room to speculate—to venture beyond the edge of the data and engage in a little "What if?" thinking. After all, the truth is still out there.

## NOTES

1. You can browse ICPSR's holdings at www.icpsr.umich.edu.
2. For excellent guidance on the meaning and measurement of political knowledge, see Michael X. Delli Carpini and Scott Keeter, "Measuring Political Knowledge: Putting First Things First," *American Journal of Political Science* 37 (November 1993): 1179–1206.
3. The cases are available from LexisNexis at www.lexis-nexis.com/academic/universe/Academic/. See also the National Center for State Courts, Court Statistics Project, at www.ncsconline.org/d_research/csp/CSP_Main_Page.html.
4. Pippa Norris of Harvard's John F. Kennedy School of Government has compiled excellent comparative and international data, which are available to the general public. These datasets are available in several formats, including SPSS. See www.pippanorris.com.
5. Examples include three books published by CQ Press: *Who's Who in Congress*, offered twice a year through 2001; *CQ's Politics in America*, published every two years; and *Vital Statistics on American Politics*, by Harold W. Stanley and Richard G. Niemi, which also appears every two years. *Vital Statistics* is an excellent single-volume general reference on American politics.
6. See www.adaction.org, www.conservative.org, and www.nationaljournal.com/njonline/.

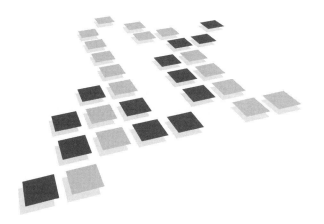

# Appendix

**Table A-1**   Descriptions of Constructed Variables in GSS2008

| Variable name and label | Source variables, GSS 1972–2008 cumulative data file[a] | Notes on construction of variable |
|---|---|---|
| abortion<br><br>How many conditions permit abortion?<br><br><br>abortion_scale<br><br>Percent of conditions permit abortion | abdefect<br>abnomore<br>abhlth<br>abpoor<br>abrape<br>absingle<br>abany | abortion: Number of permissible conditions summed.<br><br>abortion_scale: Percentage of permissible conditions calculated. |
| authoritarianism<br><br>Authoritarianism scale | thnkself<br>obey | thnkself minus obey (thnkself-obey), rescaled 0 (low authoritarianism) to 7 (high authoritarianism).<br><br>Measure based on Karen Stenner, *The Authoritarian Dynamic* (Cambridge: Cambridge University Press, 2005). |
| dogmatism<br><br>Religious dogmatism scale | punsin<br>blkwhite<br>rotapple<br>permoral | Variables coded in "more dogmatic" direction, summed and rescaled, 0 (low dogmatism) to 12 (high dogmatism).<br><br>Based on Ted G. Jelen and Clyde Wilcox, "Religious Dogmatism among White Christians: Causes and Effects," *Review of Religious Research* 33 (September 1991): 32–46. |
| egalit_scale<br><br>Economic egalitarianism | goveqinc<br>incgap<br>inequal3 | Variables recoded in most egalitarian direction and summed, rescaled, 0–12. |
| fem_role<br><br>Female role: home, work | fechld<br>fepresch<br>fefam | Variables coded in "liberal" direction, summed, and rescaled, 0 (home) to 9 (work). |

*(continued)*

**Table A-1** Descriptions of Constructed Variables in GSS2008 *(continued)*

| Variable name and label | Source variables, GSS 1972–2008 cumulative data file[a] | Notes on construction of variable |
|---|---|---|
| int_info_scale<br><br>Level of interest in current issues | intecon<br>inteduc<br>intenvir<br>intfarm<br>intintl<br>intmed<br>intmil<br>intsci<br>intspace<br>inttech | Source variables recoded 0–2, summed. Range: 0 (low interest) to 20 (high interest). |
| racial_liberal3<br><br>Racial liberalism | affrmact2<br>natrace<br>racdif1 | Variables collapsed, "most liberal" responses summed, 0 (least liberal) to 3 (most liberal). |
| relig_tol<br>relig_tol3<br><br>Religious tolerance scale | relgrpeq<br>rspctrel | relig_tol: Source variables recoded in "more tolerant" direction and summed, then rescaled, 0 through 8.<br><br>relig_tol3: relig_tol recoded (0–5 = 0; 6 = 1; 7–8 = 2). |
| science_quiz<br><br>10-item science test | hotcore<br>boyorgrl<br>electron<br>bigbang<br>condrift<br>evolved<br>earthsun<br>radioact<br>lasers<br>viruses | Number of correct responses summed. |
| soc_disconnect<br><br>Social disconnectedness | socrel<br>socomm<br>socfrend | Variables recoded and summed. Final scale ranges from 3 (low disconnectedness) to 21 (high disconnectedness). |
| social_cons3<br><br>Social conservatism | abany<br>grass<br>cappun<br>homosex<br>pornlaw | Variables recoded, "least permissive" responses summed, collapsed: 0 (most permissive) to 3 (least permissive). |
| spend10<br><br>Spending "too little" on how many programs?<br><br>spend3<br><br>Econ liberalism | natchld<br>natcity<br>natcrime<br>natdrug<br>nateduc<br>natenvir<br>natfare<br>natheal<br>natrace<br>natsoc | spend10: Number of "increase spending" responses summed.<br><br>spend3: spend10 recoded (0–1 = 0; 2–3 = 1; 4–10 = 2). |

**Table A-1** *(continued)*

| Variable name and label | Source variables, GSS 1972–2008 cumulative data file[a] | Notes on construction of variable |
|---|---|---|
| suicide_scale<br><br>No. of conditions suicide<br>okay | suicide1<br>suicide2<br>suicide3<br>suicide4 | Number of "yes" responses summed. |
| white_traits<br>black_traits<br>white_traits3<br>black_traits3<br><br>White stereotypes<br>Black stereotypes | intlblks<br>intlwhts<br>workblks<br>workwhts | 7-point source variables recoded (if necessary) in "positive stereotype" direction and rescaled, with codes of less than 4 as "negative", 4 as "neutral" (0), and codes of greater than 4 as "positive." The variables intlblks [intlwhts] and workblks [workwhts] were summed, creating a scale ranging from –6 to +6. |

[a]To find question wording for any GSS2008 variable, go to the University of California Berkeley's Survey Documentation and Analysis (SDA) Web site at http://sda.berkeley.edu/archive.htm and click the link for General Social Survey (GSS) Cumulative Datafile 1972-2008. Click Codebook → Standard Codebook and search the alphabetical listings.

**Table A-2**  Descriptions of Variables in States

| Variable name and label | Description | Source and notes |
|---|---|---|
| abort_rate<br><br>Abortions per 1,000 women (2005) | Number of legal abortions per 1,000 women aged 15–44, 2005 | Alan Guttmacher Institute, www.guttmacher.org<br><br>Rachel K. Jones, Mia R. S. Zolna, Stanley K. Henshaw, and Lawrence B. Finer. 2008. "Abortion in the United States: Incidence and Access to Services, 2005," *Perspectives on Sexual and Reproductive Health* 40(1): 6–16. |
| abortlaw<br><br>Number of restrictions on abortion (2010)<br><br>abortlaw3 | Ten restrictions coded 0 or 1, summed: (i) must be performed by licensed physician; (ii) must be performed in a hospital; (iii) second physician must participate; (iv) prohibited except in cases of life or health endangerment; (v) partial-birth abortion banned; (vi) public funds limited to life endangerment, rape, incest; (vii) providers may refuse to participate; (viii) mandated counseling; (ix) waiting period; (x) parental involvement required for minors | Guttmacher Institute, "State Policies in Brief," July 1, 2010.<br><br>*Note:* All provisions enacted into law, including legally enjoined provisions, are counted as restrictions. |
| attend_pct<br><br>% freq attend relig serv (Pew) | Percentage who frequently attend religious services | Pew Forum on Religion and Public Life<br>http://pewforum.org/How-Religious-Is-Your-State-.aspx |
| battle04<br><br>Battleground state | Coded 1 for 2004 battleground states, coded 0 for non–battleground states | U.S. Department of State<br>http://usinfo.state.gov/dhr/Archive/2004/Jul/12-250886.html |
| blkleg<br><br>Percent of state legislators who are black | Percent of state legislators who are black, 2003 | National Conference of State Legislatures<br>www.ncsl.org/programs/legismgt/about/afrAmer.htm |
| blkpct04<br>blkpct08<br><br>Percent black (2004, 2008) | Percent black or African American, 2004, 2008 | U.S. Census Bureau<br>www.census.gov |
| bush00<br><br>Percent voting for Bush 2000<br><br>gore00<br><br>Percent voting for Gore 2000<br><br>nader00<br><br>Percent voting for Nader 2000 | Number of votes cast for Bush [Gore, Nader] as a percentage of all votes cast | Federal Election Commission<br>www.fec.gov/pubrec/fe2000/2000presge.htm |

**Table A-2** *(continued)*

| Variable name and label | Description | Source and notes |
|---|---|---|
| bush04<br><br>Percent voting for Bush 2004<br><br>kerry04<br><br>Percent voting for Kerry 2004 | Number of votes cast for Bush [Kerry] as a percentage of all votes cast | www.fec.gov/pubrec/fe2004/federalelections2004.shtml |
| carfatal<br><br>Motor vehicle fatalities (per 100,000 pop) | Motor vehicle fatalities per 100,000 population, 2002 | U.S. Census Bureau www.census.gov/compendia/smadb/TableA-13.pdf |
| cig_tax<br>cig_tax_3<br><br>Cigarette tax per pack | Cigarette tax per pack, 2007 | Campaign for Tobacco-Free Kids http://tobaccofreekids.org |
| cigarettes<br><br>Packs bimonthly per adult pop | Bimonthly pack sales per adult pop, 2003 | University of California, San Diego, Social Sciences Data Collection http://ssdc.ucsd.edu/tobacco/sales/ |
| college<br><br>Percent of pop w/college or higher | Percentage of population 25 years and over who have college degrees or higher | U.S. Census Bureau www.census.gov/compendia/smadb/TableA-22.pdf |
| cons_hr06<br>cons_hr09<br><br>Conservatism score, US House delegation (2006, 2009) | Mean American Conservative Union rating of state's delegation to House of Representatives 2006, 2009 | American Conservative Union www.conservative.org<br><br>*Note:* Mean ACU ratings for each state delegation were calculated by author. |
| cook_index<br>cook_index3<br><br>Higher scores more Dem | Cook Partisan Voting Index | www.cookpolitical.com/node/4201 |
| defexpen<br><br>Federal defense expenditures per capita | Federal defense expenditures per capita, 2003 | U.S. Census Bureau www.census.gov/compendia/smadb/TableA-84.pdf |
| dem_hr09<br><br>Percent US House delegation Democratic (2009) | Number of Democratic members of U.S. House of Representatives as a percentage of state's House delegation, 2009 | American Conservative Union www.conservative.org<br><br>*Note:* Percentages were calculated by author. |

*(continued)*

**Table A-2** Descriptions of Variables in States *(continued)*

| Variable name and label | Description | Source and notes |
|---|---|---|
| demnat06<br><br>Percent US House and Senate Democratic (2006) | Number of Democratic House and Senate members as a percentage of total House and Senate delegations, 2006 | American Conservative Union www.conservative.org<br><br>*Note:* Percentages were calculated by author. |
| dempct_m<br><br>Percent mass public Democratic<br><br>reppct_m<br><br>Percent mass public Republican<br><br>indpct_m<br><br>Percent mass public Independent<br><br>libpct_m<br><br>Percent mass public Liberal<br><br>modpct_m<br><br>Percent mass public Moderate<br><br>conpct_m<br><br>Percent mass public Conservative | Pooled CBS News/*New York Times* poll party identification and ideology estimates, 1977–1999 | Gerald C. Wright, Indiana University. Data used with permission. Robert S. Erikson, Gerald C. Wright, and John P. McIver, *Statehouse Democracy* (Cambridge: Cambridge University Press, 1993). |
| demstate06<br>demstate09<br><br>Percent of state legislators who are Democrats (2006, 2009) | Number of Democratic state house and senate members as a percentage of total house and senate members, 2006 postelection/2007 preelection, 2009 | National Conference of State Legislatures www.ncsl.org |
| density<br><br>Population per square mile | Population per square mile, 2005 | U.S. Census Bureau www.census.gov/compendia/smadb/TableA-01.pdf |
| earmarks_pcap<br><br>Earmarks per capita (in dollars) | Earmark spending in dollars per capita | Lauren Cohen, Joshua D. Coval, and Christopher Malloy, "Do Powerful Politicians Cause Corporate Downsizing," Working Paper 15839, National Bureau of Economic Research, Cambridge, Mass., March 2010. |

**Table A-2** *(continued)*

| Variable name and label | Description | Source and notes |
|---|---|---|
| gay_policy<br>gay_policy2<br>gay_policy_con<br><br>Billman's policy scale | Higher scores more conservative | Jeffrey Billman, "Marriage for Some: Understanding Variation of Gay Rights and Gay Marriage Policies in the United States," University of Central Florida, M.A. thesis, 2010. |
| gay_support<br>gay_support3<br><br>Lax-Phillips opinion index | Public support for gay rights (higher scores more supportive) | Jeffrey R. Lax and Justin Phillips, "Gay Rights in the States: Public Opinion and Policy Responsiveness," *American Political Science Review* 103 (August 2009). |
| gb_win00<br>gb_win04<br><br>Did Bush win electoral vote, 2000 (2004)? | 0=Gore win/1 = Bush win<br><br>0=Kerry win/1 = Bush win | Federal Election Commission www.fec.gov |
| gunlaw_rank<br><br>Brady Campaign rank<br><br>gunlaw_scale<br><br>Brady Campaign score | Brady Campaign ranking (higher scores fewer gun restrictions) and Brady scale score (higher scores more gun restrictions) | Brady Campaign 2008 State Scorecard www.bradycampaign.org/xshare/bcam/stategunlaws/scorecard/StateRatings.pdf |
| hispanic04<br>hispanic08<br><br>Percent hispanic (2004, 2008) | Percentage of population Hispanic, 2004, 2008 | U.S. Census Bureau www.census.gov |
| hs_or_more<br><br>Percent of pop w/HS or higher | Percentage of population 25 years and over with high school degree or higher | U.S. Census Bureau www.census.gov/compendia/smadb/TableA-22.pdf |
| mccain08<br>obama08<br>obama_win08<br><br>Percent vote for McCain<br>Percent vote for Obama | Percent total vote for candidate | Dave Leip's Atlas of U. S. Presidential Elections www.uselectionatlas.org |
| over64<br><br>Percent of pop age 65 or older | Percentage of population aged 65 or older, 2004 | U.S. Census Bureau www.census.gov/compendia/smadb/TableA-04.pdf |
| permit<br><br>Percent public "Always allow" abortion | Percentage responding that abortion should "always" be permitted | Pooled American National Election Study, 1980–2004.<br><br>*Note:* Ten states have fewer than 30 cases and were set to missing values on permit. |

*(continued)*

**Table A-2** Descriptions of Variables in States *(continued)*

| Variable name and label | Description | Source and notes |
|---|---|---|
| pop_18_24<br><br>Percent age 18-24 (2004) | Percentage of population aged 18 to 24, 2004 | U.S. Census Bureau<br>www.census.gov/compendia/smadb/<br>TableA-04.pdf |
| pot_policy<br><br>Legalization policies | State legalization policy: none, pending, enacted | *USA Today*, "Slowly, Limits on Pot Are Fading," March 9, 2010. |
| prcapinc<br><br>Percapita income | Income per capita, 2004 | U.S. Census Bureau<br>www.census.gov/compendia/smadb/<br>TableA-43.pdf |
| relig_import<br><br>Percent religion "A great deal of guidance" | Percentage responding that religion provides "a great deal of guidance" in life | Pooled National Election Study, 1984–2004.<br><br>Percentage coded 3 on question #vcf0847.<br><br>*Note:* Twelve states have fewer than 30 cases and were set to missing values on relig_import. |
| religiosity<br><br>Relig-observance belief scale (Pew)<br><br>secularism<br><br>Secularism scale (Pew) | Pew index of religiosity (secularism); religiosity scale is reversal of secularism scale | Pew Forum on Religion and Public Life<br>http://pewforum.org/How-Religious-Is-Your-State-.aspx |
| seniority_sen2<br><br>Influential mem in US Sen? | Coded 1 if state does not have influential member of U.S. Senate; coded 0, otherwise | Lauren Cohen, Joshua D. Coval, and Christopher Malloy, "Do Powerful Politicians Cause Corporate Downsizing," Working Paper 15839, National Bureau of Economic Research, Cambridge, Mass., March 2010. |
| trnout00<br>trnout04 | Number of voters as a percentage of voting-age population, 2000, 2004 | Committee for the Study of the American Electorate<br>http://american.edu/ia/cdem/csae/<br>pdfs/csae041104.pdf |
| unemploy<br><br>Unemployment rate | Percent unemployed of the civilian labor force, 2004 | U.S. Census Bureau<br>www.census.gov/compendia/smadb/<br>TableA-29.pdf |
| union04<br>union07 | Percent workers who are union members (2004, 2007) | U.S. Census Bureau<br>www.census.gov |

**Table A-2**    *(continued)*

| Variable name and label | Description | Source and notes |
|---|---|---|
| urban<br><br>Percent urban population | Percentage of population in urban areas, 2000 | U.S. Census Bureau<br>www.census.gov/compendia/smadb/<br>TableA-02.pdf |
| vep00_turnout<br>vep04_turnout<br>vep08_turnout | Percent turnout of voting eligible population 2000, 2004, 2008 | Michael McDonald, United States Election Project<br>http://elections.gmu.edu/index.html |
| womleg_2007<br>womleg_2010<br><br>Percent of state legislators who are women (2007, 2010) | Percent of state legislators who are women (2007, 2010) | Center for American Women and Politics<br>www.cawp.rutgers.edu/fast_facts/<br>levels_of_office/documents/stleg.pdf |

**Table A-3** Descriptions of Variables in World*

| Variable name | Variable label | Source and notes |
|---|---|---|
| colony | Colony of what country? | CIA World Factbook |
| confidence | Confidence in institutions scale | Based on Global Indicators variable v52, Confidence in state institutions early-mid 1990s (WVS) |
| decentralization<br><br>decentralization4 | Decentralization scale | Based on the sum of Global Indicators variables Political, Fiscal, and Admin<br><br>See A. Schneider, (2003) "Who Gets What From Whom? The Impact of Decentralisation on Tax Capacity and Pro-Poor Policy," Institute of Development Studies Working Paper, No. 179 (Brighton: University of Sussex). |
| dem_other<br><br>dem_other5 | Percentage of other democracies in region | Calculated by author from region and democ_regime |
| district_size3 | Average # of members per district | Based on Global Indicators variable dm, mean district magnitude |
| durable | Number of years since the last regime transition | Polity IV 2000 |
| effectiveness | Government effectiveness scale | Kaufmann 2002 |
| enpp_3 | Effective number of parliamentary parties, June 2000 (Agora) (Banded) | Based on Global Indicators variable enpp (2000) |
| enpp3_democ | Effective number of parliamentary parties | Variable enpp3, restricted to democracies |
| eu | EU member state (yes/no) | |
| fhrate04_rev<br>fhrate08_rev | Freedom House rating of democracy (reversed)<br>fhrate04_rev (0-7)<br>fhrate08_rev (0-12) | Freedom House |
| frac_eth<br><br>frac_eth3 | Ethnic fractionalization (combined linguistic and racial) | Alesina 2003 |
| free_business<br>free_corrupt<br>free_finance<br>free_fiscal<br>free_govspend<br>free_invest<br>free_labor<br>free_monetary<br>free_overall<br>free_property<br>free_trade | Heritage Foundation ratings, 2010 (higher scores more free) | Heritage Foundation<br>www.heritage.org/index/Explore.aspx<br>www.heritage.org/index/FAQ.aspx |

**Table A-3**    *(continued)*

| Variable name | Variable label | Source and notes |
|---|---|---|
| gdp08 | GDP (billions) | WB |
| gdp_10_thou | GDP per capita in 10K US$ | UNDP 2004; income data are from 2002 |
| gdp_cap2 | GDP per capita (US$): 2 cats | UNDP |
| gdp_cap3 | GDP per capita (US$): 3 cats | UNDP |
| gender_equal3 | Gender empowerment measure | Based on Global Indicators variable GEMValue2004; UNDP 2004; http://hdr.undp.org/en/statistics/faq/question,81,en.html |
| gini04<br><br>gini08 | gini coefficient, 2004 [2008] | UNDP 2004 [2008] |
| indy | Year of independence | CIA World Factbook |
| old06<br>old2003 | Population ages 65 and above (% of total) 2006 [2003] | WB 2007 [2004] |
| pmat12_3 | Post-materialism | Based on Global Indicators variable pmat12, postmaterialism 12 cat, WVS 1995 |
| pop03<br>pop08 | Population 2003 [2008] | World Bank |
| popcat3 | Size of country by population (3-categories) | UNDP |
| pr_sys | PR system? | Based on Global Indicators variable elecpr, IDEA |
| protact3 | Protest activity | Based on Global Indicators variable protact, summary mean protest activity, WVS 1995 |
| regime_type3 | Regime types | Based on Global Indicators variable Cheibub4Type (2000); author set mixed systems to missing |
| typerel | Predominant religion | CIA World Factbook |
| unions | Union density | International Labour Office 1995; www.ilo.org |
| urban03<br>urban06 | Urban population (% of total) 2003 [2006] | WB |
| vi_rel3 | Percent saying religion "very" important | Based on Global Indicators variable vi_rel; Percent religion 'very' important, WVS 1995 |
| votevap00s<br>votevap90s | Vote/vap most recent election in 2000s<br>Vote/vap during 1990s | IDEA |
| women05<br>women09 | Percent women in lower house of parliament, 2005 [2009] | IPU; author restricted this measure to democracies only |
| womyear | Year women first enfranchised | IPU |

*(continued)*

**Table A-3**  Descriptions of Variables in World*  *(continued)*

| Variable name | Variable label | Source and notes |
|---|---|---|
| yng2003<br>young06 | Population ages 0-14 (% of total) 2003<br>[2006] | WB 2004 [2007] |

*All variables used in World were compiled and made available by Pippa Norris, John F. Kennedy School of Government, Harvard University, www.pipanorris.com. Professor Norris's dataset, Global Indicators Shared Dataset V2.0 (Updated Fall 2005 [September 13, 2005]), may be accessed at http://ksghome.harvard.edu/~pnorris/Data/Data.htm.

Alesina = Alberto Alesina, Arnaud Devleeschauwer, William Easterly, Sergio Kurlat, and Romain Wacziarg, "Fractionalization," *Journal of Economic Growth* 8 (June 2003): 155–194; www.stanford.edu/~wacziarg/papersum.html.

CIA World Factbook; www.cia.gov.

Freedom House = Freedom House, Gastil index, Annual to 2004; www.freedomhouse.org.

IDEA = IDEA, Voter turnout since 1945; www.idea.int.

IPU = Inter-Parliamentary Union; www.ipu.org.

Kaufmann = Kaufmann Governance indicators; www.worldbank.org/wbi/governance/govdata2002/.

Polity IV=Polity IV Project; www.cidcm.umd.edu/polity/data/.

UNDP = Human Development Report OUP 2003; www.undp.org.

WB = World Bank Development Indicators; www.worldbank.org.

World Bank = DPI2000 Database of political institutions; http://econ.worldbank.org/.

WVS = World Values Survey (1995–2000 waves); wvs.isr.umich.edu.

**Table A-4**  Descriptions of Variables and Constructed Scales in NES2008

| Variable name | Variable label | ANES source variables and notes[a] |
|---|---|---|
| abort_rights<br>abort_rights3 | When Should Abortion Be Permitted? | V085086 |
| abort_rights_imp | How Important Is Abortion Issue? | V085087 |
| age<br>age5 | Age Of Respondent | V083215x |
| asian_therm | Feeling Thermometer: Asian-Americans | V085064v |
| asian_trait1 | Asians Hardworking? | V083207d: +3 to –3 [recode] |
| asian_trait2 | Asians Intelligent? | V083208d: +3 to –3 [recode] |
| asian_traits | Asian Stereotype | asian_trait1 + asian_trait2<br>Range: –6 to +6 |
| atheist_therm | Feeling Thermometer: Atheists | V085065h |
| bclinton_therm_pre | Feeling Thermometer: Bill Clinton | V083041 |
| biden_therm_post | Feeling Thermometer: Joe Biden | V085063t |
| biden_therm_pre | Feeling Thermometer: Biden | V083039a |
| bigbus_therm | Feeling Thermometer: Big Business | V085064n |
| black_pres_hope | Hope That US Has African-American President | V083172 |
| black_pres_pleased | Summary: Black President Make R Pleased | V083171x |
| black_pres_ready | US Ready For African-American President | V083173 |
| black_pres_uncmfrtble | Summary: Black President Make R Uncomfortable | V083170x |
| black_trait1 | Blacks Hardworking? | V083207b +3 to –3 [recode] |
| black_trait2 | Blacks Intelligent? | V083208b +3 to –3 [recode] |
| black_traits | Black Stereotype | black_trait1 + black_trait2<br>Range: –6 to +6 |
| blacks_therm | Feeling Thermometer: Blacks | V085064y |
| blue_dog | Is R In Blue Dog District? | McCain-Dem district; source: www.cqpolitics.com/wmspage.cfm?parm1=197 |
| bush_secure | R3ax. Summary: U.S. More/Less Secure Than When Pres Took Office | V083180x |
| bush_therm_post | Feeling Thermometer: President George W. Bush | V085063a |
| bush_therm_pre | Feeling Thermometer: President Bush | V083036 |

*(continued)*

**Table A-4**   Descriptions of Variables and Constructed Scales in NES2008   *(continued)*

| Variable name | Variable label | ANES source variables and notes[a] |
|---|---|---|
| camp_acts | R Number Of Campaign Acts | V085030<br>V085031<br>V085032<br>V085033<br>V085034<br>V085035 |
| camp_attent | General Attention To Presidential Campaign News | V085012<br>V085023 |
| camp_int | Campaign Interest | Constructed from camp_intA and camp_intB |
| camp_intA | Campaign Interest (Version A) | V083001a |
| camp_intB | Campaign Interest (Version B) | V083001b |
| catholic_therm | Feeling Thermometer: Catholics | V085064c |
| christian_therm | Feeling Thermometer: Christians | V085065g |
| cindymc_therm | Feeling Thermometer: Cindy McCain | V085063d |
| cong_approval | R Approve/Disapprove Congress Handling Job | V083035x |
| cong_care_who | Care Who Wins House Election | V083034 |
| cong_incumb_approval | E2x. Summary: Approve/Disapprove House Incumbent | V085068x |
| congress_therm | Feeling Thermometer: Congress | V085064w |
| conservative_therm | Feeling Thermometer: Conservatives | V085064q |
| contact_GOTV | Anyone Talk To R About Registering Or Getting Out To Vote | V085028 |
| contact_partisan | Was R Contacted By Parties / Others? | V085025<br>V085026 |
| deathpen | R Favor/Oppose Death Penalty | V083163<br>V083163a |
| def_spend_dem | Dem Cand Position: Defense Spending | V083114a<br>V083117x |
| def_spend_imp | Importance Of Defense Spending Issue To R | V083113<br>V083116 |
| def_spend_r | R's Position: Defense Spending | V083112<br>V083115x |
| def_spend_rep | Rep Cand Position: Defense Spending | V083114b<br>V083118x |
| deficit_mil | E6c3x. Summary: Cut Deficit By Reducing Military Spending | V085076x |

**Table A-4**  *(continued)*

| Variable name | Variable label | ANES source variables and notes[a] |
|---|---|---|
| deficit_other | E6d3x. Summary: Cut Deficit By Cutting Other Programs | V085077x |
| deficit_reduce | E6a3x. Summary: Reducing The Budget Deficit | V085074x |
| deficit_taxes | E6b3x. Summary: Reduce Deficit By Raising Taxes | V085075x |
| dem_therm | Feeling Thermometer: Democratic Party | V083044a |
| drug_coverage | N3d1x. Summary: R Position On Senior Prescription Drug Cover [new] | V083122x |
| drug_coverage_imp | N3e. Importance Of Prescript Drug Coverage For Seniors [new] | V083123 |
| educ_r<br>educ_r2<br>educ_r3 | Highest Grade Of School Or Year Of College R Completed | V083217 |
| educ_spouse | Spouse: Highest Grade Or Year Of College | V083219 |
| employ_status | Employment Status | V083222 |
| enviro_emissions_dem | P4f1ax. Summary: Dem Pres Cand Fav/Opp Lower Emissions [new] | V083159x |
| enviro_emissions_imp | P4e. Importance Of Emission Std Issue [new] | V083158 |
| enviro_emissions_r | P4d1x. Summary: Favor/Oppose Lower Emission Stds [new] | V083157x |
| enviro_emissions_rep | P4f2ax. Summary: Rep Pres Cand Fav/Opp Lower Emissions [new] | V083160x |
| enviro_fuel_r | Summary: Favor/Oppose Higher Fuel Std | V083161x |
| enviro_gastax_r | Summary: Favor/Oppose Higher Gas Taxes | V083162x |
| enviro_jobs_dem | Protect Environment/Jobs: Dem Pres Cand | V083156a |
| enviro_jobs_imp<br>enviro_jobs_imp2 | Importance Of Environment/Jobs Issue To R | V083155 |
| enviro_jobs_r7<br>enviro_jobs<br>enviro_jobs_r3 | Environment Vs. Jobs Tradeoff Scale - Self-Placement | V083154 |
| enviro_jobs_rep | Protect Environment/Jobs: Rep Pres Cand | V083156b |
| environ_therm | Feeling Thermometer: Environmentalists | V085064s |
| external_eff | External Political Efficacy | V083080c [No say]<br>V083080d [No care] |
| fedgov_therm<br>fedgov_therm4 | Feeling Thermometer: Federal Government In Washington | V085064e |

*(continued)*

**Table A-4** Descriptions of Variables and Constructed Scales in NES2008 *(continued)*

| Variable name | Variable label | ANES source variables and notes[a] |
|---|---|---|
| fedspend_border | Summary: Increase Or Decrease Spending On Border | V083149x |
| fedspend_childcare | Summary: Increase Or Decrease Spending On Child Care | V083146x |
| fedspend_crime | Summary: Increase Or Decrease Spending On Crime | V083144x |
| fedspend_environ | Summary: Increase Or Decrease Spending On Environment | V083151x |
| fedspend_foreign | Summary: Increase Or Decrease Spending On Foreign Aid | V083147x |
| fedspend_highway | Summary: Increase Or Decrease Spending On Highways | V083140x |
| fedspend_poor | Summary: Increase Or Decrease Spending On Aid To Poor | V083148x |
| fedspend_schools | Summary: Increase Or Decrease Spend On Public Schools | V083142x |
| fedspend_science | Summary: Increase/Decrease Spend On Science And Techn | V083143x |
| fedspend_socsec | Summary: Increase Or Decrease Spend On Soc Security | V083141x |
| fedspend_terrorism | Summary: Increase/Decrease Spendg On War On Terrorism | V083150x |
| fedspend_welfare fedspend_welfare3 | Summary: Increase Or Decrease Spending On Welfare | V083145x |
| feminist_therm | Feeling Thermometer: Feminists | V085064d |
| fundchristian_therm | Feeling Thermometer: Christian Fundamentalists | V085064b |
| gay_adopt gay_adopt_dum | X15. Should Homosexual Couples Be Allowed To Adopt | V083213 |
| gay_discrim | X13x. Summary: Fav/Opp Laws Protect Gays Against Job Discrim | V083211x |
| gay_marriage gay_marriage3 | X16. R Position On Gay Marriage | V083214 |
| gay_mil | X14x. Summary: Favor/Opp Allowing Gays To Serve In Military | V083212x |
| gay_therm | Feeling Therm: Gay Men And Lesbians (Homosexuals) | V085064u |
| gender | R Gender | V081101 |

**Table A-4**  *(continued)*

| Variable name | Variable label | ANES source variables and notes[a] |
|---|---|---|
| global_goals | Imp Intl Policy: 10 Goals | Following variables summed: policy_amerjobs, policy_democ, policy_humanrights, policy_hunger, policy_immig, policy_markets, policy_nuclear, policy_terrorism, policy_UN; rescaled, 0–100 |
| goal_UN policy_UN | U.S. Policy Goal: Strengthen United Nations | V085078c |
| goal_amerjobs policy_amerjobs | U.S. Policy Goal: Protect American Jobs | V085078e |
| goal_democ policy_democ | U.S. Policy Goal: Bring Democracy To World | V085078f |
| goal_humanrights policy_humanrights | U.S. Policy Goal: Defending Human Rights | V085078b |
| goal_hunger policy_hunger | U.S. Policy Goal: Combat World Hunger | V085078d |
| goal_immig policy_immig | U.S. Policy Goal: Control Illegal Immigration | V085078g |
| goal_markets policy_markets | U.S. Policy Goal: Promote Market Economies Abroad | V085078h |
| goal_nuclear policy_nuclear | U.S. Policy Goal: Preventing Nuclear Weapons | V085078a |
| goal_terrorism policy_terrorism | U.S. Policy Goal: Combat Intl Terrorism | V085078j |
| govt_assist_blacks_dem | N5c1. Aid To Blacks Placement: Dem Pres Cand | V083139a |
| govt_assist_blacks_imp | Importance Of Govt Assistance To Blacks Issue To R | V083138a V083138b |
| govt_assist_blacks_r | Govt Assistance To Blacks Scale: Self-Placement | V083137 |
| govt_assist_blacks_rep | N5c2. Aid To Blacks Placement: Rep Pres Cand | V083139b |
| gunlaw gunlaw2 | Should Fed Govt Make It More Difficult To Buy A Gun | V083164 |
| gunlaw_imp | Importance Of Gun Issue To R | V083164a V083164b |
| gunown | Does R Have A Gun In His Or Her Home Or Garage | V083165 |
| hh_adults | Number Of Adults In HH | V081108a |

*(continued)*

**Table A-4** Descriptions of Variables and Constructed Scales in NES2008 *(continued)*

| Variable name | Variable label | ANES source variables and notes[a] |
|---|---|---|
| hh_kids | Number Of Children In HH | V081109 |
| hh_kids10 | Y31a. How Many Children In HH Age 10 And Younger | V083265a |
| hh_kids17 | Y31b. How Many Children In HH Age 11-17 | V083265b |
| hillary_therm | Feeling Thermometer: Hillary Clinton | V085063v |
| hillary_therm_pre | Feeling Thermometer: Hillary Clinton | V083040 |
| hindu_therm | Feeling Thermometer: Hindus | V085065f |
| hispanic | Hispanic Ethnicity | V081103 |
| hispanic_therm | Feeling Thermometer: Hispanics | V085064a |
| hispanic_trait1 | Hispanics Hardworking? | V083207c +3 to −3 [recode] |
| hispanic_trait2 | Hispanics Intelligent? | V083208c +3 to −3 [recode] |
| hispanic_traits | Hispanic Stereotype | hispanic_trait1 + hispanic_trait2 Range: −6 to +6 |
| hlth_insured | Does R Have Health Insurance | V083060 |
| illegal_citizenshp_dem | Summary: Dem Pres Cand On Illegal Immigrant Citizenship[new] | V083135x |
| illegal_citizenshp_imp | Importance Of Citizenship Illegals Issue [new] | V083134 |
| illegal_citizenshp_r | Fav/Opp Citizenship Process For Illegals | V083133x |
| illegal_citizenshp_rep | Summary: Rep Pres Cand On Illegal Immigrant Citizenship [new] | V083136x |
| illegal_wrk | Fav/Opp 3yr US Work Period For Illegals [V083131x] | V083131x |
| illegal_wrk_imp | N4e. Importance Of Illegal Immigrant Work Period [new] | V083132 |
| illegals_therm | Feeling Thermometer: Illegal Immigrants | V085065a |
| income_HH | Household Income | V083248 |
| income_r income_r2 income5 | R Income | V083249 |
| internal_eff | Internal Political Efficacy | V083079a [Politics complicated] V083079b [Good understanding] |
| internet | Does R Have Internet Access? | V083018 |
| internet_news_new | Days Reviewed News Online | V083023 |
| internet_news_old | Days Read Online Newpaper | V083021b |
| iraq_cost | Was Iraq War Worth The Cost | V083103 |

**Table A-4** *(continued)*

| Variable name | Variable label | ANES source variables and notes[a] |
|---|---|---|
| iraq_terror | Iraq War Increased Or Decreased Threat Of Terrorism | V083104 |
| israel_therm | Feeling Thermometer: Israel | V085065d |
| iw_age_group | Interviewer Age Group | V082258 |
| iw_education | Interviewer Education | V082253 |
| iw_ethnicity | Interviewer Hispanic? | V082255 |
| iw_experience | Interviewer Years Of Experience | V082257 |
| iw_gender | Interviewer Gender | V082252 |
| iw_language | Interviewer Language Used | V082256 |
| iw_race | Interviewer Race | V082254 |
| jew_therm | Feeling Thermometer: Jews | V085064f |
| jobs_imp | Importance Of Guaranteed Job-Income Issue To [old] | V083129 |
| jobs_scale_dem | Guar Job Scale: Dem Pres Cand [old] | V083130a |
| jobs_scale_r | Guaranteed Job-Income Scale: Self-Placement [old] | V083128 |
| jobs_scale_rep | Guar Job Scale: Rep Pres Cand [old] | V083130b |
| know_brown | Recog Gordon Brown | V085122a |
| know_cheney | Recog Dick Cheney | V085121a |
| know_hr | Know Party Control HR | V085066 |
| know_pelosi | Recog Nancy Pelosi | V085120a |
| know_roberts | Recog John Roberts | V085123a |
| know_scale know_scale3 | R Recog Pelosi, Cheney, Brown, Roberts? | Variables summed: know_pelosi, know_cheney, know_brown, know_roberts |
| know_sen | Know Party Control Sen | V085067 |
| libcon7_demparty | Liberal/Conservative Dem Party | V083071a |
| libcon7_mccain | Liberal/Conservative - Rep Pres Cand | V083070b |
| libcon7_obama | Liberal/Conservative - Dem Pres Cand | V083070a |
| libcon7_probe | E1b. If R Had To Choose Liberal Or Conservative Self-Placement | V083069a |
| libcon7_r libcon7 libcon3_r | R Self-Placement Liberal-Conservative Scale | V083069 V083069a |

*(continued)*

**Table A-4**  Descriptions of Variables and Constructed Scales in NES2008  *(continued)*

| Variable name | Variable label | ANES source variables and notes[a] |
|---|---|---|
| libcon7_repparty | Liberal/Conservative Rep Party | V083071b |
| liberal_therm<br>liberalism3 | Feeling Thermometer: Liberals | V085064g |
| limbaugh_therm_pre | Feeling Thermometer: Rush Limbaugh | V083043 |
| marital | Summary: Marital Status | V083216x |
| mccain_therm | Feeling Thermometer: Mccain | V083037b |
| mccain_therm_post | Feeling Thermometer: Republican Presidential Candidate | V085063c |
| medinsur_dem | Insurance Scale: Dem Pres Cand | V083121a |
| medinsur_imp | Importance Of Govt Insurance Issue To R | V083120 |
| medinsur_r | Govt/Private Medical Insur Scale: Self-Placement | V083119 |
| medinsur_rep | Insurance Scale: Rep Pres Cand | V083121b |
| memnum<br>memnum2 | No. Of Memships | V085127 |
| michelleo_therm | Feeling Thermometer: Michelle Obama | V085063e |
| midclass_therm | Feeling Thermometer: Middle Class People | V085064h |
| military_therm | Feeling Thermometer: The Military | V085064m |
| muslim_therm | Feeling Thermometer: Muslims | V085065e |
| native_parents<br>native2 | Native Status Of Parents | V083257 |
| natl_btr_wrs_future | F2a1x. Summary: Economy Better/Worse Next 12 Mos [Version E] | V083084x |
| natl_btr_wrs_past<br>nat_econ_past2<br>nat_econ_past3 | F1ax. Summary: Economy Better Worse In Last Year | V083083x |
| newspaper | Number Of Days R Reads Newspaper | V083021a V083025 |
| obama_repub | Is R In Pro-Obama District W/GOP Rep? | Obama-Rep district; www.cqpolitics.com/wmspage.cfm?parm1=197 |
| obama_therm | Feeling Thermometer: Obama | V083037a |
| obama_therm_post | Feeling Thermometer: Democratic Presidential Candidate | V085063b |
| old_new_filter | Old Or New Question Battery | V082402 |
| optimism_US | Summary: Optimism/Pessimism About The U.S. | V085062x |
| optimism_personal | Summary: Personal Optimism | V085061x |

**Table A-4** *(continued)*

| Variable name | Variable label | ANES source variables and notes[a] |
|---|---|---|
| own_rent | Does R Family Own/Rent Home | V083281 |
| palin_therm_post | Feeling Thermometer: Sarah Palin | V085063u |
| palin_therm_pre | Feeling Thermometer: Palin | V083039b |
| party_btr_econ | Which Party Better: Handling Nations Economy | V083093 |
| party_btr_war | Which Party Better Keeping Out Of War | V083094a<br>V083094b |
| partyid7<br>partyid3 | Summary Party ID | V083098x |
| poor_therm | Feeling Thermometer: Poor People | V085064k |
| pres_Iraq | R Approve/Disapp Pres Handling War In Iraq | V083033x |
| pres_approval | R Approve/Disapp Pres Handling Job | V083028x |
| pres_care_who | Care Who Wins Presidential Election | V083072 |
| pres_econ | R Approve/Disapp Pres Handling Economy | V0839x |
| pres_environ | R Approve/Disapp Pres Handling Environment | V083031x |
| pres_foreign | R Approve/Disapp Pres Handling Foreign Rel | V083030x |
| pres_hlthcare | R Approve/Disapp Pres Handling Health Care | V083032x |
| r_btr_wrs_future | R Better/Worse Off 1 Year From Now | V083058x |
| r_btr_wrs_past<br>r_econ_past3 | R Better/Worse Off Than 1 Year Ago | V083057x |
| race<br>race2 | R Race | V081102 |
| region<br>south | Region Of Residence | V081204 |
| relig_attend<br>relig_attend3<br>relig_attendHi | How Often R Attends Religious Services | V083186 |
| relig_bible<br>relig_bible_word | Bible Is Word Of God Or Men | V083184 |
| relig_born_again | R Consider Self Born Again | V083203 |
| relig_denom | Summary: Initial Mention Major Religious Group | V083188x |

*(continued)*

**Table A-4** Descriptions of Variables and Constructed Scales in NES2008 *(continued)*

| Variable name | Variable label | ANES source variables and notes[a] |
|---|---|---|
| relig_imp | Importance Of Religion To R | V083181<br>V083182 |
| relig_pray<br>relig_pray_freq | How Often Does R Pray | V083183 |
| relig_transub | R Belief In Transubstantiation | V083204 |
| relig_transub_imp | Importance Of Belief In Transubstantiation | V083205 |
| rep_therm | Feeling Thermometer: Republican Party | V083044b |
| rice_therm_pre | Feeling Thermometer: Condoleezza Rice | V083042 |
| rich_therm | Feeling Thermometer: Rich People | V085065b |
| roots | How Long Lived In This Community: Years | V083266a |
| serv_spend_dem | Dem Cand Position: Services And Spending | V083107a<br>V083110x |
| serv_spend_imp | Importance Of Spending-Services Issue To R | V083106<br>V083109 |
| serv_spend_r | R's Position: Services And Spending | V083105<br>V083108x |
| serv_spend_rep | Rep Cand Position: Services And Spending | V083107b<br>V083111x |
| sexl_orient_famfrnds | GLB Orientation Among Family And Friends? | V083210 |
| sexl_orient_r | Sexual Orientation Of R | V083209 |
| social_trust_new | Can People Be Trusted [Version H] | V083092b |
| social_trust_old | Can People Be Trusted [Version G] | V083092a |
| southerner_therm | Feeling Thermometer: Southerners | V085064z |
| spend10 | Incr Federal $:10 Programs | Number "increase": fedspend_childcare, fedspend_crime, fedspend_environ, fedspend_foreign, fedspend_highway, fedspend_poor, fedspend_schools, fedspend_science, fedspend_socsec, fedspend_welfare |
| state | State Of Residence | V081201a |
| stock_investor | Does R Or Spouse Have Any Money Invested In Stock Market | V083059 |
| supremecrt_therm | Feeling Thermometer: The U.S. Supreme Court | V085064t |
| tax_corp | Big Companies Pay Larger Or Smaller Percent Of Profits | V083153 |

**Table A-4**  *(continued)*

| Variable name | Variable label | ANES source variables and notes[a] |
|---|---|---|
| tax_income | Statement Best Agrees With R About Graduated Tax | V083152 |
| trust_media | How Often Trust The Media To Report News Fairly | V085024 |
| tv_news | Number Of Days R Watches TV News | V083019<br>V083020a<br>V083020b<br>V083024 |
| union_HH | Y19. Anyone In HH Belong To Labor Union | V083245 |
| union_therm | Feeling Thermometer: Labor Unions | V085064j |
| universal_hlthcr_dem | N3h1ax. Summary: Dem Pres Cand Posn On Univ Healthcare [new] | V083126x |
| universal_hlthcr_imp | N3g. Importance Of Universal Health Coverage [new] | V083125 |
| universal_hlthcr_r | N3f1x. Summary: R Position On Universal Health Care [new] | V083124x |
| universal_hlthcr_rep | N3h2ax. Summary: Rep Pres Cand Posn On Univ Healthcare [new] | V083127x |
| us_stay_home | Country Would Be Better Off If We Just Stayed Home | V083096 |
| veteran | Is R Or Has R Ever Been In The Military | V083221 |
| voted | Did R Vote? | V085036x |
| voted_04 | Did R Vote In 2004? | V083007 |
| voted_absentee | C5a1. Did R Vote In Person Or By Absentee Ballot | V085043 |
| voted_early | C5. Did R Vote On Election Day Or Before Election Day | V085042 |
| voted_primary | Did R Vote In The Presidential Primary Or Caucus | V083077 |
| voted_primary_who | For Which Candidate Did R Vote In Presidential Primary | V083077a |
| voter0408 | Voter '04, '08, Both? | Calculated from voted and voted04 |
| welfare_therm | Feeling Thermometer: People On Welfare | V085064p |
| white_therm | Feeling Thermometer: Whites | V085065c |
| white_trait1 | Whites Hardworking? | V083207a +3 to −3 [recode] |
| white_trait2 | Whites Intelligent? | V083208a +3 to −3 [recode] |

*(continued)*

**Table A-4**  Descriptions of Variables and Constructed Scales in NES2008  *(continued)*

| Variable name | Variable label | ANES source variables and notes[a] |
|---|---|---|
| white_traits | White Stereotype | white_trait1 + white_trait2<br>Range: −6 to +6 |
| who08 | Pres Vote 2008 | V085044a |
| who08_strong | Preference Strong For Candidate For Whom R Voted | V085044b |
| who_04 | R's 2004 Vote Choice | V083007a |
| wom_role_dem | Women's Role Scale: Dem Pres Cand | V083168a |
| wom_role_imp | Importance Of Women's Role To R | V083167 |
| wom_role_r<br>wom_role_r7 | Womens Role Self-Placement | V083166 |
| wom_role_rep | Women's Role Scale: Rep Pres Cand | V083168b |
| wrkingclass_therm | Feeling Thermometer: Working Class People | V085064r |
| yob | Y1a. Birthdate Year | V083215a |

[a]To find source variable question wording, go to the University of California Berkeley's Survey Documentation and Analysis (SDA) Web site at http://sda.berkeley.edu/archive.htm and click the link for American National Election Study (ANES) 2008. Click the Codebook tab and select Alphabetical Variable List.